PLANNING CO-EXISTENCE: ABORIGINAL ISSUES IN FOREST AND LAND USE PLANNING

Marc G. Stevenson and David C. Natcher
(editors)

Research and Insights from the Aboriginal Research
Program of the Sustainable Forest Management
Network: Volume II

CCI Press
2010

Library and Archives Canada Cataloguing in Publication

Planning co-existence: aboriginal issues in forest and land use planning /
Marc G. Stevenson and David C. Natcher, editors.

(Occasional publication series, 0068-0303; no. 65)
"Research and insights from the Aboriginal Research Program
    of the Sustainable Forest Management Network: volume II".
Co-published by Sustainable Forest Management Network.
Includes bibliographical references.
ISBN 978-1-896445-49-6

        1. Native peoples—Land tenure—Canada. 2. Native peoples—Canada
—Government relations. 3. Aboriginal title—Canada. 4. Land use—Canada—
Planning. 5. Forest management—Canada. I. Stevenson, Marc II. Natcher,
David C., 1967– III. Sustainable Forest Management Network . IV. Series:
Occasional publication series (Canadian Circumpolar Institute) ; no. 65.
KE7709.P53 2010   346.7104'3208997   C2010-904006-6
KF5660.P53 2010

CCI Press Occasional Publication No. 65
ISSN 0068-03030   ISBN 978-1-896445-49-6
© 2010 CCI Press and Sustainable Forest Management Network (SFMN)
Cover photos courtesy editors and contributors.
Cover design by CCI in collaboration with art design printing, inc.
Printed in Canada by art design printing, inc.

About the Cover
The two-row wampum symbolically and graphically illustrates the principles for
an ethical research relationship between peoples of different cultures; it is a
theory of international cooperation, setting out the basis of relationships between
nations. It is also an ethical theory when it talks of respect, dignity, honesty and
kindness as the principles for relationships…, and "we ought to remember that
the space between the rows is a place of conversation, discussion and debate"
(David Newhouse in SSHRC 2003:18).
        -quoted from M.G. Stevenson (2008), p. 201 in 'Negotiating research
    relationships with Aboriginal communities: Ethical Considerations and
    Principles,' Chapter 12 in the companion volume *Changing the Culture of
Forestry in Canada: Building Effective Institutions for Aboriginal Engagement
in Sustainable Forest Management.* Edmonton: CCI Press and SFM Network,
                Occasional Publication No. 60.

Planning Co-existence: Aboriginal Issues in Forest and Land Use Planning

Table of Contents

Acknowledgements

Planning Co-existence: Aboriginal Issues in Forest and Land Use Planning is the last of a two-volume set highlighting the research and insights of the Aboriginal Research Program of the Sustainable Forest Management (SFM) Network. While not all the contributors to this volume report on research funded by the SFM Network, most have been affiliated with the Network for many years and felt that this venue was an appropriate place for the publication of their research. We note with a sense of melancholy that the SFM Network ended its 14 year life as a National Network Centre of Excellence on March 31st of this year (2010). Yet, we point out with satisfaction that the contributing authors to this volume include a mixture of researchers (Aboriginal and non-Aboriginal) and Aboriginal partners in research. Gone are the days when research on Aboriginal peoples and issues was the exclusive domain of university-based academics, though we acknowledge with pride that three (and soon-to-be four) contributors to this volume are First Nations' authors with PhDs in Forestry or related disciplines (see Notes on Authors).

Both this volume and its predecessor, *Changing the 'Culture' of Forestry in Canada: Building Effective Institutions for Aboriginal Engagement in Sustainable Forest Management* (Stevenson and Natcher 2009) would not have been possible without the support of many people associated with Sustainable Forest Management Network including: Marv Abugov, Vic Adamovicz, Dale Bischoff, Alison Boddy, Harry Bombay, Robert Charlie, Shirley Devries, Margaret Donnelly, J.P. Gladu, Jim Fyles, Cliff Hickey, Cynthia Kaufman, Gary Lipinski, Bruce MacLock, Bruce MacNab, Jim McGrath, Morris Monias, Lucille Partington, Janet Pronovost, Johnson Sewepagaham, John Stager, Don Sharp, John Turner, Peter Usher, Terry Veeman, Barry Waito and countless others. We would also like to recognize Elaine Maloney and Cindy Mason of the Canadian Circumpolar Institute for their tireless efforts in producing both this volume and the first, and Cliff Hickey's expert and sage review of this volume, which improved the final draft considerably, though we (and by this we mean all the contributors to this volume) alone assume full responsibility for any errors or omissions (which we apologize for in advance).

Frideres and Rowe (Chapter 1) would like to thank the SFM Network for funding support as well as Jim Webb, Marc Stevenson, Monique Passelec-Ross and Chief Roland Wilson for their contributions to and continuing encouragement of their work. Stevenson (Chapter 2) would like to thank Heather Kennedy (Land Use Secretariat), Chief Morris Monias and Betty Kennedy of the Heart Lake First Nation, and Dwayne (Sonny) Nest and Dave Scott of Tribal Chief Ventures Incorporated (Treaty 6) for moral and other support during his time on the Lower Athabasca Regional Advisory Council. Webb (Chapters 3 and 4) would like to acknowledge Cliff Hickey and Dave Natcher, and especially Marc Stevenson, for undertaking countless reviews of his chapters over the years, and the SFM Network for finally publishing this body of work. Passelec-Ross (Chapter 5) and Johnson and Trosper (Chapter 6) acknowledge the support of the SFM Network in their research. Johnson's sincere thanks also goes to Sarah Weber for her cooperation and team support while working together and the many intellectual knowledge exchanges they had during this period. Finally, Johnson and Trosper wish to thank the Stellaquo for their willingness to share their knowledge, time and vision with researchers, Aboriginal peoples and others on their proposal for change in the governance of forestland and resources in British Columbia and Canada.

Deutsche and Davidson-Hunt (Chapter 7) acknowledge the SFM Network and Fikret Berkes and David Natcher (co-applicants) for supporting their research, and wish to thank the community and elders of Pikangikum for their teachings and critiques throughout the research process. In particular they would like to thank Paddy Peters who helped them to understand 'a little bit' of what was being taught, and Alex Peters who made their research partnership so rewarding in many ways. Kneeshaw et al. (Chapter 8) would like to thank Marc Stevenson for editorial advice and assistance. Wyatt et al. (Chapter 9) acknowledge the insights and contributions of Keith Carlson, Philippe Chartier, George Duckworth, Liam Haggarty, Martin Pelletier, Nadia Saganash, Marc Stevenson, Terry Tobias, Jim Webb and Brad Young to the workshop, *Mapping Aboriginal Land Use and Occupancy in Forestry*, held in Saskatoon in January 2009. (This workshop was jointly organised by an SFM Network funded project and by the Indigenous Land Management Institute at the University of Saskatchewan, with significant participation by a number of First Nations and Métis organisations in Saskatchewan.) McGregor (Chapter 10) would like to thank the editors for their editorial advice and direction, as do Smith and her co-authors (Chapter 11). Stevenson and Perreault acknowledge the support of the SFM Network, National Aboriginal Forestry Association and Aboriginal Capacity Working Group, in producing the research synthesis upon which their chapter (Chapter 12) is based. Kuefler and his co-authors would like to thank the Cree of Moose Factory for their support of and participation in Kuefler's research.

Introduction

Planning Co-existence

Marc G. Stevenson and David C. Natcher

Introduction

Planning Co-Existence: Aboriginal Issues in Forest and Land Use Planning is the second of two volumes produced by members of the Aboriginal Research Group (Fig. 1) of the Sustainable Forest Management Network (SFM Network). Comprising some of Canada's top academic and Aboriginal researchers, Aboriginal community partners and professional affiliates, this 'community of practice' has evolved and expanded over the life of the SFM Network (1995-2010) to address a broad spectrum of issues on the Aboriginal forestry agenda. Notwithstanding the fact that members of our group are accountable to different 'masters,' we share a strong commitment to create more equitable arrangements and space for Aboriginal peoples in land use planning and environmental decision-making processes.

Figure 1. *Pikangikum elders and land stewards with SFMN researchers, graduate students and staff in the Whitefeather Forest.* (Photo credit: Irena Creed and Iain Davidson-Hunt)

The first volume, *Changing the Culture of Forestry in Canada: Building Effective Institutions for Aboriginal Engagement in Sustainable Forest Management* (Stevenson and Natcher 2009), sought to examine Aboriginal values, institutions, models and approaches to achieving social, cultural, economic and environmental sustainability outcomes for Canada's forest-dependent Aboriginal communities. While *Changing the Culture of Forestry in Canada* was aimed most directly at Canada's forest sector, the issues raised (rights, cultural survival, equity and social justice) extend well beyond that sector and should prove relevant to all those who have an interest in lands where Aboriginal peoples assert constitutionally protected rights—that is, all of Canada!

While it was always our intention to produce a second volume, it was not until we reflected more carefully upon the issues raised in Volume I that we came to see a unifying, yet unanswered question: How do we begin to reconcile the land and resource use rights and interests of Canada's Aboriginal peoples while finding common ground for co-existence with others who have come to occupy these shared spaces?

With this potential theme in mind, we began to consult with members of our small research community. This involved phone and conference calls, emails, opportunistic meetings in airports and at conferences, and so on. While all seemed to agree that 'co-existence' was an appropriate theme for Volume II, we came to realize, quite explicitly, that before co-existence can be achieved, before effective institutions and policies can be developed and put into place, reconciliation must occur.

Despite heartfelt attempts, we can offer no concrete definitions for 'co-existence.' Co-existence is likely to be context-specific, and what may be considered as co-existence to one party, may not be to another. Embedded in the use of the concept, however, is the notion that co-existence requires the recognition and peaceful, equitable and respectful accommodation of each others' rights and interests on the same land base—a distinctly Canadian and Aboriginal trait if there ever was one (Saul 2008). Co-existence thus necessitates a shared commitment to respectful negotiations, of setting aside past and historic grievances that have for too long characterized relations between Canada and its first peoples, to achieve outcomes for the common good. Reconciling the rights and interests of both the Canadian state and its Aboriginal peoples through negotiation, of finding the tough compromise, will not be an easy task, but the investment promises to pay social, economic, cultural and environmental dividends for generations to come.

Reconciliation also necessitates an objective and critical self-examination by all parties of what went wrong and why, of learning from the past, and correcting mistakes made collectively, together, without blame or malice. Yet, reconciliation requires political will—by all indications a rare commodity in Canada, despite direction offered by the Supreme and lower courts, not to mention the *Report of the Royal Commission on Aboriginal Peoples* (Government of Canada 1996). More often than not, it has been the threat of litigation, conflict, bad press and lost profits and revenues from natural resource extraction that has prompted government action toward the accommodation of Aboriginal rights and interests on traditional lands. Too often ignored in the acrimonious debate over land and resource rights in 'Indian Country'[1] are the costs of keeping Aboriginal peoples marginalized, dependent and impoverished. One would think that a full-cost accounting of the 'Aboriginal dependency project,' especially as natural resource development expands into areas where Aboriginal peoples continue to exercise their land-based rights, would be motivation enough for governments to engage seriously in reconciliation. But rarely have such issues been considered, let alone acted upon. The failure of Canada to sign the *Declaration on the Rights of Indigenous Peoples* (2007) indicates that reconciliation may at best be a very distant hope.

Despite the challenges—and they are considerable—we have come to realize that finding solutions to overcome these barriers are critically important if we are to make any advancement toward reconciliation and ultimately find the space needed for co-existence. As our discussions unfolded, it became clear that, at a minimum, reconciliation and co-existence will require the equitable participation of Aboriginal peoples in land use planning and environmental decision-making. It is from this point of departure that *Planning Co-existence* was launched.

State Land Use Planning: Curse or Opportunity?

As the breadth and complexity of resource management issues increase, governments have recognized the need to engage the public in participatory forms of management and decision-making. Recognizing that sustainability can only be realistically achieved through the active involvement of a wide array of interests groups, government agencies are taking a more inclusive approach to environmental decision-making and land use planning (Natcher 2001). More particularly, and specific to Aboriginal peoples, these participatory processes are often required due to unique legal and political relationships between 'state' and Indigenous governments, as set out by various treaties, statutes and court decisions. For some, the direct involvement of Aboriginal peoples in land use planning processes represents a significant social and political achievement (Lane 2006; Langton 2004). By providing a platform or institutional space for Aboriginal involvement, land use planning can effectively advance Aboriginal claims of sovereignty in social, political and economic realms while protecting the lands and resources that have long sustained their cultures and economies. Langton (2004:25) argues that, although Indigenous peoples may, in many ways, continue to be disadvantaged by colonial legacies, they and state governments can forge new relationships through consensual land use planning arrangements, that provide them with a genuine role in decisions that affect their lives and territories. By implication, the recognition of Aboriginal co-jurisdiction over lands and resources can redefine future interactions between Indigenous peoples and state governments by leaving open new avenues for negotiated improvements on social and political inequities (Langton 2004:22). In this way, land use planning and other modern forms of environmental

---

[1] We use the term 'Indian Country' to denote those lands within Canada where Aboriginal peoples retain constitutionally protected rights and interests based on long-term use and occupancy.

agreement-making can advance the accommodation of Aboriginal interests, and provide restitution for historical and continuing injustices against Indigenous peoples (Langton 2004:25).

An alternative position, however, suggests that Aboriginal involvement in state-sponsored land use planning processes serves only to undermine Aboriginal peoples' claims, and ultimately their quest for economic self-sufficiency and political self-determination, while entrenching the state's authority and control over Aboriginal peoples, lands and resources. Lane (2006) identifies three reasons why land use planning has often failed to meet the needs of Aboriginal peoples:

1) the power of European developmentalism as an ideology has a tendency to marginalize Indigenous perspectives in planning and decision-making;
2) Aboriginal peoples generally lack capacity to participate in state planning initiatives; and
3) the rational-comprehensive paradigm tends to marginalize Indigenous cultural perspectives.

Lane (2006) and others (Chase 1990; Escobar 1992; Howitt 2001) argue that the power of developmentalism, with its uncritical emphasis on the benefits of development and capital accumulation, affects all aspects of the planning process. Shared largely "by planning practitioners, private sector proponents of resource development, and bureaucratic and elected officials" (Lane 2006:386), developmentalism tends to marginalize Aboriginal peoples and their perspectives and contributions to planning through its reliance on the dominant rational–comprehensive paradigm of planning, planners' claims to scientific rationality and uncritical dependence on positivist science as the authoritative and often only legitimate discourse in contemporary planning processes (Lane 2006). Thus, land use planning more often than not recasts these projects to be a political process, reflecting the ideologies and interests of dominant actors (Lane 2006). In this way, land use planning is used simply as a tool by the state to further subjugate Aboriginal territorial rights and interests. While Aboriginal involvement in land use planning, including the use of traditional land use information, may create the impression that Aboriginal concerns are being addressed, in reality they constitute a vehicle for continuing dispossession of Aboriginal interests (O'Faircheallaigh 2004:304).

Despite the risks involved, many Aboriginal communities enter into land use planning processes with the state and/or other resource sector industries knowing that, in light of other alternatives (e.g., litigation and civil disobedience), they may be able to derive short-term benefits that can be channeled to community members. In other cases, Aboriginal involvement in collaborative land use planning initiatives is used as a strategy of self-defense, whereby Aboriginal communities form external alliances with other actors to ward off threats of environmental degradation and socio-cultural impoverishment (Borrini-Feyerabend *et al.* 2004:xxxiii). In still other instances, Aboriginal participation in such initiatives is viewed as way to track and develop effective strategies to respond to Crown activities (Dwayne Nest, personal communication to Stevenson, January 2010).

However, by participating in the processes and institutions of the state planning process, Aboriginal peoples risk distortion of their own territorial vision (Hirro and Surralles 2005:10), while providing the appearance of tacit approval and legitimacy to such initiatives. For all these reasons, many academics, Aboriginal leaders and legal advisors prefer to see the negotiation, reconciliation and accommodation of Aboriginal and Treaty rights take place outside the planning process and at the highest of political levels.

Planning Co-existence does not seek to determine which path will best serve the rights and interests of Canada's Aboriginal peoples, and ultimately all Canadians. That task is best determined by the parties involved on a case-by-case basis. Rather, it is our intention to explore the current state of land use planning initiatives in Canada, what may be required to meet the Crown's legal and fiduciary obligations in these processes, and a variety of issues of critical importance to Aboriginal peoples that need to be addressed in the design and implementation of land use plans. In so doing, it is our intention to lay the groundwork for a more informed discussion about reconciliation and co-existence in the context of land use planning in Indian Country in the hope of achieving social and environmental justice sooner rather than later.

Co-existence 2010

Section One (Chapters 1 and 2) of this volume explores the natural resource management and Aboriginal rights policies of the Alberta provincial government. Specifically, Alberta government policies with respect to cumulative impacts assessment and the accommodation of Aboriginal rights in the context of natural resource development and land use planning are discussed. While Alberta may be unique in certain respects, how it has handled and is dealing with, these issues is not incongruent with the policies of other Canadian provinces, and in fact may be considered fairly typical at the time of publication (Summer 2010). However, real institutional and policy reform that creates a solid foundation for true reconciliation and co-existence continues to elude us all.

Jim Frideres and Cash Rowe, in *Alberta Government Policy, First Nations Realities: Barriers to Land Use Policy and Cumulative Impact Assessment* (Chapter 1), review Alberta's policies with respect to resource development, cumulative impact assessment and accommodation of Aboriginal rights and interests. Informed by interviews conducted with government officials (federal and provincial), industry representatives and First Nations leaders, these authors seek to understand more fully the positions and actions taken by various actors with regard to land use, economic development and First Nations relations in the Treaty 8 region of Alberta. They identify the decision boundaries of both the provincial and federal governments, and analyze the actions taken by the former over the past three decades with regard to resource development. Until recently, the province's policy with respect to cumulative impacts assessment and consultation with First Nations may be summarized as reactive, piecemeal and rigid—critiques that are shared by many today. Policy gaps in relation to cumulative impacts assessments and the cultural sustainability of Alberta's First Nations peoples are identified, and will require greater consideration of valued social and ecosystem components at all levels of decision making. If resource development in Alberta is ever going to be truly sustainable, the government will need to develop 'best practices,' draw on Indigenous ways of knowing and build a 'made in Alberta' policy to ensure that cumulative impacts are properly assessed, monitored and mediated.

Marc Stevenson's chapter, *Trust Us Again, Just One More Time: Alberta's Land Use Framework and First Nations* (Chapter 2), describes the current initiative by the Government of Alberta to institute regional land use plans in order to manage the multiple and cumulative impacts of resource development, while striving to "balance the constitutionally protected rights of Aboriginal communities with the interests of all Albertans." The creation of Conservation Areas where Aboriginal traditional uses are supported, consultation plans specific to First Nations and the encouragement of Aboriginal peoples in land-use planning provide some hope that an era of improved Aboriginal-Provincial relations may be dawning. Even so, Alberta's First Nations are being asked, once again, to trust that the provincial government will act with their interests in mind. Stevenson's chapter questions whether the proposals being put forward by the province under *Alberta's Land Stewardship Act* and *Land Use Framework* are sufficient to warrant such expressions of faith. Stevenson's chapter advances recommendations with respect to improving relations between Alberta and its Aboriginal peoples in the context of regional land-use planning and the management of the natural resource development and oil/gas sectors in northeastern Alberta.

Setting the Planning Stage for Co-existence

Section Two (Chapters 3-6) sets the stage for significant transformational change in the nature of the relationship between Canada and its Aboriginal peoples through an exploration of the legal foundations for accommodating Aboriginal and Treaty rights in British Columbia and Alberta. Chapters 3 and 4 deal with the unfulfilled Treaty obligations of Alberta and Canada to protect the livelihoods of Treaty 8 First Nations peoples in Alberta, while constructing an interpretative framework for a shared understanding of 'treaty,' and how this understanding may be applied in the contemporary context to achieve co-existence. Chapter 5 reviews jurisprudence relating to Aboriginal rights and examines the legal context for the development of Aboriginal forest tenures in BC, and how they might be built to achieve co-existence. The final chapter in this section describes a new institutional model of co-existence that focuses on forested lands and resources, while combining elements of both Aboriginal traditional and contemporary Canadian systems of governance. Together, these chapters lead the way and forge a path for reconciliation and co-existence.

Chapter 3, *Unfinished Business: The Intent of the Crown to Protect Treaty 8 Livelihood Interests (1923-1939)*, by Jim Webb examines federal Crown commitments to safeguard Treaty 8 (1899) Indians' 'livelihood' interests and post-Treaty Crown government policy initiatives to implement these commitments. This chapter documents historic, post-Treaty efforts of the

Dominion government to safeguard and protect the 'usual vocations' of Treaty 8 peoples, incident to the transfer of provincial lands and resources under the *Natural Resources Transfer Act* (1930). Webb argues that these unfulfilled commitments, specifically the creation of 'special reserves,' provide a *sui generis* basis for the conduct of Crown-First Nations negotiations focused on the need for equitable reallocation of natural resources within Treaty 8 traditional territories, and the commercial development of a broad range of natural resources to create and sustain First Nation economies. As recently recognized by the courts, Treaty 8 (1899) was no more than a first step in a long process of reconciliation—a fact that has been effectively ignored by the federal and provincial Crowns. Webb proposes that the record of Crown negotiations related to establishment of 'special reserves' provides a catalyst for recognition of Crown obligations to revisit Treaty obligations and possibly act as the nexus for principled negotiations and meaningful consultations regarding natural resource development and the creation of First Nation economies in Treaty 8 territory.

The Canadian courts have developed an impressive array of interpretive principles and guidelines related to the reconciliation of Aboriginal land and resource interests with respect to the sovereignty of the Crown. Webb's second contribution to this volume, *On Vocation and Livelihood: Interpretive Principles and Guidelines for Reconciliation of Treaty 8 Rights and Interests* (Chapter 4), outlines how and why these interpretive principles can be used to reframe Aboriginal-Crown relations related to land and resource development within the Treaty 8 territory of northern Alberta. This chapter, which builds upon the previous chapter, constructs an interpretative framework for understanding what the respective parties to Treaty 8 may have meant by the terms 'livelihood' and 'vocations.' Webb proposes that these historical understandings provide a firm foundation for Crown and Treaty 8 First Nations negotiations of a mutual understanding as to the contemporary scope of Treaty-protected livelihood interests, which must now be accommodated in the context of ongoing provincial Crown allocation of lands and resources to third party interests.

The research described in Monique Passelac-Ross's chapter, *Aboriginal Rights and Aboriginal Forest Tenures: Towards 'Reconciliation' in British Columbia* (Chapter 5), explores some of the key attributes of Aboriginal forest tenures, and the legal framework upon which they might be built. Specifically, she examines the legal context in BC for the development of Aboriginal forest tenures and the provincial government's steps to address recent court decisions in an effort to determine what needs to occur legally for the development and implementation of culturally appropriate Aboriginal forest tenures. Background information on the negotiation of treaties and a brief review of the jurisprudence on Aboriginal rights, title and governance allow Passelac-Ross to draw implications from case law for the management of forest lands and resources in the BC context. Specifically, Passelac-Ross discusses the Carrier Sekani Tribal Council's (Fig. 2) efforts to obtain access and regain some measure of control over its forested lands in order to highlight issues of importance and identify options in relation to the development of Aboriginal forest tenures in BC.

Based on research with the Stellaquo First Nation, Eddison Lee-Johnson and Ronald Trosper, in *Designing a New Governance Structure: Analysis of a Stellat'en Model for Implementing Forest Management Devolution in British*

Columbia (Chapter 6), describe a new institutional model of governance that focuses on forested lands and resources, while combining elements of both Aboriginal traditional and contemporary Canadian systems of governance. Driven by the need to transform the current system of government, the Stellaquo governance model is based on the recognition and implementation of Aboriginal rights and sovereignty, and the institutionalization of Stellat'en self-governance within a large Canadian polity. The Stellaquo proposal envisions a co-governance status with Canada and BC in all aspects relating to the management of forestland and resources within Stellat'en territory, and considers forest and land use planning as a governance process interconnected with major aspects of Aboriginal community well-being. The model provides a detailed description of the structural roles and relationships of different arms of government, calls for the institutionalization of informal governance structures and demonstrates that reconciliation between Aboriginal peoples and the Crown is possible under the province's *New Relationship Policy*. The process of transforming First Nation governments must be realized according to First Nations-defined plans, priorities and processes, and be sustainable over a sufficient period of time so as to achieve a constitutionally defined mechanism of devolving authority.

Planning Tools and Considerations for Co-existence

Co-existence just doesn't happen on its own. Strategies, approaches and planning tools need to be considered, designed and implemented to achieve reconciliation and to build firm foundations for co-existence. Section Three (Chapter 7-11) examines several issues that are critical to achieving these goals in the context of land use planning and management. Specifically, this section explores:

- the delineation of Aboriginal planning areas;
- the impacts of roads on Aboriginal communities and their lands;
- the benefits and pitfalls of Aboriginal land use and occupancy studies;
- the ethics and efficacy of Aboriginal values mapping; and
- the development of appropriate criteria and indicators for sustainable forest management in Indian Country.

While these planning considerations and approaches certainly do not exhaust the range of tools that will be required to achieve reconciliation and co-existence, they do at least inform the discourse and point us in the right direction so that we can initiate the conversations we need to have.

Planning boundaries and areas are categorical imperatives in state-sponsored regional land use planning processes. How these concepts resonate with First Nation traditional and contemporary perceptions of territory and space, however, remains unclear. Nathan Deutsch and Iain Davidson-Hunt in *Pikangikum Family Hunting Areas and Trapline: Customary Lands and Aboriginal Land Use Planning in Ontario's Far North* (Chapter 7) describe the processes by which Ontario's registered trapline system articulated with the pre-existing system of family hunting territories, and the approach to land use planning utilized by the Pikangikum First Nation. Their experience leads these authors to query whether Aboriginal peoples can participate in regional land use planning initiatives without being subordinated by government planning processes, and whether the Pikangikum people can maintain their relationship to the land, while being able to negotiate land use boundaries, implement their own forms of management and create new empowering relations with the state. They caution against denying Aboriginal peoples the agency to devise new solutions to meet their contemporary circumstances in ways that maintain continuity between their past and future. In conclusion they suggest that, where First Nations are willing to engage in land use planning and forge new relations with the state, the approach utilized by the Pikangikum First Nation is consistent with a regional planning approach and their own understanding of who should speak for and steward their lands. They note, however, that the northern Ontario case has been marked by cooperative relations between the Ontario Government and the First Nations for decades, including the setting up of the trapline system, which is being used as the basis for land use planning. This kind of cooperation has not always been the case in other jurisdictions.

Land use planning in Canada's northern regions requires careful consideration of the ecological and social impacts of roads and other linear disturbances. In Chapter 8, *Road Rash: Ecological and Social Impacts of Road Networks on Aboriginal Communities,* Kneeshaw and colleagues examine the effects of forest road networks on 'natural' ecosystems and discuss the related effects of road development on remote Aboriginal communities. They address the impact of roads on ecosystem processes and wildlife populations, focusing on game species important to traditional livelihoods of Aboriginal peoples, and elaborate on both the positive and negative effects of roads on Aboriginal communities and their socio-environmental dynamics. Although roads provide access to otherwise inaccessible tracts of land and affect new travel patterns, they have far more complex social and cultural repercussions for remote Aboriginal communities. These effects contrast with the expected benefits of roads (e.g., enhanced mobility and increased access to valued goods and services, and potential economic opportunities outside of isolated communities), and underscore the fact that the impacts of different types of roads are not the same. The objective of this chapter is to identify and discuss the range of impacts that land use and forest planners may expect, and need to consider, with the introduction of road networks into forested environments that are home to many of Canada's remote Aboriginal communities.

Although their origins can be traced to the turn of the 20th century, Aboriginal land use and occupancy studies have only become a component of forest management in the last decade or so. Aboriginal land use studies are now used commonly as tools in land use planning and forest management processes, and are being employed by government and industry as a means to comply with legal obligations to consult with Aboriginal peoples when Crown actions and land use decisions may adversely affect Aboriginal land use rights. As these studies have become more common, they have also attracted criticism, from both forest managers who are concerned about getting accurate and scale-relevant information for effective planning, and from Aboriginal peoples who contend that such studies are incapable of capturing and conveying the complex and dynamic relationships they have with the land. The end result is misrepresentation and devaluation of the cultural significance of traditional use. In *Aboriginal Land Use Mapping: What Have We Learned From 30 Years of Experience?* (Chapter 9), Stephen Wyatt *et al.* consider some of the challenges and issues associated with Aboriginal land use studies, and the ways in which Aboriginal land use information is being used in forest management and land use planning in Canada.

Aboriginal participation in environmental decision-making is increasingly being recognized as vital to environmental and cultural sustainability in Canada's forests. In Ontario, the Ontario Ministry of Natural Resources has assumed a lead role in enhancing Aboriginal participation in forest management, whereby such involvement now constitutes a significant component of Ontario's forest management planning system. The key method used to achieve Aboriginal participation in Ontario forest management is through mapping Aboriginal traditional knowledge and values via a process known as 'Aboriginal Values Mapping.' Based on interviews conducted with representatives from the provincial government, industry and First Nations, Deborah McGregor's chapter, *Aboriginal Values Mapping in Ontario's Forest Management Planning Process* (Chapter 10), questions the ethics and efficacy of such projects. While the results indicate ongoing Aboriginal dissatisfaction with the process, positive lessons were learned. The Aboriginal concept of 'co-existence' is suggested as a framework for improving Aboriginal/non-Aboriginal relations, and as a necessary aspect of achieving greater sustainability in Ontario forest management. The author concludes with a list that characterises the most successful of such cooperative undertakings to date.

Criteria and Indicator (C&I) frameworks for sustainable forest management have been developed, both nationally and internationally, to guide forest planning and forestry operations. Peggy Smith, Erin Symington and Sarah Allen's *First Nations' Criteria and Indicators of Sustainable Forest Management: A Review* (Chapter 11) questions whether such frameworks are sufficient to accommodate First Nation needs, rights and interests, while examining the processes involved in the development of six First Nation's C&I frameworks: the Nuu-chah-nulth Nation's Iisaak Forest Resources Ltd. (British Columbia), Tl'azt'en Nation (British Columbia), Little Red River Cree Nation (Alberta), the Algonquins of Barriere Lake (Quebec), Waswanipi Cree Nation (Quebec) and the Innu Nation (Labrador). They ask how local-level First Nation C&I frameworks are applied, and whether they are an effective tool for First Nations to protect their values in forest development and land use planning. These authors conclude that Aboriginal C&I development has been an important tool for formalizing and introducing Aboriginal values and worldviews into a forest management system that is only beginning to understand and implement the sustainable development paradigm. Further, they propose a framework for developing, understanding and analyzing Aboriginal C&I sets, which may also have relevance to other land use planning processes and initiatives.

Capacities for Co-existence

True reconciliation and co-existence cannot be achieved in the absence of building the capacities of both parties to visualize and realize these outcomes. The final two chapters in this volume argue that current 'top down' approaches to Aboriginal capacity building must be met in equal or greater measure with 'bottom-up' approaches, which put Aboriginal peoples at the centre of determining and realizing their capacity needs; then, only then, will reconciliation and co-existence be possible.

Capacity building has become a rallying call for government, industry as well as Aboriginal peoples, communities and nations in the context of natural resource development. Much of the time, money and energy invested in 'Aboriginal capacity building projects' has focused on enhancing the capacities of Aboriginal peoples to participate in economic opportunities and institutionalized government frameworks and processes relating to natural resource extraction and management. While admirable, such initiatives too often overlook the need for Aboriginal peoples to sustain and build capacities rooted in their cultures, collective identities and fundamental connections and stewardship responsibilities to the land. In other words, most capacity building efforts ignore the need to maintain and enhance those values, activities and ways of life that are integral to who Aboriginal peoples are, and give rise to the protection of their rights under the *Canadian Constitution Act*.

In *Capacity for What? Aboriginal Capacity and Canada's Natural Resource Development and Management Sectors* (Chapter 12), Marc Stevenson and Pamela Perreault explore Aboriginal capacity building in the natural resource development and management sectors from multiple perspectives, scales and dimensions. Theoretically grounded in a growing body of literature on Aboriginal and indigenous empowerment, where local communities drive the design and delivery of capacity building programs, their approach builds on and situates existing Aboriginal capacity building initiatives within a conceptual framework that, if new institutions, arrangements, resourcing and political will are found, should facilitate Aboriginal peoples aspirations to become architects of their future and true partners in Confederation. Essentially, what is being advocated by these authors is greater attention to building the capacities of Aboriginal peoples in order to develop a firm foundation in both worlds, and to navigate successfully between them—a tall order indeed, but one that can no longer be ignored.

In *Capacity Building Moose Cree Style: Moose Cree Strategies for Becoming a Goose Hunter* (Chapter 13), Brent Kuefler, Adrian Tanner and David Natcher demonstrate that, for the Moose Cree, capacity building most importantly involves the learned ability to be Moose Cree, i.e., to embrace and advance the values, needs, rights and interests of the Moose Cree people. For the Moose Cree First Nation, enhancing traditional land-based skills, knowledge, and the social values and obligations that attend them, and by which they are constituted, can no longer be dismissed at the expense of building the skills and acumen of its individual members to participate in what the outside world has brought to them. Both are required. Indeed, it might even be argued that without the former, the abilities of Aboriginal peoples to successfully navigate the complexities of the 'modern' world are diminished, set adrift, untethered to a solid anchor. We can think of no better chapter to end this volume than one that turns its attention to building the capacity of a First Nation's member to survive on the land and within a social context that necessarily involves obligations and responsibilities to the environment and the larger community upon which they depend. Only by building capacities in both worlds will reconciliation and co-existence, and survival of Aboriginal cultures in Canada, be realistic and achievable goals.

References Cited

Borrini-Feyerabend, G., M. Pimbert, M.T. Farvar, A. Kothari and Y. Renard (2004). *Sharing Power. Learning by Doing in Co-management of Natural Resources Throughout the World.* IIED and IUCN/ CEESP/ CMWG, Cenesta, Tehran.

Chase, A.K. (1990). Anthropology and impact Assessment: Development pressures and Indigenous interests in Australia. *Environmental Impact Assessment Review* 10(1/2): 11-25.

Declaration on the Rights of Indigenous Peoples, 2007. http://www.iwgia.org/sw248.asp

Escobar, A. (1992). 'Planning,' pp. 132-145 in Wolfgang Sachs (ed.), *The Development Dictionary.* London, UK: Zed Press.

Government of Canada (1996). Report of the Royal Commission on Aboriginal Peoples (five volumes). http://www.collectionscanada.gc.ca/webarchives/20071115053257/http://www.ainc-inac.gc.ca/ch/rcap/sg/sgmm_e.html

Hirro, P. G. and A. Surralles (2005). 'Introduction,' pp. 8-21 in *The Land Within: Indigenous Territory and the Perception of Environment.* Copenhagen: International Work Group for Indigenous Affairs.

Howitt, R. (2001). *Rethinking Resource Management: Justice, Sustainability, and Indigenous Peoples.* Routledge Press, New York.

Lane, M.B. (2006). The role of planning in achieving indigenous land justice and community goals. *Land Use Policy* 23: 385-394.

Langton, M. (2004). 'Introduction: Unsettling Sovereignties,' pp. 29-33 in Langton, M., M. Tehan, L. Palmer and K. Shain, eds., *Honour Among Nations? Treaties and Agreements with Indigenous People.* Carlton, Victoria: Melbourne University Press.

Natcher, D.C. (2001). Land use research and the duty to consult: A misrepresentation of the Aboriginal landscape. *Land Use Policy* 18(2): 113-122.

O'Fairchaellaigh, C. (2004). 'Evaluating agreements between Indigenous peoples and resource developers,' pp. 303-328 in Langton, M., M. Tehan, L. Palmer and K. Shain, eds., *Honour Among Nations? Treaties and Agreements with Indigenous Peoples.* Carlton, Victoria: Melbourne University Press.

Saul, J.R. (2008). *A Fair Country: Telling Truths about Canada.* Penguin, Toronto.

Stevenson, M.G. and D.C. Natcher, eds. (2009). *Changing the 'Culture' of Forestry in Canada: Building Effective Institutions for Aboriginal Engagement in Forest Management.* Edmonton, Alberta: Canadian Circumpolar Institute Press.

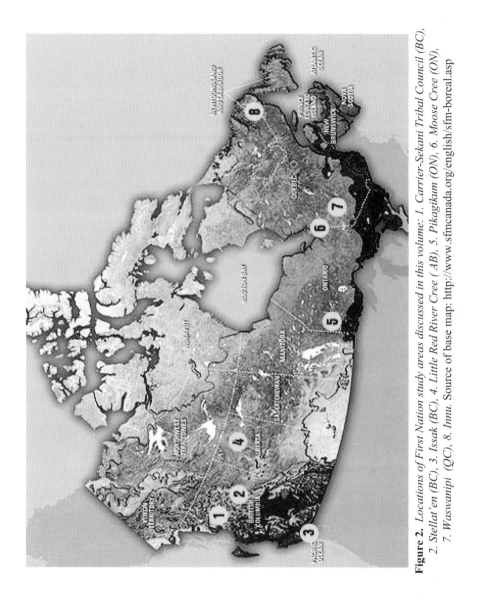

Figure 2. *Locations of First Nation study areas discussed in this volume: 1. Carrier-Sekani Tribal Council (BC), 2. Stellat'en (BC), 3. Issak (BC), 4. Little Red River Cree (AB), 5. Pikagikum (ON), 6. Moose Cree (ON), 7. Waswanipi (QC), 8. Innu. Source of base map:* http://www.sfmcanada.org/english/sfm-boreal.asp

SECTION 1:
CO-EXISTENCE 2010

Chapter One

Alberta Government Policy, First Nations Realities: Barriers to Land Use Policy and Cumulative Impact Assessment

Jim S. Frideres and Cash Rowe

Introduction

Over the past three decades, provincial and territorial governments and resource developers have focused their attention on the North in their quest for new economic opportunities (Passelac-Ross and Potes 2007; Sallenave 1994). During this time, the expansion of fossil fuel extraction, mining and forestry in the Treaty 8 region of Alberta has proceeded unabated and at a rapid rate. Moreover, the Alberta Government intends to accelerate the development of natural resources in the Treaty 8 region in the coming years (McGuigan 2006; Ross 2003). At the same time, it is clear that First Nations communities in the Treaty 8 region have not been the beneficiary of resource developments (Stevenson and Perreault 2008). Not only has natural resource extraction in and around their traditional lands not improved the quality of their lives, it would seem that, for many First Nations peoples, it has had a negative impact on their quality of life. As such, First Nation governments have raised concerns about the pace and scope of development, the nature of First Nation involvement, as well as the flow of benefits from these development projects. These concerns have been interpreted by the private sector as attempts to 'stop' development in the region and have been downplayed, if not ignored, by government. Given that the potential for further natural resource development in the Treaty 8 region (Fig. 1), and northeastern Alberta in particular, is significant, there is some recognition by governments and the private sector that First Nations involvement, albeit not consent, will be required if these future developments are to go forward.

However, this realization has not been translated into any substantive policy changes that would directly address the concerns of First Nations peoples. Only after the courts rendered their decisions in *Delgammukw, Taku River, Haida,*[1] and other Aboriginal rights cases was Alberta forced to take into consideration First Nation concerns through the development of a specific consultation policy, known as the *Government of Alberta's First Nations Consultation Policy on Land Management and Resource Development* (Government of Alberta 2005).[2] However, the Assembly of Treaty Chiefs of

---

[1] *Delgamuukw v. British Columbia,* [1997] 3 S.C.R. 1010. *Taku River Tlingit First Nation v. British Columbia* [2004] 3 S.C.R. 550, 2004 SCC 74. *Haida Nation v. British Columbia (Minister of Forests),* 2004 SCC 73, [2004] 3 S.C.R. 511.

[2] http://www.aboriginal.alberta.ca/images/Policy_APROVED_-_May_16.pdf

Alberta has unanimously rejected Alberta's First Nation Consultation Policy (*see* Stevenson this volume). In the end, the Government of Alberta remains mired in an ethos that suggests that First Nations peoples do not have the right to impose conditions on development, stop development (even temporarily) or have the capacity to participate in such developments once projects have been approved.

Figure 1. *Typical aerial view of northern Alberta with multiple disturbances (roads, seismic lines, clear cutting).*

Moreover, it is clear that the interests of the province and the private sector do not always align with those of First Nations with regard to resource development in the Treaty 8 region. For example, government and the private sector have a foremost interest in forestry, oil and gas extraction, pipeline and construction, mining and agricultural projects for obvious financial reasons. While it is true that First Nations peoples also have an economic interest in these activities, they are also concerned with issues such as hunting, trapping, fishing, ecological impacts, health, Treaty and Aboriginal Rights, cultural and environmental sustainability, and spiritual practices. While there is some overlap in the interests of these parties, it is evident that their priorities are not well aligned. The end result is considerable tension and disagreements as to the nature and pace of resource development in the Treaty 8 region. This suggests that issues defined as important by First Nation communities may not garner attention by or support from the provincial government.

Notwithstanding the constitutionally protected rights of Alberta's First Nations and Court pronouncements that the Crown has the duty to consult with First Nations, Alberta has argued against these decisions, and continues to undertake unilateral actions that seem contrary to Supreme Court findings (Potes *et al.* 2006). Conflicts between the province and First Nations cost money and time, and the provincial government, in comparison with First Nations, has a full supply of both. Moreover, even if the courts rule against the actions of the provincial government, its liability costs are postponed for several years as the case winds its way through the courts. Meanwhile, the development has gone forward and only 'compensation' for damages incurred is awarded, while First Nations negotiate impacts and benefits agreements with industrial proponents. This ethos is particularly well-articulated in Alberta where the private sector has had 'sway' over natural resource development decisions for over fifty years and government officials have been complicit in allowing business to set the agenda as well as policy for natural resource development in the province (Kennett 2007; Wenig and Poschwatta 2008).

Methods and Data Collection

In gathering information for this paper, a dual data collection procedure was utilized. First, we reviewed existing documentation related to the topic. Having completed that task, we then began to interview government officials (federal, provincial) and private sector and First Nations leaders to more fully understand the contents of the documents related to land use and cumulative impact as well as to why certain decisions were made. Individuals responded to our request for an interview—some by phone, some by internet and some face-to-face. In addition, retired individuals from various departments and the private sector were interviewed. The interviews were semi-structured and ranged in scope depending upon individual experience and knowledge of the subject. Our goal was to understand more fully the actions taken by the stakeholders with regard to land use, economic development and First Nations relations in the Treaty 8 region.

To better understand the rights and obligations of the government, we begin by providing a brief description of the Treaty 8 region and then by outlining the decision boundaries of both the provincial and federal governments with regard to resource development. This is followed by an analysis of the actions taken by the provincial government over the past three decades, the gaps in these decisions, reactions by First Nations and the policies currently being proposed by the government.

The Boreal Region (Treaty 8 Area)

The Boreal forest region of Canada, which includes the Treaty 8 area, is sparsely populated with less than ten percent of the Canadian population residing within this area. At the same time, it is estimated that nearly half of the Aboriginal population of Canada live in the Boreal region. Overall, the National Aboriginal Forestry Association claims that more than 80 percent of the First Nations in Canada live in productive forest areas. Major oil and gas development also occurs in this region of Canada, and, in northern Alberta we find that it contributes nearly half of the conventional oil production and over one third of the gas produced in Alberta (Ross 2003). In addition, oil sands developments are found in Cold Lake, Peace River and Fort McMurray with a supporting infrastructure of roads, pipelines and power corridors traversing the Treaty 8 region. Even though much of the area is covered by treaty, the provincial and federal governments have taken very different views of the terms of the treaty and what rights it confers on each party.

The extensive development that has taken place in the Boreal forest across Canada has impacted both the social and environmental circumstances of Aboriginal peoples. Moreover, the many developments in the Boreal region fall under multiple levels of government jurisdiction and an array of planning processes. Impacts have ranged from a sudden influx of workers into isolated Aboriginal communities to the destruction of wildlife habitat to air pollution to the cultural deterioration of Aboriginal communities. With regard to northern Alberta, Huff (1999) and Geertsema (2008) have identified the detrimental impacts of oil and gas exploration on the environment and social structure of First Nations that reside in the area.

Jurisdictional Boundaries

Alberta Provincial Government
We begin by noting that natural resource development by the provinces has been granted broad authority. The *Natural Resources Transfer Agreement* (1930) gave provincial jurisdictional control over the lands and resources within its boundaries. Moreover, a quick review of existing legislation reveals that there are many different components of the *Canadian Constitution* (1982) that allow the province unfettered access and control to developing Crown lands. Sections 92(5), 92(A), 92(10) and 92(13), for example, grant the provinces other powers in relation to property and civil rights, local works and undertakings and all matters of a merely local or private nature in the province. These provisions allow the provinces to specify the nature and pace of resource development within their provincial boundaries. Provincial governments have interpreted the law as giving them sole and uncontested rights to develop Crown lands without having to engage in consultations with other parties.

Federal Government
At the national level, there are a number of constitutional provisions that engage the federal Crown in natural resource development in Canada. Subsection 91(2) of the *Canadian Constitution* (1982) gives the federal government exclusive jurisdiction over the regulation of trade and commerce, inter-provincial and international trade, including the export of crude oil from the provinces (Vlavianos 2007a). It also has other regulatory powers with regard to resource development within provinces, e.g., regulating emissions of toxic substances and legislating on matters that are of national concern such that one province cannot adequately address issues such as water pollution and greenhouse gas emissions. Other federal pieces of legislation, e.g., the federal *Fisheries Act, Navigable Waters Protection Act, Canadian Environmental Protection Act* (1999), *Canadian Environmental Assessment Act* also limit activities by provincial governments. However, Vlavianos (2007b) notes that it is difficult to "demystify the role of the federal government" in resource development because of uncertainties regarding the roles and responsibilities of federal departments

themselves. Some can compel action while others can only 'oversee' and still others can only play a consultative role. It would seem that no federal department has well-delineated boundaries that specify where their powers end and another's begins. As such, federal agencies have been ineffective in developing or influencing policy on natural resource development throughout most of Canada. Even though the federal government is, under the *Canadian Constitution* (1982), responsible for 'Indians and Indian lands,' it is clear that there continues to be uncertainty as to where federal responsibility ends and provincial responsibility begins.

Provincial Land Use Policies
The early basis for decision-making regarding resource development in Alberta resulted in a policy that Low (2009b) characterizes as one of 'neglect.' Later the *Energy Resources Conservation Board and the Alberta Energy Utilities Board* became the regulatory bodies to assess the impacts of natural resource development on land, water and air, and to approve resource development projects. As early as the 1970s the Environment Council of Alberta produced the *Environmental Effects of Forestry Operations in Alberta: Report and Recommendations* that focused on the implications of resource development and land use conflicts and regional cumulative effects. A quarter of a century later, the Government of Alberta published a document *Ensuring Prosperity—Implementing Sustainable Development* (1995) that acknowledged the need for land and resource management. "Despite an integrated resource planning process being put into place in the early 1970s it was soon eviscerated by budget cuts and the emergence of the 'anti-regulation and anti-planning' ideology within the provincial government" (Kennett 2007). In the end, these factors stalled the implementation of a balanced land use policy and the development of a cumulative impact assessment requirement. Moreover, as increasing resource development of Crown lands in the Treaty 8 region continued during the 1980s and 1990s, there was no attempt to establish effective means to address resource-use conflicts. As Pratt and Urquhart (1994) point out, although some administrative integration was attempted, it was not sufficient and subsequent changes in the legislation, planning and regulatory regimes of forestry, oil and gas operations made it impossible to manage. The result was that by the late 1990s, land use conflicts were common throughout the province and no vision for finding a solution could be found. Perhaps this was due to the fact that, for the most part, government officials did not see the issues as a 'problem' (Kennett 2002).

In the mean time, the province continued to argue that it was following an integrated resource plan (AFLW 1991). The policies of such a plan were supported in various policy statements such as the *Alberta Public Lands* document in 1988. This document and others steadfastly argued that integrated resource management was the fundamental assumption underlying Alberta's policy for resource development. The main consideration of such a policy was the maximization of economic benefits to Albertans along with careful land and resource management. In addition, the policy supported community consultation and consideration of future needs of Albertans. In reality, the integrated resource planning process was never provided legal status, thus was never fully implemented by the provincial government. Used as a political tool, it enabled resource development in the Treaty 8 area to proceed despite the objections of most First Nations.

Notwithstanding these objections, Alberta argued it was supporting development initiatives for First Nations in the region through their signing the *National Forest Strategy* (Government of Canada 1998a) and *Canada Forest Accord* (Government of Canada 1998b), both of which make provisions for Aboriginal and Treaty rights. The *Canada Forest Accord* (Government of Canada 1998b:23), for example, recognizes and makes provision for:

> ... Aboriginal and treaty rights ensuring the involvement of Aboriginals in forest management and decision making, consistent with these rights, supporting the pursuit of both traditional and modern economic development activities and achieving sustainable forest management on Indian Reserve Lands.

One year later the government issued a policy entitled *Alberta's Commitment to Sustainable Resource and Environmental Management* (1999). The general thrust of this document focused on the need for integrated regional planning and other laws, policies and regulations reflective of the principles of integrated resource management in the province—a task that fell to the Department of Alberta Environment through the development of a number of policy papers. However, resistance by other government departments mandated to pursue resource development, inadequate support from senior officials, and the lack of political leadership thwarted the creation of any broad encompassing policy. By 2007, as Kennett (2007) points out, no real 'transformation' of the Government of Alberta's resource and land management policy had occurred.

The Government of Alberta released *Strengthening Relationships: The Government of Alberta's Aboriginal Policy Framework* in 2000 (Government of Alberta 2000). The major focus of this policy document is on how to integrate Aboriginal peoples into the industrial economy, and with only peripheral consideration given to First Nation Treaty rights and little acknowledgment of the deterioration of Aboriginal livelihoods and traditional use activities where extensive development is taking place. While this policy document accepts the Treaty rights of First Nation peoples in the province to hunt, trap and fish on lands not 'taken up' by the province, it takes the position that ever since treaties were signed, only the provincial Crown has the legal right of ownership and management of provincial lands and resources. Again, this document acknowledges the need to support 'traditional land use' studies with regard to Aboriginal peoples, but evidence from provincial officials indicate that little has been accomplished in the past nine years on this front. Moreover, there is considerable confusion as to the value and use of Traditional Land Use documents. In interviews conducted by the lead author, senior government officials were unclear as to how these documents could be used in the resource development planning process, if at all.

The Integrated Resource Management Division's proposed framework for regional resource and land management was never pursued, and by 2001, the division was closed with only the Regional Sustainable Development Strategy for the Athabasca Oil Sands policy remaining. Out of this development, the Cumulative Environmental Management Association was developed. Unfortunately, this partnership was not organized properly, essentially ignored the input of First Nation communities (knowledge and process), lacked effective decision-making procedures and treated First Nations as just one more 'stakeholder' in the process of decision-making. In particular, it refused to accept Aboriginal traditional knowledge as a legitimate source of knowledge to be used in decision-making (Tanner 2008). From the perspective of Aboriginal groups, the structure of this committee undervalued traditional knowledge and failed to recognize the broader political, social, cultural, economic and spiritual contexts which needed to be taken into consideration when making decisions on development proposals (Stevenson 2006; Tanner 2008). As a result, the committee has made only one major decision in its life and most First Nations organizations have withdrawn their membership (Tanner 2008).

In 2006, the Alberta government implicitly acknowledged that a vision and strategy for resource development in Alberta was needed and proceeded to conduct a public consultation process in the development of a new policy on sustainable resource and environmental management. The result of this 'consultation' has been the recently produced *Land Use Framework* (Government of Alberta 2008) that outlines how the provincial government will address economic growth and cumulative effects in the province. It is clear that the proposed framework seeks to manage economic growth, not slow or stop it (however temporarily), providing underlying support for Alberta's philosophy that any growth is good. However, two elements in this document are new and relevant to the issue at hand. First of all, the government acknowledges that the issue of 'cumulative impacts' must be more systematically dealt with, and secondly, it recognizes a need for consultation with First Nations with regard to resource development projects (*see* Stevenson, Chapter Two this volume).

The second new aspect of this framework is the inclusion of Aboriginal peoples in land use planning. The new policy will "strive for a meaningful balance that respects the constitutionally protected rights of Aboriginal communities and the interests of all Albertans," while "encouraging the participation of Aboriginal peoples in land use planning" (Government of Alberta 2008). This document is the latest expression of what the province believes its responsibilities are to Aboriginal peoples with regard to natural resource development. When the policy was made public, Aboriginal peoples across the province rejected the proposed framework. The Alberta Government, however, has effectively ignored their objections and concerns, and is moving forward to implement the policy. Whether the *Land Use Framework* represents more of the same, or is a harbinger of a new relationship between the province and its First Nations peoples is considered at length in Stevenson (Chapter Two, this volume).

Environmental and Social Impact Assessment

The requirement of an environmental impact assessment emerged in the early 1970s out of the realization of the inadequacy of the *Coal Policy* to make informed decisions about the impacts of resource developments. Until the 1970s, provincial conservation boards took a very cooperative and consensual approach to their mandate, spending considerable time and effort working with industry and developing a collaborative approach that is still evident today (Low 2009a). Under the Energy Resources Conservation Board model, the Government of Alberta determined that maximizing production and encouraging investment in all resource development projects was in the public interest and that became the basis for approving proposals. The overall policy of the province was to 'orderly develop Alberta's oil and gas resources' and it would not be until the late 20th and early 21st century that demands from environmental protection and public interest groups began to challenge the province's position.

Generally speaking, energy and utility development in Alberta are subject to review by the Alberta Energy and Utilities Board, while natural resource development projects are reviewed by the National Resources Conservation Board.[3] Environmental impacts assessments provide relevant bodies with a description of potential positive and negative environmental, social, economic and cultural impacts of a proposed activity, *including cumulative regional, temporal and spatial considerations* (emphasis ours). However, there is no prescription as to how the proponent is required to address cumulative impacts. As such, proponents are at liberty to select the approach and methodology for assessing cumulative impacts. Moreover, there are no rules as to how far in the past or the future cumulative impacts assessments should extend. A similar absence of guidance with regard to temporal and spatial considerations can be noted.

Cumulative Impact Assessments

Cumulative impact is defined as an effect on the environment that has an incremental impact on the action when added to other past, present and reasonably foreseeable future actions (Canter 1999; Hegmann *et al.* 1999; Kennett 1999; McGuigan 2006). These effects may be small, short-term or extend over sizeable geographic areas and entail significant time lags. Generally, cumulative impacts are caused by a large number of small incremental changes, that individually may have little or no major impact on the environment, community or people. In short, cumulative impacts reveal problems created by a combination of individually minor impacts of multiple actions over time (Council on Environmental Quality 1997). Moreover, cumulative impacts may be additive, multiplicative, synergistic or antagonistic. As noted by the United States Council on Environmental Quality (1997:1, 7):

---

[3] An Environmental Impact Assessment report is required for certain energy, utility and natural resource developments as defined by the *Alberta Environmental Protection and Enhancement Act*.

...evidence is increasing that the most devastating environmental effects may not result from the direct effects of a particular action, but from the combination of individually minor effects of multiple actions over time....the impact on the environment which results from the incremental impact of the action when added to other past, present and reasonably foreseeable future actions regardless of what agency or person undertakes such actions. Cumulative impacts can result from individually minor but collectively significant actions taking place over a period of time.

Since the 1990s project evaluators working in the United States have required that all direct, indirect and cumulative impacts of proposed developments be assessed and disclosed, e.g., both the *National Environmental Policy Act* and the *California Environmental Equality Act* require cumulative impacts be assessed with regard to developments (Ma *et al.* 2009).

Environmental assessment legislation, such as the *Environmental Protection and Enhancement Act* (1992) was enacted in Alberta during the 1990s, although social scientists had admonished the provincial government two decades earlier that cumulative impacts should be noted in natural resource development projects.[4] The above *Act* has requirements that proponents of development projects consider cumulative effects. At the same time, those government agencies that were to investigate cumulative impacts were subjected to major budget cuts and retained little capacity to conduct baseline studies or to collect data with regard to cumulative impacts of existing resource development projects.

In addition, the private sector's resistance to 'planning' ensured that officials in the government would not re-implement the Integrated Resource Planning Program. Finally, the attempt to deal with cumulative impacts within a legal and bureaucratic process built for specific projects on a 'one-off' basis means that the implementation of such a policy would not be successful. For example, the traditional environmental assessment process is highly restrictive in terms of spatial and temporal parameters of a specific project. Not only have social science and First Nations communities' voices been silenced by government, but so too has the voice of decision-making bodies such as the Energy Resources Conservation Board.[5] Notwithstanding the recent introduction of the *Alberta Land Stewardship Act* (2009), which anchors the *Land Use Framework* (yet remains to be implemented), the policy of the Alberta Government with regard to land use planning or cumulative impacts assessment has been one of minimal interference. As a result, project proponents are the 'drivers' of the impacts assessment, and because of the complexity of the

---

[4] For a more detailed chronology of Canadian environmental impact assessments, the reader should consult Hanna (2005).

[5] The Alberta Energy Utility Board was created in 1995 when the provincial government merged the Public Utilities Board and the Energy Resources Conservation Board. Then, in 2007, it eliminated the Alberta Energy Utilities Board and created the Alberta Utilities Commission and a 'new' Energy Resources Conservation Board (Low 2009a).

cumulative impact assessment process, this conflict of interest has gone unchallenged, thereby leaving proponents free to proceed with project development without addressing the issue of cumulative impacts.

The outdated process of evaluating one project, one at a time, with no consideration of cumulative or interactive effects is no longer sustainable or socially acceptable. Decisions with regard to resource development and management need to fit within one policy that coordinates each of these processes. In the end, what few policies Alberta has developed are inconsistent, lacking in specifics and fail to balance resource development with social, cultural and environmental protection. Even the Energy Utilities Board has commented on the fact that it has had to make decisions without being able to assess in a proper manner proposed developments since there is no regional development strategy that includes cumulative effects limits and thresholds. As Neil McCrank, Chairman of the Alberta *Energy Utilities Board* noted in a speech in Calgary, Alberta on 14 March 2007:

> We should look at the possibility of regional hearings...where we examine the broader issues, the broader societal and environmental issues... We would hope that in the course of this kind of approach there would be standards developed that would look at the cumulative impact.

Today, as in the past, cumulative effects resulting from natural resource developments are noted, but not managed in certain areas of the country. On the other hand, in the Northwest Territories, Yukon and Nunavut, cumulative impacts are part of the overall Environmental Impact Assessment legislation. As such, the *Mackenzie Valley Environmental Impact Review Board*, an administrative tribunal in the Northwest Territories, has required and evaluates cumulative impact assessments as part of their mandate (Ehrlich and Sian 2002). Under provincial/federal regulations, environmental assessments are intended to become the basis by which cumulative impacts are addressed. In reality, they are difficult to deal with as legislation for conducting environmental impact assessment is limited to specific projects. It is only because of litigation by Aboriginal peoples and increased public attention by environmental advocacy groups that governments have begun to think about dealing with cumulative impacts when evaluating the impacts of resource developments. In the Boreal region of Canada, where few people live (and most are Aboriginal), the issue of cumulative impacts has been further marginalized. On the other hand, First Nations peoples have argued for years that the cumulative impacts of the numerous developments in the Treaty 8 region have an impact on their communities and the exercise of their Treaty rights; many can no longer hunt, fish or trap as was guaranteed to them under treaty.

For example, the Treaty 8 Tribal Association of Alberta had, for many years, pursued a cooperative approach with government to address its concerns with regard to land and resource development. However, because of the scope, scale and pace of resource development in the Treaty 8 region, the cumulative impacts of all such developments, and the unwillingness of the provincial government to address these concerns, it has petitioned Canada and Alberta to undertake a comprehensive and collaborative multi-party assessment of the cumulative environmental, economic, health and social impacts linked to resource developments in the Treaty 8 region. This request follows decisions of the Supreme Court with regard to the efficacy of integrated land use planning processes and environmental assessment processes for conduct of meaningful consultation with First Nation peoples. Today, over three years later, no official response has yet been received, and it remains to be seen whether the Alberta's *Land Use Framework* will fulfill Alberta's treaty and fiduciary obligations in these regards.

Barriers to Cumulative Impact Assessments
Nearly 20 years ago, New Zealand introduced the condition of 'cumulative impact' as part of its environmental impact assessment legislation, while the United States introduced its cumulative impact policy in the early 1990s. In Canada, the requirement of cumulative impacts are included in the investigation phase of the environmental impact assessment process, although there seems to be insufficient descriptions of what cumulative impacts are (Warnback and Hilding-Rydevik 2009; Lawrence 2007a, 2007b). However, as Duinker and Greig (2006:153) noted "Cumulative effects assessment in Canada … has not lived up to its glowing promises of helping to achieve sustainability of diverse valued ecosystem components…." On the other hand, Baxter *et al.* (2000) argue that, in some cases, assessing the cumulative impacts of major resource developments is now carried out as part of the environmental impact assessment. Smith (2006) reviewed court decisions regarding challenges to the adequacy of federal agency cumulative impact analyses and found that nearly 75% of the cases during 2001-2004 were deemed inadequate by the courts. All agree that cumulative impacts assessments and their methodologies are limited.

We live in many different systems simultaneously, e.g., social, ecological and technological. Moreover, each of these systems has many inter- and intra-related components. Adding to the complexity, both internal and external events can impact the 'balance' of the system(s) and their resilience (i.e., the ability to absorb disturbance or shocks and reorganize while undergoing change) may differ in time and setting. Governments have been hesitant in developing comprehensive development strategies because the more holistic and inclusive a strategy becomes, the greater the complexity and magnitude of systems involved. In addition, because of this complexity, issues of cause and effect will emerge (Wenig and Poschwatta 2008). And, in an attempt to unravel these complexities, new political will, new policies, new institutions, new decision-making procedures and new expertise will have to be dedicated to addressing these issues and developing a new paradigm for assessing the acceptability of proposed natural resource development projects.

Decision making will become fettered with a variety of standards and guidelines, e.g., environmental, social, ethical, and informed by a broader understanding of the impacts of past, present and future projects. Thus, additional resources and focus will have to be given to these issues by both the proponent of a project as well as provincial government agencies (e.g., Environment, Energy, Sustainable Resource Development). All this also means that the time required for an agency, e.g., *Energy Resources Conservation*

Board, to make decisions may well exceed the timelines of the private sector. Hence, there is a reluctance of government to embark on a new way of thinking about and assessing resource development impacts, even as new legislation has been introduced to facilitate these outcomes. As such, no agency has developed a comprehensive plan or a decision-making process based on this new paradigm. Nevertheless, scholars and policy analysts alike still argue, as they did two decades ago, that there are severe limits in taking a reductionist and mechanistic approach to addressing the relevant concerns in natural resource development projects. The complexity of society does not mirror the *status quo* (one event at a time) approach and the failure to replace policies that reflect this kind of thinking will mean that basic social and environmental problems confronting us will not be resolved anytime soon.

Few applied integrated resource and environmental management policies have been introduced in Alberta that have persisted for an extended time period. However, we found that, while provincial government officials interviewed acknowledged a commitment to such a strategy, enactment and application have yet to be delivered. For example, the provincial government published *Alberta's Commitment to Sustainable Resources and Environmental Management* (Government of Alberta 1999) a decade ago, but since its introduction it has been unwilling to develop a comprehensive province-wide land use plan. While plans were developed for the Fort McMurray–Athabasca oil sands region (Government of Alberta 1996b) and the Cold Lake region (Government of Government Alberta 1996a), these plans are outdated, fail to provide adequate guidance for development in the regions, lack legal enforceability and are subject to change by the government as it sees fit. For example, in 2002, the Alberta Government amended the *Fort McMurray Integrated Regional Plan* to allow for oil sands development in a previously protected wetland complex. Even the *Energy Utilities Board* noted that the plan had no legal status and was subject to revisions or review at the discretion of the Minister. In summary, control of natural resource development in the Treaty 8 region and the management of cumulative impacts of those developments appear not to be a government priority. Moreover, even if policies are developed, the issue of maintenance and enforcement come to the foreground.

It is not currently required that cumulative impact assessments are included in project proposals because if the specific project is considered to have little direct impact, it is quickly approved without any serious consideration of cumulative impacts (e.g., the request to drill two exploration wells within a region would be seen as having a negligible impact, even though over time and space their footprints may combine to produce impacts greater than might be anticipated by the sum of the two projects). As Kennett (1999) points out, "the management of activities that are individually insignificant on a landscape or social system is problematic even though they can nonetheless produce significant cumulative effects (e.g., 'the tyranny of small decisions')." These impacts, 'small' as they may be, can be cumulative and produce major impacts in the future. For example, the individual impacts of a small scale forestry operation, a short road, a seismic line, an electric power line and a pipeline individually may produce minimal impacts. However, viewed holistically and incrementally, they may have a great impact on the environment, both social and physical. The government has taken the stance that if specific projects are minor or 'insignificant' in terms of impact (direct and/or immediate), they would not warrant the use of an expensive, time consuming project review process.

Under the Federal Canadian Environmental Assessment Process, cumulative impacts assessments are required for developments that occur over a large region that may include cross-jurisdictional boundaries. Moreover, the need for defining a 'baseline' of impacts that encompasses both time and space is required. Unfortunately, the provincial government does not have such requirements and the federal government has not attempted to rectify this deficiency. As a result, the practice of using 'current' conditions as 'baselines' negates the effects of past actions (Stakhiv 1988; Pauly 1995).

In our interviews with government officials and individuals from the private sector, all exhibited a rudimentary understanding of what a 'cumulative impact assessment' was and what it entailed. They also agreed that it was an important issue and needed to be addressed. At the same time, few suggested that 'interaction' effects of single impacts could produce unanticipated and/or substantively different impacts than would be expected from simply 'adding' the impacts of individual projects. Nearly all of our respondents noted that they 'followed what is required in the legislation' and few could provide examples of recent reports that actually investigated cumulative impact assessments. They also noted that the issue was new and thus had not been subjected to intense discussion or dialogue, nor was there a body of knowledge available for them to review on the subject. Both the private sector and consultants noted that government officials did not really 'require' cumulative impact assessments. They also noted the issue of 'responsibility' has to be resolved before any action could be taken on their part.

Government officials observed that cumulative impact statements were required by legislation. However, they also noted that they felt constrained in enforcing these requirements because of project time constraints and trepidation about asking proponents for additional information—information that may turn out to be expensive to acquire or unnecessary. They also were fearful about being labelled 'the bureaucracy' and interfering with responsible development, thus incurring the wrath of superiors and being accused of simply lacking the necessary knowledge to evaluate any cumulative impacts that the proponent might set forth in their Environmental Impact Assessment report.

While researchers and policy makers agreed on the need for cumulative impact assessments, arguments for not undertaking them include: (1) they don't know enough about many causal relationships; (2) aggregate influences on natural systems are difficult to predict with accuracy; (3) temporal and spatial considerations are unclear; and, (4) there are constraints imposed by the current environmental impact assessments.

Others concomitantly noted that carrying out a cumulative impact assessment is impossible since there is a lack of time and resources to assess effectively past, present and future developments. They further noted that there is a lack of sufficient data to analyze some, if not all of the impacts thoroughly. In addition, they argued that adequate ecological baseline data in northern regions do not exist and thus impacts, and particularly cumulative impacts, due to resource developments cannot be assessed. They also commented that there is no agreed-upon methodology to assess the cumulative impacts of resource development (Smith 2006). As a result, while the issue of cumulative impacts is acknowledged, there has been little appetite by government to require proponents of resource developments to carry out cumulative impact assessments. Proponents of developments also argue that the inclusion of cumulative impact assessments will drive up the cost and time it takes to carry out an impact assessment and thus delay the project. Finally, there is reluctance among proponents and consultants alike to address the issue because definitional questions have yet to be dealt with through case law. As a consequence of the above arguments, the issue of cumulative impact assessment has been quietly ignored by both the private sector and government officials.

Adding to the debate is the question 'Whose problem is it?' For example, developers argue that they are responsible for dealing with the impacts of their own developments, but they should not be responsible for managing or dealing with the effects of other developments. Those who are first to develop a project in a region argue that their direct impact is their responsibility, but that they should not be responsible for future activities by others. On the other hand, those who arrive later argue that they should not be penalized for the impacts of their development since earlier developments have contributed to the cumulative impacts. To date, three solutions have been proposed.

The first is simply to allow only a certain amount of development in a region so that the total cumulative impact is restricted. This approach has been rejected by both the provincial government and the private sector, although in some regions communities have favoured such an approach. A second solution is to have a coordinated mitigated activity undertaken by all the developers in the region so as to minimize the cumulative impact (Asselin and Parkins 2009). A third solution is found in the *Espoo* (1991) and the *Aarhus* conventions (1998). While these have been developed for transnational boundary projects, the provisions within these agreements could easily be implemented to cover multiple regional impacts within one nation (Fitzpatrick and Sinclair 2009; Therivel and Ross 2007).

The lack of action by government in measuring cumulative and regional impacts means there is little baseline data by which to compare the impact (direct or cumulative) of a proposed resource development. Moreover, we find that information collected to establish baseline data has been collected or stored in such a way that it cannot be used to make comparisons over time.

Attempts to involve First Nations people in establishing a variety of baseline data points has not been undertaken and attempts to incorporate traditional ecological knowledge have likewise been rejected by scientists, government and the private sector as many harbour a scepticism about the credibility of Indigenous ways of knowing. Another barrier to the inclusion of traditional knowledge is the political obstacle. If traditional Aboriginal knowledge were to be incorporated into the impact assessment process, changes would have to be made in the decision making process of impact assessments, and such alterations do not appear to be palatable to policy makers.

Conflicts and Resolution

Major areas of disconnect between government and First Nations centre on the nature and 'scope' of concerns identified by the First Nations communities with regard to natural resource development projects. Government officials note that the issues identified by First Nations (e.g., spiritual concerns, cumulative impacts) were often broader than the mandates of the agencies conducting the consultations. As such, the participating government agencies are unable to deal with the concerns of Aboriginal peoples and thus negotiations become fruitless. We also found that government officials noted that the capacity of First Nations to participate in the negotiations about land use, impact assessment and resource development varied considerably. However, the capacity issue was always couched in terms of western ways of knowing, doing business, or relating to the natural world; Indigenous ways were considered irrelevant in discussing the capacity needs of community members (*see* Stevenson and Perreault 2008; this volume).

Treaty First Nations support a position of 'bioregionalism' which involves the decentralization of environmental governance so that decisions about development would have considerable input from local communities. This approach is diametrically opposed to the current centralized-state approach to environmental management (Sale 2001) championed by the provincial and federal governments. Aboriginal leaders argue that societies, economies and politics should be organized around local bioregions rather than the reverse. These bioregions would then become the focal point for human activity and consciousness since this sense of place is obtained as humans orient their lives and livelihoods to their local bioregion and its natural limits (Davidson and Hatt 2005). Also included in this way of thinking is that economic interactions should be focused on co-operation rather than competition and the development of governance should be based on respectful self-sufficiency rather than solely on expansionism. As such, governance would be based on decentralization whereby local First Nation communities would become engaged in development issues through a participatory democratic framework. In turn, the values, knowledge and experience of First Nations would be employed in decision-making. Alberta's *Land Use Framework,* which separates land use planning into seven bioregions, provides a glimmer of hope that this approach may yet still be possible in Alberta.

In the end, however, no one in government suggested that the 'rules' need to change to fit the circumstances of those communities who live in the area. Overall, government officials and proponents of development hold a view of the North as a 'frontier' to be exploited and to enrich Canadians, while First Nations peoples in the region advocate a view of the North as their 'homeland' that must be treated with respect and with their rights and interests in mind.

There is no process or policy currently operating in Alberta that ensures adequate management of the cumulative impacts of development, regardless of whether or not it is directed to First Nations or non-First Nations communities. Nor, might we add, has there been adequate assessment of cumulative impacts framed from the First Nations' perspective (McGuigian 2006). In this respect it is important to note that Aboriginal peoples are at the vanguard of forcing government and the private sector to address this concern. To be sure, while the practice of employing Indigenous Knowledge to assess and monitor social/environmental impacts has become more common (e.g., the Traditional Knowledge in the Kache Tue Study Region, 2002), the importance of this local knowledge is generally ignored once collected. This means that there has been a failure by the government to establish and collaboratively identify ecosystem values and relationships (Stevenson 2006) prioritized by First Nation communities in the region.

The Crown *a priori* assumes that First Nations peoples want to stop resource development of any sort. The reality is quite the contrary; many also look to economic opportunities associated with natural resource development as a way to improve their economic and social conditions. Like others, First Nations are being driven to development decisions with regard to local resources that are valued on a national and international market. Second, there is the assumption that First Nations peoples do not have the expertise or skills to make decisions, e.g., there is a 'capacity deficit model' that is always applied to First Nations communities (Stevenson and Perreault 2008, this volume). Third, there is an unwillingness of the Crown to allow First Nations communities to be involved in the decision-making process. What is particularly troubling is that the Crown often makes resource development decisions in the absence of understanding their impacts on First Nations communities. The private sectors' involvement with First Nations peoples focuses on promising jobs/employment so the project can move forward, although in reality, this seldom is a result. The Crown is clear that it DOES NOT need to consult with First Nations peoples prior to the disposition of Crown rights, and consultation is not a condition imposed on resource developers of acquiring or renewing mineral agreements. Thus, the Alberta Government has retained an 'arms length' relationship with First Nation communities, preferring to delegate consultation with First Nations to the private sector with regard to their project.

Challenges for the Future

Alberta needs to adopt a comprehensive energy policy complemented by a detailed province-wide land use framework. Until recently, there was a lack of both in the province (Vlavianos 2007b). Today, *Alberta's Land Stewardship Act* and *Land Use Framework* hold some promise. However, it is important to add that once land use plans are developed, they become legally binding on all parties. Without legal status, these plans, and any ecological limits and thresholds they set, can be modified at the will of the government or simply ignored by regulators when making decisions (Vlavianos 2007b).

The major concern with current environmental impact assessments and social impact assessments is that they are limited to one project at a time and they are not required to address cumulative impacts of resource development projects. However, if several resource developments take place in a single region or even 'up-stream' from an 'impacted' community, these small, non-major impacts may accumulate to produce a major impact. However, they have never been addressed and no one appears to want to accept responsibility for the 'cumulative' impact. In addition, when cumulative impact is raised as an important issue, there is quibbling over what the baselines, scales and definitions should be. In the end, there is a general rejection of addressing the cumulative impacts of resource development as defined by the perspectives and objectives of impacted First Nations' communities.

In reaction to the government's resistance to carrying out cumulative impacts as well as baseline studies, specific First Nation communities in the region have undertaken their own land use and planning strategies to see how developments can fit into their culture and lifestyle. They have implemented their own methodology, their own data collection and, using Traditional Environmental Knowledge techniques, have developed their specific community land use integrated plans (West Moberly First Nation 2006, Little Red River Cree Nation 2008). In various policy documents, the government of Alberta has encouraged these activities, but it remains to be seen how the government will react when there is a conflict between their proposed plans and the plans adopted by the First Nation communities. They also have begun to engage the courts to resolve social and environmental issues that remain contested.

Current regulations have introduced the notion of 'ecological threshold'—a concept that represents the point at which the impacts of development exceed acceptable parameters and remedial action is required. However, there is no consideration for invoking a concept of 'cultural threshold' which identifies the point where culture is irrevocably impacted and the resilience of the community has been exceeded. Issues of cultural sustainability and cultural viability are not part of the 'impact' assessment scenario identified by any level of government. In other cases, the tangible aspects of culture, e.g., employment/income are part of the social impact assessment of resource development. However, the intangible factors (e.g., quality of life values), are not considered relevant to social impacts of development. More recently, under the American *National Environmental Policy Act*, the issue of 'environmental justice' must now be dealt with by proponents of development. This concept refers to the fair and ethical treatment and meaningful involvement of all people regardless of race, colour, national origin or income with respect to a proposed development. And it means that minority and low-income groups should not bear a disproportionate share of the negative environmental impacts of developments.

The result of our analysis demonstrates that there are very different world views that exist between western scientific and Aboriginal approaches to what most people would label 'resource and environmental management.' At one level, it is easy for researchers, policy makers, and bureaucrats to view these differences as intractable. However, Stevenson (2006) and McGregor (this volume) note that strategies such as the 'two-row wampum' approach to co-management are possible, whereby government and Aboriginal communities undertake land use planning and management in ways that respect the ethical spaces and knowledge contributions of both parties. In the end, a mutually respectful relationship that integrates the understanding inherent in both systems will bring better, more lasting and acceptable decisions to all parties.

Discussion

In 1990, Justice Cory of the Supreme Court of Canada noted in. *Horseman*,[6] that the *Natural Resources Transfer Agreement* (NRTA) was a unilateral decision made by the federal government that fundamentally altered treaty relations with First Nations people without any consultation with them. Section 12 of the NRTA, however, retains the subsistence rights of Treaty Indian peoples:

> Indians shall have the right, which the Province hereby assures to them, of hunting, trapping and fishing game and fish for food at all seasons of the year on all unoccupied Crown lands and on any other lands to which the said Indians may have a right of access.

The extensive development of the Treaty 8 region over the years and its cumulative impact has now begun to infringe upon that right. Moreover, by severing the relationship between First Nations people and their land, it denies their physical, spiritual and economic right to make a living by limiting their access to traditional lands and resources.

The *Northwest Territories Cumulative Impact Monitoring Program* represents some 'best practices' with regard to assessing land use as well as cumulative impacts. It was established to develop inventories of all relevant data that could contribute to the assessment of cumulative impacts and it includes aspects of the environment that have importance for communities in a given geographical area. For example in the NWT, the McKenzie Valley Boards (Land and Water Board, McKenzie Valley Environmental Impact Review Board) have been established to regulate resource development in the region. All decisions made by these boards are subject to review by the Supreme Court of the NWT and the Federal courts of Canada. In other cases in the NWT, the Deh Cho and Canada have signed a *First Nations Interim Measures Agreement* (2001) that will allow the First Nations to develop a land use plan for lands outside the border of local governments and the Nahanni National Park. Recently, the *Mackenzie Valley Environmental Impact Review Board* announced the rejection of the U-R Energy proposal for a uranium mine near the Lutsel K'e Dene First Nation (2007). It rejected the proposal based on a broad understanding of cultural and spiritual concerns about the landscape and about the social, cultural, economic and environmental impacts of the proposed development.

The Government of Alberta has ignored these best practices and quickly points to the 'guidelines' they have developed over the years with regard to impact assessments (Government of Alberta 2006a,b,c). The Alberta Government also has ignored the Precautionary Principle that addresses the dilemma of what to do when scientific knowledge is incomplete, but there is a threat of serious adverse consequences due to a development (Lawrence 2007a). Thus far, the province refuses to accept the ideas embodied in the Precautionary Principle and rejects its incorporation into impact assessment evaluation.

If a systems analysis approach is never undertaken, then our knowledge about relationships and interconnectedness of real world phenomena will not be

---

[6] *R. v. Horseman* [1990] 1 S.C.R. 901.

improved, and uncertainty will prevail. Dixon and Montz (1995) argue that the key characteristics of a cumulative impact assessment must have the following: some representation of interaction, the incorporation of impacts as they occur over space and time, and the ability to trace impacts through from first order direct impacts to second and greater order indirect impacts. This approach can assess interaction effects as well as differentiate between individual, additive and synergistic interactions. It can also show what outputs affect which components of the region. While this approach is not comprehensive (e.g., it will not allow for determination of magnitude of effects), it may serve as the basis for workable methods with practical application to policy makers, government officials and decision-makers with regard to proposed developments. These techniques have the advantage of simplifying complex relationships and provide a framework for introducing a workable cumulative impact assessment process.

A first step in dealing with the cumulative impact assessments will have to involve the identification of valued social and ecosystems components with greater emphasis being placed on regional and local level decision-making regarding resource management issues. As other countries have discovered, cumulative impact analysis is not easy, but it is possible and advisable. However, if sustainable developments are to become a reality, the government will need to draw upon these best practices, utilize Indigenous ways of knowing and build a 'made in Alberta' policy to ensure that cumulative impacts are assessed, monitored and mediated.

Until recently, the province's policy with regard to cumulative impacts assessment and management as well as consultation with First Nations may be summarized as reactive, piecemeal and rigid. While concern for First Nations may be important to government, the fact is that the province has been able to develop natural resources in the region without the consent or support of First Nations. A lack of land use and cumulative impact policies has emerged over time and continues to impact the process of resource development in the Treaty 8 region. However, the Boreal area of Canada is changing fundamentally due to resource development, globalization and cumulative impact. Active interventions by First Nations today about the cumulative impacts of natural resource development and the decisions that have emerged from the courts have forced governments to address the issues, even though they may prefer to avoid dealing with such concerns.

There is not much talk in provincial government circles about engaging in sustainable development that would involve trade-offs. Nor has the provincial government actually thought about what it means, how it would be put into operation, or the implications for the future if the paradigm is not put in place. Further, the Crown does not see itself as responsible for the support of cultural sustainability. While the Crown argues that development activity with regard to natural resource covers just over 5% of the Treaty 8 area, unrecognized is the fact that such activities have an impact on an area many times that size (e.g., a road placed in the area takes up little space, but its impact may change the ecology for many miles around). The Crown needs to acknowledge its duty to consult with First Nations in a manner consistent with its treaty and fiduciary obligations, and then do something meaningful about it. Biological and cultural diversity may be impacted long before 'the shovel hits the dirt' as even the prospect of a development will impact many dimensions of their way of life.

References Cited

AFLW—Alberta Forestry, Lands and Wildlife (1991). *Integrated Resource Planning in Alberta.* Edmonton, AB: Resource Planning Branch.

Aarthus Convention (1998). *Convention on Access to Information. Public Participation in Decision Making and Access to Justice in Environmental Matters.* Aarthus, Denmark.

Asselin, J. and J. Parkins (2009). Comparative case study as social impact assessment: Possibilities and limitations for anticipating social change in the Far North. *Social Indicators Research* 94(3): 483-497.

Baxter, W., W. Ross, and H. Spaling (2000). Improving the practice of cumulative effects assessment in Canada. *Impact Assessment Project Appraisal* 4: 253-262.

Canter, L. (1999). 'Cumulative effects assessment,' pp. 405-440 in J. Petts, ed., *Handbook of Environmental Impact Assessment.* Oxford, UK: Blackwell Science Ltd.

Council on Environmental Quality (1997). *Considering Cumulative Effects under the National Environmental Policy Act.* Washington, DC: Council on Environmental Quality.

Davidson, D. and K. Hatt (2005). 'Towards a sustainable future,' pp. 228-255 in D. Davidson and K. Hatt, eds., *Consuming Sustainability.* Halifax: Fernwood Publishing.

Deh Cho and Government of Canada (2001). *First Nations Interim Measures Agreement.* (www.dehchofirstnations.com/documents/agreements/dehchoirda_e.pdf)

Dixon, J. and B. Montz (1995). From concept to practice: Implementing cumulative impact assessment in New Zealand. *Environmental Management* 19(3): 445-456.

Duinker, P. and L. Greig (2006). The impotence of cumulative effects assessment in Canada: Ailments and ideas for redeployment. *Environmental Management* 37 (2): 153-161.

Ehrlich, A. and S. Sian (2002). 'Cultural Cumulative Impact Assessment in Canada's Far North.' Paper presented at the 24th Annual Conference International Association for Impact Assessment. Vancouver, BC.

Espoo (1991). *Convention on Environmental Impact Assessment in a Transboundary Context.* Espoo, Finland.

Fitzpatrick, P. and A. Sinclair (2009). Multi-jurisdictional environmental impact assessment: Canadian experiences. *Environmental Impact Assessment Review* 29: 252-260.

Geertsema, K. (2008). *Nakatehtamasoyahk Ote Nekan Nitaskenan—Ecological Monitoring of the Lesser Slave Lake Cree, Edmonton.* Masters Thesis, University of Alberta.

Government of Alberta (1992). *Environmental Protection and Enhancement Act.* Edmonton, AB: Queens Printer.

Government of Alberta (1996a). *Cold Lake Subregional Integrated Resource Plan.* Edmonton, AB: Queens Printer.

Government of Alberta (1996b). *Fort McMurray-Athabasca oil Sands Subregional Integrated Resource Plan.* Edmonton, AB: Queens Printer.

Government of Alberta (1999). *Alberta's Commitment to Sustainable Resources and Environmental Management.* Edmonton, AB: Queens Printer.

Government of Alberta (2000). *Strengthening Relationships: The Government of Alberta's Aboriginal Policy Framework.* Edmonton, AB: Queens Printer.

Government of Alberta (2005). *The Government of Alberta's First Nations Consultation Policy on Land Management and Resource Development*. Edmonton, AB: Queens Printer.

Government of Alberta (2006a). *The Government of Alberta's First Nations Consultation Guidelines on Land Management and Resource Development*. Edmonton, AB: Queens Printer.

Government of Alberta (2006b). *Framework for Consultation Guidelines*. Edmonton, AB: Queens Printer.

Government of Alberta (2006c). *First Nations Consultation Policy on Land Management and Resource Development*. Edmonton, AB: Queens Printer.

Government of Alberta (2008). *Draft Land-Use Framework*. Edmonton, AB.

Government of Alberta (2009). *Alberta Land Stewardship Act*. Statues of Alberta, Chapter A-26.8. Edmonton, AB: Queens Printer.

Government of Canada (1998a). *National Forest Strategies-1998-2003. Sustainable Forest: A Canadian Commitment*. Ottawa.

Government of Canada (1998b). *Canada Forest Accord*. Ottawa.

Hanna, K. ed. (2005). *Environmental Impact Assessment*. Toronto: Oxford University Press.

Hegmann, G., C. Cocklin, R. Creasey, S. Dupuis, A. Kennedy, L. Kingsley, W. Ross, H. Spaling, and D. Stalker (1999). *Cumulative Effects Assessment Practitioners Guide*. Ottawa: Canadian Environmental Assessment Agency.

Huff, A. (1999). Resource development and human rights: A look at the case of the Lubicon Cree Indian Nation of Canada. *Colorado Journal of International Environmental Law and Policy* 10: 161-194.

Kennett, S. (1999). *Towards a New Paradigm for Cumulative Effects Management*. Calgary: Canadian Institute of Resources Law.

Kennett, S. (2002). *Integrated Resource Management in Alberta: Past, Present and Benchmarks for the Future*. Calgary: Canadian Institute of Resources Law.

Kennett, S. (2007). *Closing the Performance Gap: The Challenge for Cumulative Effects Management in Alberta's Athabasca Oil Sands Region*. Calgary: Canadian Institute of Resources Law

Lawrence, D. (2007a). Impact significance determination—Back to basics. *Environmental Impact Assessment Review* 27: 755-769.

Lawrence, D. (2007b). Impact significance determination—Pushing the boundaries. *Environmental Impact Assessment Review* 27: 770-788.

Little Red River Cree Nation (2008). *Critique of the Draft Alberta Land Use Plan*. North Peace Tribal Council.

Low, C. (2009a.) The AEUB: A Short chapter in Alberta's long history of energy and utilities regulation. *Resources* No. 105. Calgary, Canadian Institute of Resources Law.

Low, C. (2009b). *Energy and Utility Regulation in Alberta: Like Oil and Water?* Occasional Paper, # 25, Calgary: Canadian Institute of Resources Law.

Lutsel K'e Dene Band (2007). *Traditional Knowledge in the Kache Tue Study Region*.

Ma, Z., D. Becker, M. Kilgore (2009). Assessing cumulative impacts within state environmental review frameworks in the United States. *Environmental Impact Assessment Review* 29: 390-398.

McGuigan, E. (2006). *Of Moose and Man: Collaborating to Identify First nations' Priorities for Cumulative Impact Assessment in Northeast British Columbia*. Vancouver: University of British Columbia.

Passelac-Ross, M. and V. Potes (2007). Consultation with Aboriginal peoples in the Athabasca Oil Sands Region: Is it meeting the Crown's legal obligations? *Resources* No. 98. Calgary: Canadian Institute of Resources Law.

Pauly, D. (1995). Anecdotes and the shifting baseline syndrome of fisheries. *Trends in Ecological Evolution* 10: 423-1056.

Potes, V., M. Passelac-Ross, and N. Bankes (2006). *Oil and Gas Development and the Crown's Duty to Consult.* Calgary: Institute for Sustainable Energy, Environment and Economy, University of Calgary.

Pratt, L and I. Urquhart (1994). *The Last Great Forest: Japanese Multinationals and Alberta's Northern Forests.* Edmonton, AB: NewWest Publishers, Ltd.

Ross, M. (2003). Aboriginal Peoples and Resource Development in Northern Alberta Canadian Institute of Resources Law. University of Calgary.

Sale, K. (2001). There's no place like home: Bioregionalism. *The Ecologist* 31(2): 23-35.

Sallenave, J. (1994). Giving traditional ecological knowledge its rightful place in environmental impact assessment. *Northern Perspectives* 22(1): 1-7.

Smith, M. (2006). Cumulative impact assessment under the *National Environmental Policy Act*: An analysis of recent case law. *Environmental Practice* 8(4): 328-240.

Stakhiv, E. (1988). An evaluation paradigm for cumulative impact analysis. *Environmental Management* 12: 725-748.

Stevenson, M. (2006). The possibility of difference: Re-thinking co-management. *Human Organization* 65(2): 167-180.

Stevenson, M. and P. Perreault (2008). *Capacity for What? Capacity for Whom?* Edmonton, AB: Sustainable Forest Management Network.

Tanner, T. (2008). *Rights vs. Resources: Why the First Nations Left the Cumulative Environmental Management Association.* M.A. Thesis, Royal Roads University.

Therivel, R. and B. Ross (2007). Cumulative effects assessment: Does scale matter? *Environmental Impact Assessment Review* 27: 365-383.

Vlavianos, N. (2007a.) Key shortcomings in Alberta's regulatory framework for oil sands development. *Resources*, No. 100. Calgary, Canadian Institute of Resources Law.

Vlavianos, N. (2007b). *The Legislative and Regulatory Framework for Oil Sands Development in Alberta: A Detailed Review and Analysis.* Calgary, Canadian Institute of Resources Law.

Warnback, A. and T. Hilding-Rydevik (2009). Cumulative effects in Swedish EIA practice—Difficulties and obstacles. *Environmental Impact Assessment Review* 29: 107-115.

Wenig, M. and J. Poschwatta (2008). *Developing a "Comprehensive Energy Strategy" with a Capital 'C'.* Calgary, Canadian Institute of Resources Law.

WMFN—West Moberly First Nations (2006). *The Peace Moberly Tract Draft Sustainable Resource Management Plan.* British Columbia: Saulteau First Nations.

Chapter Two

Trust Us Again, Just One More Time:
Alberta's Land Use Framework and First Nations

Marc G. Stevenson

Introduction

Chapter 1 by Frideres and Rowe was completed prior to the introduction and enactment of legislation required to support the implementation of Alberta's *Land Use Framework* (ALUF) (Government of Alberta 2008a). While critical of Alberta's policies with respect to Treaty 8 Indian peoples and cumulative impacts management, their claims cannot be said to be without substance; Alberta's track record on both fronts is envied by few across Canada, or the world for that matter. At the same time, Chapters 3 and 4, respectively, by Webb suggest that there is an unfulfilled federal/provincial Crown treaty obligation to provide 'hunting reserves' to Treaty 8 First Nations, while offering interpretive principles and guidelines to achieve this outcome. This chapter is relevant to arguments and positions put forward in all three chapters. Specifically, it discusses the recent initiative by the Government of Alberta to create regional land use plans under ALUF in order to manage the multiple and cumulative impacts of resource development, while striving for "a meaningful balance that respects the constitutionally protected rights of aboriginal communities and the interests of all Albertans" (Government of Alberta 2008a:41).

Initiatives currently being contemplated by the province under ALUF and the *Terms of Reference for Developing the Lower Athabasca Regional Plan* (Government of Alberta 2009c) suggest that an era of improved Aboriginal-Provincial relations may soon be upon us.[1] However, Alberta's First Nations are being asked, once again, to trust that the provincial government will act with their interests in mind. This chapter questions whether the proposals being put forward by the province under ALUF are sufficient to warrant such expressions of faith, and advances some recommendations with respect to improving relations between the province and its Aboriginal peoples in the context of regional land use planning, and the stated objective of balancing the rights of Alberta's Aboriginal peoples with the interests of other Albertans.

---

[1] These include: the creation of conservation areas where Aboriginal traditional uses are supported, consultation plans specific to First Nations, and the encouragement of Aboriginal peoples in land-use planning.

Alberta's Land Use Framework

Under development since 2006 and introduced in late 2008, ALUF seeks, through the development of regional land use plans, to manage land use in Alberta on a regional scale. In particular, ALUF intends to address the cumulative effects of all land-use activities from recreation to municipal expansion to large scale resource extraction. The emphasis on managing the cumulative impacts of industrial development resonates with Alberta's Aboriginal communities (e.g., HLFN 2008; LRRCN 2008) and environmental communities (e.g., Water Matters Society of Alberta 2009). In the view of the province (Government of Alberta 2008a:31), cumulative effects management...

> ...recognizes that our watersheds, airsheds and landscapes have a finite carrying capacity. Our future well-being will depend on how well we manage our activities so that they do not exceed the carrying capacity of our environment. Alberta's current regulatory system is based on a project-by-project approval and mitigation of the adverse effects of each project. Until now, the approach has been to control the impact of each project. While this may be acceptable for low levels of development, it does not adequately address the cumulative effects of all activities under the current pace of development. Cumulative effects cannot be managed as an 'add-on' to existing management approaches; nor is it about shutting down development. It is about anticipating future pressures and establishing limits; not limits on new economic development, but limits on the effects of this development on the air, land, water and biodiversity of the affected region. Within these limits, industry would be encouraged to innovate in order to maximize economic opportunity. The Government of Alberta will develop a process to identify appropriate thresholds, measurable management objectives, indicators and targets for the environment (air, land, water and biodiversity), at the regional levels and, where appropriate, at local levels. Land-use planning and decision-making will be based on balancing these environmental factors with economic and social considerations.

Anchored by the *Alberta Land Stewardship Act* (Government of Alberta 2009a), which received Royal assent on 1 October 2009, ALUF provides the legislative structure and authority to create a regional level of planning that does not currently exist in the province. ALUF will affect many laws and policies that currently guide decisions by provincial ministries, municipalities and land users, including First Nations. Indeed, 27 existing provincial acts have been amended to ensure consistency under *Alberta Land Stewardship Act* (ALSA). Decision-making bodies will be required to comply with regional plans once developed, and because regional land use plans are approved by Cabinet, they are government policy (Government of Alberta 008a:27). Whereas the ALSA relies mainly on existing appeal processes, new appeal processes will also be created under the *Forests Act* and *Public Lands Act* (Government of Alberta 2008a:9). It has been recognized that, while the advantage of this approach is that it does not create new bureaucratic structures, it also means that land use decisions will not be subject to one consistent appeal process, and that individuals appealing relevant rules and availability of costs, will vary depending on the subject matter of the decision being appealed and the applicable legislation (Chiansson 2009).

ALUF envisions a system of regional plans for each of seven new land use regions established by the province on the basis of watershed characteristics (e.g., Lower Athabasca, South Saskatchewan, etc.). These land use plans will set out regional objectives, integrate provincial policies and bind decision-makers, such as municipalities and regulatory tribunals. Regional land use plans are to consider a 50 year time horizon, and are subject to amendment at least once every 10 years as new information, issues and planning needs emerge. A new governance structure will include a Land Use Secretariat to oversee implementation of the framework and the development of regional land use plans. ALUF also creates Regional Advisory Councils (RACs) for each land use region to provide multi-stakeholder input into the development of regional plans. Included within the make-up of some RACs are seats for Aboriginal representatives. Ultimate authority for the scope, content and implementation of regional plans, however, rests with the provincial cabinet. Poschwatta-Yearsley and Zelmer (2009) regard the decision-making power of both the Land Use Secretariat and Regional Advisory Councils as limited under the ALSA, as the Lieutenant Governor in Council retains the authority to implement this *Act*. In particular,

> ..the role of the RACs under ALSA is purely a consultative one. The LGIC (Lieutenant Governor in Council) is not required to consider, or follow, the advice of a RAC. The ALSA does not require the LGIC to provide any reasons, written or otherwise, if it disregards the advice of a RAC (Poschwatta-Yearsley Zelmer 2009:3-4).

Planned attributes of the new framework include cumulative effects management, development of a suite of conservation and stewardship tools (including Conservation Areas), establishment of an information/monitoring /knowledge system, and the inclusion of First Nations in land use planning. Under the *Terms of Reference for Developing the Lower Athabasca Regional Plan* (Government of Alberta 2009c), the Regional Advisory Council will consider three oil sands development scenarios modeled after the status quo (1.5-2.0 million barrels per day), mid-range growth (4.0-4.5 million barrels per day) and high-end production (6 million barrels per day or more), and predict the environmental, social and economic implications of each.

ALUF is to be implemented via pursuit of seven strategic initiatives. Strategy 7, *The Inclusion of Aboriginal Peoples in Land-use Planning,* commits the province to "strive for *a meaningful balance that respects the constitutionally protected rights of aboriginal communities and the interests of all Albertans*" (Government of Alberta 2008a:21, emphasis added). The Province (Government of Alberta 2008a:41) recognizes that:

...the aboriginal peoples of Alberta have an historic connection to Alberta's land and environment... ...and that those First Nations and Métis communities that hold constitutionally protected rights are uniquely positioned to inform land-use planning. The Government of Alberta has the constitutional mandate to manage lands in the province for the benefit of all Albertans. However, the Government of Alberta will continue to meet Alberta's legal duty to consult aboriginal communities whose constitutionally protected rights under section 35 of the *Constitution Act, 1982 (Canada)* are potentially adversely impacted by development. To support meaningful consultation in the province, Cabinet approved *The Government of Alberta's First Nations Consultation Policy on Land Management and Resource Development* in 2005. This policy is a key step towards engaging First Nations in land management decision making. Ongoing review and monitoring of the policy with the intent of changing and improving it will ensure that it meets the needs of Albertans, First Nations and industry. To address specific implementation challenges, Alberta has created a 'trilateral process' involving senior representatives from industry, First Nations and government. Efforts to build First Nations capacity have been underway for several years and include programs such as the Traditional Use Studies Program and the First Nations Consultation Capacity Investment Program, which are administered by the Ministry of Aboriginal Relations. By investing in the gathering and maintenance of information on First Nations land uses, Alberta has also helped prepare First Nations for increased dialogue in regional planning.

Recent critical commentary of the ALSA (Poschwatta-Yearsley and Zelmer 2009:9-10) however, suggests that meaningful Aboriginal engagement in land use planning and implementation in Alberta will remain more a dream than a reality for Alberta's First Nations peoples:

Most of the goals and policies outlined in the LUF (Land Use Framework) have translated themselves in some form into the ALSA: however, the vast majority are optional or *discretionary*. The *LGIC (Lieutenant Governor in Council) retains unconstrained power to independently create, amend and implement the planning regions and regional plans*. Although the LUF specifies that the framework will not create a heavy-handed centralized bureaucracy that appears to be what is created by ALSA. The discretionary nature of the legislation limits the public's (including Aboriginal peoples) ability to participate in the development of the regional plans. There is no right or mechanism in the legislation for the public to contribute to, or to challenge, the content of a regional plan. The right to participate in the complaint procedure appears to be a toothless right without the ability to influence the regional plans themselves. The structure of ALSA makes it very difficult to predict the outcome of the legislation on actual land use planning. With few guiding principles and mandatory requirements, the ALSA only adds to the uncertainty surrounding land use in the province. (emphasis added)

Of more than passing interest, and perhaps a harbinger of how committed the province is in regards to engaging Aboriginal peoples in land use planning is the reference in ALUF (Government of Alberta 2008a:21) to balancing 'rights' with 'interests.' These are not the same thing, nor are they of the same order. The 'constitutionally protected rights' of one cannot be balanced with the 'interests' of others.[2] Conflation of these terms, whether deliberate or not, places the constitutionally protected 'rights' of Alberta's Aboriginal communities on par with the 'interests' of all Albertans. The more appropriate phrasing would have been 'to strive for a meaningful balance that respects the constitutionally protected rights of aboriginal communities and all Albertans.'

Nevertheless, two actions are proposed to achieve this balance. Specifically, the province will consult with Aboriginal communities "whose constitutionally protected rights, under section 35 of the *Constitution Act, 1982* (Canada), are potentially adversely impacted by development," and "encourage Aboriginal peoples to participate in the development of land-use plans" (Government of Alberta 2008a: 41).

Planned Consultations with Alberta Treaty First Nations: Are They Sufficient?

Aboriginal peoples are specifically mentioned in ALSA as persons that may be appointed to RACs.[3] However, in the opinion of Poschwatta-Yearsley and Zelmer (2009:7), this is inadequate to meet the constitutionally protected rights of Alberta's Aboriginal peoples. To some extent, the Land Use Secretariat has acknowledged this fact by proposing to undertake a regional consultation process for the Lower Athabasca Regional Plan (LARP) specific to affected First Nations communities (Government of Alberta 2009b) (see below).

The *Terms of Reference for Developing the Lower Athabasca Regional Plan* (Government of Alberta 2009c) (LARP TOR) acknowledges that "land use must be managed to include Aboriginal traditional use activities" (Government of Alberta 2009c:11), and commits the province to taking into "account Aboriginal issues with respect to Aboriginal consultation, environmental protection and human development" (Government of Alberta 2009c:9) in the

---

[2] An alternative interpretation of this policy statement is that the province intends to balance the interests of all Albertans, including its First Nation peoples, with the constitutionally protected rights of its Aboriginal communities, thus elevating the latter to a status higher than mere interests alone. However, this interpretation would be out of character with the province's track record, and certainly a reversal of decades of provincial policy (*see* Frideres and Rowe, this volume).

[3] ALSA s. 52(2)(a).

development of the regional land use plan. However, whether ALUF and LARP will have largely negative or positive impacts on the treaty and Aboriginal rights of Alberta's Treaty 6 and 8 First Nations remains to be seen, and will depend on the extent to which the latter are consulted and involved in the land use planning process, as well as the implementation and modification of regional plans.

In the fall of 2007, the provincial government sought input from Alberta's First Nations and Métis communities regarding ALUF. During these consultations, Aboriginal peoples provided their views on the future of land use in the province and their concerns about protecting their cultures and local environments under the new policy framework. The following spring and summer, Aboriginal communities reviewed the Draft Land-use Framework advocating for, among other things:

- Recognition and accommodation of their Aboriginal and Treaty rights,
- Parallel RACs comprising Aboriginal representatives to provide unfettered input into regional land-use plans that would be considered on par with input received from the proposed multi-stakeholder RACs (LRRCN 2008);
- First Nation representation at higher levels of decision-making; i.e., within the Land Use Secretariat (HLFN 2008);
- Regional First Nation land use planning co-ordinators, who would be responsible for working with individual First Nations to provide land use planning information to RACs within each region (HLFN 2008); and
- Enough time and capacity funding to permit First Nations to undertake their own land use planning processes in order to provide input into the development of regional land use plans (HLFN 2008).

However, none of these concerns or recommendations were considered or addressed in the Final Land-Use Framework,[4] which changed very little from the Draft Land Use Framework. Alberta Aboriginal Affairs and Northern Development (2009b) nevertheless reported in a subsequent consultation plan developed specifically for First Nations in the Lower Athabasca region that, during the course of these consultations, First Nations identified the following common themes and perspectives with respect to regional land use planning:

- Participation in land use planning,
- Consultation and rights,
- Land use planning and traditional use,[5]
- Emphasis on environmental and social outcomes, and
- Cumulative effects.

---

[4] I have avoided using, against great temptation and in consideration of better judgement, the acronym for 'Final Land Use Framework.'

[5] The linkage between land use planning and traditional use was not explained.

Relevant statements that made it into the final ALUF include references to the facts that Albertans wanted "enhanced conservation and stewardship on both private and public lands to promote ecological sustainability" and "increased consultation with First Nations and Métis communities, stakeholders and the public to ensure a fair opportunity to influence new policies and decisions" (Government of Alberta 2008a). In the opinion of the province, feedback received from First Nations in connection with the ALUF consultation process has been considered by Alberta and incorporated into the LARP TOR (Government of Alberta 2009c).

The Lower Athabasca Region First Nation Consultation Plan
In developing the *Lower Athabasca Regional Plan* (LARP)—the first regional plan to be created under ALUF—the Land Use Secretariat intends to undertake *The Lower Athabasca Region First Nation Consultation Plan* (Government of Alberta 2009b), once a draft regional plan has been produced by the Regional Planning Team.[6] This is not to say that there has not been any Aboriginal input thus far into the development of the LARP. Treaty 8, Treaty 6 and the Métis Settlements each have a seat on the LARP Regional Advisory Council, and have provided Aboriginal input, advice and direction on such issues as Aboriginal and treaty rights, conservation areas, parks and recreation, transportation and utility corridors, and environmental and social conditions, indicators and thresholds. To what extent this input will be included in the final LARP is uncertain and dependent on many factors, not least of which is the discretion of Cabinet, the Lieutenant Governor in Council, and the Ministerial Working Group.[7] Regardless, individuals occupying Aboriginal seats (including the author, who was appointed to the Treaty 6 seat) do not represent the views and opinions of those Treaty organizations whose seats they occupy. Rather, their input into the LARP RAC deliberations is their own, and their views and opinions are not necessarily concordant with those of Treaty 6, Treaty 8 or the Métis Settlements. Both tribal organizations responsible for placing individuals on the LARP Regional Advisory Council—Tribal Chief Ventures Inc. for Treaty 6 and the Athabasca Tribal Council for Treaty 8—have submitted letters to the Land Use Secretariat to that effect (Roy Vermillion, Dave Scott, personal communications 2009). However, at the same time, there is an expectation, held by the Alberta Government and both tribal organizations that these individuals will report back on the deliberations and work of the LARP RAC. The Regional Planning Team has also provided the LARP RAC with information held by the province on Aboriginal values within the Lower Athabasca planning region. Nonetheless,

---

[6] Regional Advisory Teams are composed of government personnel and private consultants—in the case of the Lower Athabasca Regional Plan, *Stantec Consultants*—who, with input and feedback received from the Regional Advisory Councils, Cabinet (via the Land Use Secretariat) and public and Aboriginal consultation processes, are charged with drafting regional land use plans.

[7] A Ministerial Working Group provides direction to and receives input from the Land Use Secretariat in the design of the LARP (Heather Kennedy 2009, pers. comm.).

neither input discharges the province's duty to consult directly with affected First Nations regarding the potential of the LARP to adversely impact the Aboriginal and Treaty rights of the region's First Nations, a fact acknowledged by the province.

Under the *First Nation Consultation Plan for the Lower Athabasca Region* (Government of Alberta 2009b), each First Nation having either a reserve or land use practices within the region will be consulted. The proposed consultation process encompasses eight steps from initial notification to final reporting from Cabinet back to each First Nation. The Land Use Secretariat will "keep records of all correspondence and other forms of successful and attempted contact throughout the consultation process to help establish whether any applicable legal duties or policy commitments to consult have been fulfilled" (Government of Alberta 2009c).

There is little question that ALUF, and any regional land use plans developed and implemented under its authority, has the potential to adversely impact the constitutionally protected Aboriginal and treaty rights of Alberta's Aboriginal peoples. It is also clear from the Supreme Court of Canada's (SCC) decisions in *Haida*[8] and *Mikisew*[9] that the duty to consult is that of the Crown's, and the Crown's alone. However, the nature and extent of this duty required for regional land use plans places a heavy burden on the provincial Crown that it seems unwilling to shoulder seriously. The SCC directed in *Haida* that when governments have real or constructive knowledge of the potential existence of an Aboriginal right, including title, and contemplates conduct that may adversely affect that right, it is under a duty to consult with Aboriginal peoples and accommodate their interests[10] (*see also* Passelac-Ross and others in this volume). With reference to the SCC rulings in *Haida* and *Taku*,[11] the degree of consultation owed to First Nations is not contingent upon court-ordered recognition, but directly proportional to:

1) The strength of the right claimed, and
2) The degree or extent of the infringement being contemplated on the right claimed.

For example, industrial proponents undertaking activities within a core traditional area of a First Nation where it can meet the test of Aboriginal title as outlined by the courts will require a more involved or engaged process of consultation than other areas of its traditional territory where Aboriginal and treaty rights are characterized by harvesting rights and other interests. The greater the strength of the right claimed and the greater the potential impact on that right, the greater the degree of consultation required (Fig. 1). In other

---

[8] *Haida Nation v. British Columbia (Minister of Forests)*, 2004 SCC 73, [2004] 3 S.C.R. 511.
[9] *Mikisew Cree First Nation v. Canada (Minister of Canadian Heritage)*, 2005 SCC 69, [2005] 3 S.C.R. 388.
[10] *Haida* at para. 35
[11] *Taku River Tlingit First Nation v. British Columbia (Project Assessment Director)* [2004], 3 S.C.R. 511.

words, it is at that point where the lines intersect on both axes in Figure 1 that will determine the degree of consultation required. For example, an instance where the strength of the right claimed is great, but the infringement is not, might require the same level of consultation where the strength of the right claimed is weaker, but the infringement is greater.

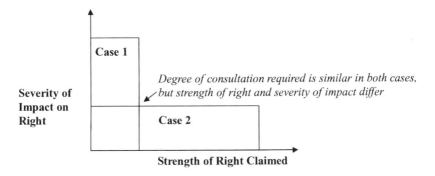

Figure 1. *Spectrum of consultation* (adapted from Statt 2008).

Following the Supreme Court of Canada's decision in *Delgammukw* (1997),[12] the goals of consultation are well-defined: 1) to minimize the infringement of the proposed activity/development on, and to accommodate, the exercise of the constitutionally protected rights of Aboriginal peoples, and 2) where infringement cannot be avoided, agreement as to the appropriate forms of mitigation and compensation for the infringement. As outlined in *Delgammukw,* the overriding objective of consultation where resource development has the potential to adversely impact Aboriginal and Treaty rights is to "substantially address the concerns of Aboriginal peoples." Following this, it is obvious that none of the efforts undertaken by the province prior to releasing ALUF, or currently being contemplated by the province, even marginally addresses the concerns of Alberta's Aboriginal peoples, and thus cannot be considered as having discharged Alberta's duty to consult.

That said, there is light, however faint, on the horizon. First, it is the understanding of the author that representatives of Alberta's Treaty Nations and the province are about to enter into discussions to amend provisions of Alberta's Aboriginal Consultation Policy (Government of Alberta 2005) to reflect the Supreme Court of Canada's decisions in *Haida* and *Mikisew.* What, if any, amendment(s) will be made to provincial land use policies (e.g., ALSA, ALUF, LARP TOR, etc.) as a result of these discussions is a matter of conjecture. Secondly, depending on the outcome of the province's consultations with First Nations and Métis regarding the LARP (Government of Alberta 2009b), and the extent to which their concerns are addressed in both this consultation process and the LARP, "a meaningful balance that respects the constitutionally protected rights of aboriginal communities and the interests of all Albertans" (Government of Alberta 2008a:21) may yet be achieved.

---

[12] *Delgamuukw v. British Columbia,* [*1997*] 3 S.C.R. 1010.

Complicating the issue of Aboriginal consultation regarding the development of the LARP, and other regional land use plans in the province, is the fact that the Alberta Treaty Chiefs have rejected both the ALUF and Alberta's Aboriginal Consultation Policy,[13] preferring to take a treaty-based approach to reconciling their Treaty and Aboriginal rights with the resource rights of the provincial Crown, which it acquired in 1930 through the *National Resource Transfer Agreement*. This makes it particularly problematic for individual First Nations who see their participation in regional land use planning initiatives as a way to protect their rights and interests in the face of development.[14] By way of avoiding this conflict of interest, Treaty Nations may wish to consider developing a coordinated and strategic approach to advancing their constitutionally protected rights and interests at both levels. This dualistic approach would allow Alberta First Nations to participate, and engage the provincial Crown in existing land use policies, instruments, institutions and planning initiatives without negating or prejudicing nation-to-nation discussions with the province over the recognition and accommodation of their Treaty rights. In other words, both 'interest-based' and 'rights-based' strategies are needed, and one should not be pursued in the absence of, or at the expense of, the other.

Encouraging Participation in Land Use Planning: Is it Enough?
Alberta's Land Use Framework (Government of Alberta 2008a), *Terms of Reference for the Lower Athabasca Regional Plan* (Government of Alberta 2009c), and *First Nation Consultation Plan for the Lower Athabasca Region* (Government of Alberta 2009b) commit the province to "encourag(ing) Aboriginal peoples to participate in the development of land-use plans." What this means is not exactly clear. To date, information sent out by the Land Use Secretariat to Treaty Nations has not met the information needs of First Nations communities, although the Land Use Secretariat has made it clear that it has and will continue to meet with any First Nation at its request to discuss matters relating to ALUF, the LARP TOR and the regional land use planning process in advance of implementing the proposed First Nation consultation plan (Heather Kennedy 2009, pers. comm.).

Further, no specific programs or resources have been identified by the province to facilitate Aboriginal participation in the development of regional land-use plans. While there are government programs and sources of funding available to First Nations to undertake traditional land use studies and to develop consultation capacity to engage industry and other third parties granted

---

[13] Assembly of Alberta Treaty Chiefs Resolutions, 13 February 2009.
[14] According to the province (Government of Alberta 2006), 23 of 47 Alberta chiefs signed a resolution on September 14, 2006 rejecting Alberta's Consultation Guidelines. However, as the province (Government of Alberta 2006 acknowledges, "although all parties agreed to continue the dialogue on improving the Guidelines, the limited participation of First Nations in Land Use Framework discussions is inextricably linked to the release of the Guidelines."

access/use rights by the province to traditional lands,[15] these programs are not specific to the development of regional land use plans *per se*. There is simply no funding in the overall Land Use Secretariat budget for meaningful Aboriginal participation in the LARP consultation or planning processes, and First Nations are simply left to their own devices should they choose to participate. Meaningful Aboriginal participation in this context would involve, in addition to other things, having sufficient time, opportunity, and information and data sets to document and support the concerns of Aboriginal peoples so that they can be adequately addressed by the Crown.

The Land Use Secretariat will, however, provide funding to First Nations for venues, hosting and attendees' travel expenses to engage in consultations with the province (Government of Alberta 2009b). Viewing this policy as deficient and unable to meet their input requirements, several Treaty 6 and 8 First Nations in the Lower Athabasca region have forwarded proposals to the Land Use Secretariat for funding to build the capacity and data sets necessary to effectively engage the province in consultations regarding the development of the LARP (Betty Kennedy and Roy Vermillion, personal communications 2009). The province formally responded to these requests in January 2010 acknowledging that it would give each First Nation $100,000 to provide input into the LARP and to undertake consultations with the province after the plan is approved by Cabinet, regardless of the size of the First Nation, its traditional territory or its original funding request (Dave Bartesko, person. comm. 2010). Conditions attached to the approvals stipulate that the funding is to be used not just for the LARP planning region, but all planning regions in the province where each First Nation possesses Aboriginal and treaty rights.[16]

Traditional Land Use Information: Is it Sufficient to Discharge the Duty of the Crown?
It is reasonable to conclude that land use planning by a First Nation would be a critical first step in its quest for social, cultural and economic sustainability, and a desired use of its traditional land use study (TLUS) information. In an ideal world, First Nation communities would have developed their own specific land use plans, based on their data sets, traditional use information, resource inventories, full cost accounting of current and future economic development opportunities, current and future human capital and infrastructure needs, etc., in order to provide meaningful input into the development of regional land use plans. However, few, if any, First Nation communities in northern Alberta have undertaken the necessary research and local-level planning to achieve this outcome. In contrast, other stakeholders in the region (e.g., municipal

---

[15] *Traditional Use Studies Program* and the *First Nations Consultation Capacity Investment Program* administered by the Ministry of Aboriginal Relations, Government of Alberta.

[16] In the case of the Heart Lake First Nation and other Treaty 6 First Nations, this means that significant portions of the $100,000 would have to be allocated to regional land use planning outside of the Lower Athabasca region (e.g., the North Saskatchewan, Lower Peace and other regions) where they retain constitutionally protected Aboriginal and treaty rights.

governments, forest companies, oil/gas companies, etc.), many of whom are represented on LARP Regional Advisory Council, come to the table with their land use plans well-formulated and approved by the province.

    To date, funding requests by some First Nations (e.g., Heart Lake First Nation) to use TLUS data to inform land use planning have been rejected by the province (Betty Kennedy, HLFN 2009, pers. comm.). The view of the province is that "by investing in the gathering and maintenance of information on First Nations land uses, Alberta has ...prepare(d) First Nations for increased dialogue in regional planning" (Government of Alberta 2008a:41). While this may or may not be true, the province's access to TLUS information produced by First Nations, which is a condition of funding, may produce the opposite result, effectively excluding Aboriginal peoples from regional land use planning while using their TLUS data to whatever ends the province deems appropriate. At the same time, the province's narrow materialistic view of the contributions of Aboriginal peoples to land use planning hamstrings the whole planning process by limiting their information contributions to the identification of specific traditional land use locales, so that they might be avoided should resource development proceed. As others have made clear, this 'dots on a map' (Webb *et al.* 2009) or 'small spots' (Passelac-Ross, this volume) approach to traditional land use is untenable, and ignores the fact that most land use sites are part of, and integral to, a much broader complex of related activities that take place over a much larger landscape.[17] In other words, the connections and linkages between/among traditional use locales are as important as the sites themselves; together, they form the cultural landscapes of many First Nation communities. At the same time, the province's impoverished view ignores valuable information and knowledge that Aboriginal peoples might have to contribute toward the setting of social and environmental indicators, thresholds and outcomes, among other things.

Alberta Treaty First Nations and Conservation Areas

A significant, encouraging, if not the watershed, component of ALUF is the creation of 'Conservation Areas' to maintain ecosystem integrity and habitat. Conservation Areas would be established to offset oil sands mining, oil/gas exploration and development, agriculture, forestry and other industrial activities, and their cumulative impacts, while protecting biological and cultural diversity. The LARP TOR (Government of Alberta 2009c:15, 17-18) asks the RAC, among other things, to:

---

[17] Government and industry appear to value TLUS information primarily as means to identify specific locales that resource development needs to avoid. Moreover, any rights that Aboriginal peoples may hold are confined to these areas. This view of traditional land use relegates Aboriginal peoples to 'postage-stamp' existences, which Judge Vickers of the BC Supreme Court soundly rejected in *Tsilhqot'in Nation v. British Columbia* (2007 BCSC 1700).

- explore the feasibility of meeting a conservation scenario higher than 20 percent while achieving the stated economic objectives,
- demonstrate how the conservation scenario can be met, while minimizing and limiting any negative impacts, including mineral tenure and fiscal implications, and
- observe the key criteria for conservation lands... and assess and advise which lands in the region could meet the following four criteria:
  o little or no industrial activity,
  o supportive of Aboriginal traditional uses,
  o representative of the biological diversity of the area, and
  o sufficient size, i.e., 4,000 to 5,000 km².

While the size criterion is/was undoubtedly influenced by the Cumulative Effects Management Association (CEMA 2008) and Canadian Boreal Initiative (CBI) proposals,[18] whatever economic development scenario is ultimately approved and implemented by Cabinet, Conservation Areas will undoubtedly be the biggest 'stick' against which the success of ALUF will be 'measured' in the court of public opinion. The second and last criteria are discussed here.

Use and Access Rights to Conservation Areas
The idea that Aboriginal traditional use is compatible with conservation objectives, while familiar to many First Nation elders, land users and researchers, is novel for a provincial government, and should be duly recognized. Incidentally, no mention in the LARP TOR is made about supporting traditional uses by non-Aboriginal peoples. Nevertheless, there is an emerging expectation among some LARP RAC members, and their constituents, that Conservation Areas will be freely available to all Albertans for hunting, fishing and other resource uses. To date, this issue has been among the most contentious and debated within the LARP RAC. Whatever the outcome of this discourse, it needs to be guided by the reality that First Nations and Métis peoples and communities in the Lower Athabasca region experience levels of poverty, social pathologies, health problems, unemployment and population growth that are many times the provincial and national averages. Currently, over 12,000 Aboriginal people live in nine First Nation and four Métis communities in the region, with this number expected to double with each passing generation. At the same time, as more land is 'taken up' for oil/gas, agriculture, forestry and other forms of development, the ability of Aboriginal peoples to exercise their constitutionally protected rights is diminished. Ultimately, there will come a time when promises made by the Crown under Treaties 6 and 8 will become impossible to keep. For the Fort McKay First Nation and the Beaver Lake First

---

[18] Conservation areas of between 20 and 40% have been recommended by CEMA (2008)—a multi-stakeholder group including the oil sands industry to study the cumulative environmental effects of industrial development in the region—while the CBI is an organization of NGO conservation interests, First Nations, industry and other parties dedicated to preserving 50% of the Boreal forest across Canada.

Nation, which is pursuing legal action against the Provincial and federal Crowns for breach of treaty and fiduciary obligations, this time has come.[19] For other First Nation communities, this time is fast approaching.

The position of both levels of government up to this point, as deduced from their ongoing failure to protect the traditional livelihood rights of its Treaty Indian peoples (*see* Webb, both chapters, this volume) is clear: Despite the pace and extent of industrial development in the Lower Athabasca region, Aboriginal peoples still have access to a large enough land base where they can exercise their rights, interests and traditional livelihoods guaranteed to them under treaty. While this may or may not be true, ensuring that Aboriginal peoples have a large enough land base is only half the equation. Ensuring that there is a sufficient supply and quality of game, fur, fish, plants and medicines as well as adequate water and air quality on that land base to exercise their rights and livelihoods is the other half. The cumulative effects of a host of industrial activities (seismic lines, road construction, oil sands mining, well pads, pipelines, clear cutting, agricultural expansion and other use activities) in the Lower Athabasca region has so fragmented Crown land and so degraded the environmental integrity and biodiversity of the region as to render the promises made under treaty virtually 'meaningless' and without effect in many areas. Thus, First Nation peoples in the Lower Athabasca region, despite the political position of some of their chiefs (see above), are looking to participate in the development of the province's regional land-use plan, and particularly the establishment of 'Conservation Areas' as a measure to protect their rights and interests.

ALUF (Government of Alberta 2008a:51) defines 'conservation' as "the responsible preservation, management and care of our land and of our natural and cultural resources." The reference to cultural resources is noteworthy. Biological diversity and cultural diversity in most parts of the Indigenous world are intimately related, and Conservation Areas must ultimately accommodate and preserve this dual purpose (and their inter-related functions). Although the LARP TOR does not list non-Aboriginal traditional use of 'Conservation Areas' as one of its four criterion, it would be outside the spirit and intent of treaties to exclude non-Aboriginal traditional use. It would also undermine one of the primary objectives of 'conservation:' the preservation of cultural diversity. In this regard, the existing hunting, trapping, gathering and fishing rights of non-Aboriginal peoples in Conservation Areas who can prove that such practices/uses are 'integral to their identity and culture' (through demonstration, for example, of long-term use and occupancy), should not be denied. Co-existence and sharing is what treaty, as understood and advocated by Treaty Indian elders and chiefs, was all about, and the management of land use in Conservation Areas must reflect this reality.

The LARP TOR also makes no reference to supporting the modern or non-traditional uses of Conservation Areas by Aboriginal peoples. As elaborated upon in *Delgammukw,* Aboriginal rights are not frozen in time, and modern expressions or manifestations of such rights are protected under *section 35.1,*

---

[19] *Beaver Lake First Nation v. the Queen in Right of Alberta and the Attorney General of Canada,* 2008 Statement of Claim, Court of Queen's Bench, Edmonton, Alberta.

Canadian Constitution Act, as long as they do not compromise the ability of future generations of Aboriginal peoples to use the land in ways that gave rise to the constitutional protection of their rights in the first place, i.e., 'are integral to their identity and culture.' This would likely be a small sacrifice for many First Nations, provided <u>no new</u> industrial tenures (oil/gas, forestry, etc.) or land-based tenures, leases or licenses (registered traplines, commercial fishing licenses, etc.) be granted in Conservation Areas. Neither should provincial hunting, angling or other renewable resource harvesting licenses of general application apply in Conservation Areas. This will ensure the protection of biological and cultural diversity that such areas are intended to preserve. Unrestricted public access to and use of Conservation Areas would quickly undermine the conservation outcomes and objectives they were designed to achieve in the first place.

Balancing the 'constitutionally protected rights of Aboriginal communities with the interests of all Albertans,' however one interprets this statement, means that neither the province nor its First Nations/Métis communities has a veto over what can be done to the land, in this case, in Conservation Areas. As the SCC stated in *Haida*,[20] "what is required is a process of balancing interests, give and take;" in other words, finding a compromise that both can live with. However, in order to achieve this balance, tough choices and compromises must be made, and the *status quo* must be rejected. In other words, there can be no ultimate losers or winners, and all Albertans must benefit.

We must also bear in mind that the rights of Aboriginal Albertans are not absolute. They can justifiably be infringed, providing that there is: 1) a 'compelling legislative objective' (such as conservation or the taking up of lands in the national best interest), and 2) that such infringement upholds the honour of the Crown by consulting with the affected Aboriginal parties, mitigating any impacts arising from the infringement, and adequately compensating the infringement, where necessary. However, excluding non-Aboriginal hunters, fishers and trappers from Conservation Areas, who can prove that their use of such areas is integral to their culture and identity, is not the solution. As stated above, the latter have rights which should be retained in Conservation Areas under the conditions outlined above.

In stating that it will "strive for a meaningful balance that respects the constitutionally protected rights of aboriginal communities and the interests of all Albertans," the province is reaffirming its position in Confederation and support of *section 35* of the *Canadian Constitution Act* (1982). The nature and scope of Aboriginal and treaty rights protected under the Canadian Constitution are continually being clarified and refined by Canada's judicial system, and the SCC in particular. In the landmark decision in *R. v Sparrow*,[21] the SCC identified 'conservation' as a compelling and substantial objective that may justifiably infringe Aboriginal rights. Thus, where resources are limited and there is a conservation issue, conservation trumps Aboriginal rights. However, beyond conservation and public safety, the SCC has determined that Aboriginal rights to specific resources have priority over commercial rights, while

---

[20] *Haida,* at para. 46-47.
[21] *R. v. Sparrow,* [1990] 1 S.C.R. 1075.

commercial rights trump recreational interests. This may be an important and constitutionally-grounded guiding principle that will help establish resource use rights and priorities of renewable resources in Conservation Areas, and indeed all zones within each region. However, resource allocations cannot be determined without undertaking the appropriate research and consultations with First Nation and Métis communities in regard to matching Aboriginal and non-Aboriginal needs with resource availability. If we are to get this correct, we must acknowledge and plan for the research workload before us.

Size Matters!
In the eyes of Albertans, the requirement that Conservation Areas be 4,000-5,000 km^2 in size will surely be the linchpin of regional land use plans and ALUF. In the Lower Athabasca region, this requirement is driven, in large part, by the ongoing loss of biodiversity and fragmentation of the land base by industrial development, in addition to the spatial habitat requirements of woodland caribou—a keynote indicator species of Boreal forest health and biodiversity if there ever was one. To date, however, few Conservation Areas identified by the LARP Regional Planning Team appear to fulfill this size requirement. Moreover, most proposed Conservation Areas are located in the northern half of the region in Canadian Shield country far from bitumen-producing geological formations and oil/gas activity. Indeed, only 0.2% of the total combined Conservation area overlaps with what the province calls the 'Bitumen Zone,' an area comprising over half the region and a critical habitat for endangered woodland caribou. Further, as one proceeds from north to south, proposed Conservation Areas diminish in number and become smaller in size. By the time Treaty 6 traditional lands are encountered, less than 5% of the total area set aside for conservation remains.

While this current proposal for Conservation Areas will undoubtedly find disfavour among Treaty 6 First Nations, it is important to point out that the LARP Regional Advisory Council is considering meeting the size requirement by placing Conservation Areas adjacent to parks, recreation areas and other areas where conservation values can be sustained and there is little or no potential for industrial development. Sleight of hand, perhaps, but it does make sense, and is perhaps the only realistic option for meeting the size criteria given that Conservation Areas in the Lower Athabasca region are contingent upon reaching economic objectives. In this regard, several types of Conservation Area designations or classes, based on their degree of connectivity, may be warranted, each with its own management regime, in order to reach the 4,000-5,000 km^2 size requirement (Table 1).

In order that land use in Conservation Areas are properly managed in the name of preserving and protecting biological and cultural diversity, appropriate management policies, regimes and institutions need to be implemented. In this regard, co-management boards composed of provincial representatives, and local Aboriginal and non-Aboriginal land-users should be formed to manage land use and conservation activities within Conservation Areas. Such boards would, among other things, establish sustainable and allowable takes of game, fish, fur and other resources, and oversee remedial habitat restoration and environmental monitoring activities in Conservation Areas. Ensuring Aboriginal representation on Conservation Area Co-management Boards would not only place Aboriginal peoples in a position to once again assume their roles as stewards and caretakers of their traditional lands, but it would go a long way to balancing their constitutionally protected rights with those of other Albertans. The above proposal, in all likelihood, is the minimum that First Nations peoples and communities in the Lower Athabasca region would be willing to accept and support.

Table 1. Proposed conservation classes to meet size criteria.

Conservation Class/Category	Degrees of Jurisdictional Connectivity	Conservation and Co-management Jurisdictions
A	0	Provincial Crown lands only
B	1	Federal Crown lands (National Parks, Military Ranges, Indian Reserves)
C	1	Existing Provincial Crown lands (provincial parks, recreation areas, environmental reserves, etc.)
D	2	Federal Crown lands, Provincial Crown lands
E	3	Federal Crown lands, Provincial Crown lands, 'Best practices' industrial/ municipal/other use lands

Conclusions

Much is at stake for Alberta Treaty First Nations, the province and indeed all Albertans in implementing ALUF and designing regional land use plans that incorporate the constitutionally protected rights and interests of all Albertans. ALUF may not be able to correct past wrongs, but it cannot ignore the fact that achieving its promise is, as Brian Slattery (2006:282-83) suggests, contingent upon 'recognition and reconciliation,' i.e., "striking a balance that remedies past injustices with the need to accommodate contemporary interests." In the absence of these guiding principles, ALUF will prevent us from moving forward together in designing a future that is environmentally, socially, culturally and economically sustainable for all Albertans.

Currently ALUF, and the processes that it sets in motion, are far from perfect, and Alberta would do well to look at other provinces and their efforts to meaningfully incorporate Aboriginal peoples into land use planning. Ontario's *Far North Act*,[22] for example, establishes a First Nations-led land use planning process across Ontario's far north, while mandating a balanced approach to conservation and economic development (*see* Deutsche and Davidson-Hunt, this

---

[22] *Far North Act*, Bill 191, 2009. http://www.ontla.on.ca/bills/bills-files/39_Parliament/Session1/b191_e.htm

volume). The *Act* enables First Nations to work with the Minister to prepare land use plans with funding from the province. In particular, this community-based land use planning process allows First Nations and the province to jointly determine areas to be protected and to identify areas for economic development that benefit First Nations communities and consider their ecological and cultural values.

Alberta must take a proactive, not a disengaged and/or reactive, approach in dealing with its Aboriginal peoples in the context of resource development and land use planning. While resource revenue sharing with Treaty First Nations is something that the province is at least willing to entertain sometime in the future,[23] no longer tenable are its current practices of:

- viewing historic Treaties as land surrenders, rather than agreements to co-exist and share lands and resources,
- doing little or nothing to address First Nations concerns until they file claims in provincial or federal courts for breach of Treaty and fiduciary obligations, and
- failing to draw the line, and develop attendant policies, with respect to delineating and enabling the exercise of the rights of all its citizens.

At the same time, First Nation Elders and Chiefs can no longer rely exclusively on treaty-based positions to force Alberta to recognize and accommodate their constitutionally protected Treaty rights and interests. While this strategy has yet to produce the desired outcomes, there are signs, such as the *Protocol Agreement on Government to Government Relations* (Government of Alberta 2008b), that indicate that the province is willing to move, in measured ways, in this direction.[24] For these and other reasons, these high level nation-to-nation/government-to-government discussions should not only not be abandoned, they should be 'fast-tracked.'

Nonetheless, First Nation leaders ignore immediate opportunities and challenges at their peril. Treaty Indian peoples may not have set the rules of engagement, or even like the 'game' in which they are obliged to participate, but they cannot disregard them, as their rights and interests can be advanced under existing provincial policies and institutions, such as ALUF and the Regional Advisory Councils that the *Alberta Land Stewardship Act* creates. However, this necessitates the development of well thought-out and coordinated approaches that integrate both strategies. It also requires a commitment to engage provincial representatives and other stakeholders at all levels to educate and inform them about the spirit and intent of the Treaties. Existing Regional Advisory Councils

---

[23] Notes prepared by Tribal Chief Ventures Incorporated from meeting between Alberta Grand Chiefs and the Hon. Ed Stelmach, Premier, 20 May 2009.

[24] Other examples include: annual Premier-Grand Chief meetings, joint sub-table to revise Alberta's First Consultation Policy and specific First Nation consultation plans in the context of regional land use planning.

and First Nations specific consultation processes developed under ALUF may not be best the forums to effect this outcome. Even so, with adequate funding to support Aboriginal input, they at least provide an opportunity to reconcile the rights and interests of Alberta's First Nations with those of other Albertans at the 'ground level' in the contexts of resource development and land use planning. If small, multi-stakeholder groups of well-meaning and informed citizens from the 'grass-roots,' including representatives of Alberta's Aboriginal communities, cannot reach agreement and consensus on what needs to be done and how to do it, our hope is misplaced if we think that these issues are going to be resolved any time soon at higher political levels where discussions are lead by politicians, senior bureaucrats and their legal advisors.

References Cited

Chiasson, C. (2009). New legislation continues changes to Alberta's land use planning. *Environmental Law Centre News Brief* 24(2): 5-7.

CEMA—Cumulative Effects Management Association (2008). *Terrestrial Ecosystem Management Framework for the Regional Municipality of Wood Buffalo.* http://www.cemaonline.ca/index.php/cemarecommendations/terrestrial-ecosystem

Government of Alberta (2005). *The Government of Alberta's First Nations Consultation Policy on Land Management and Resource Development.* www.aboriginal.alberta.ca/.../Policy_APPROVED-May 16.

Government of Alberta (2006) *Aboriginal Input on the Provincial Land Use Framework Initiative: Summary Report.* Edmonton: Alberta Aboriginal Affairs and Northern Development http://caid.ca/AlbAborConLand111506.pdf November 2006

Government of Alberta—Alberta Aboriginal Affairs and Northern Development (2008a). *Land Use Framework.* http://www.landuse.alberta.ca/Default.aspx

Government of Alberta (2008b). *Protocol Agreement on Government to Government Relations.* Edmonton: Alberta Aboriginal Affairs and Northern Development http://www.google.com/searchq=Protocol+Agreement+on+Government+to+Government+Relations+Alberta&sourceid=ie7&rls=com.microsoft:en-US&ie=utf8&oe=utf8&rlz=1I7GGLL_en

Government of Alberta (2009a). *Alberta Land Stewardship Act*, Bill 36. Edmonton: Alberta Aboriginal Affairs and Northern Development http://www.qp.alberta.ca/574.cfm?page=A26P8.cfm&leg_type=Acts&isbncln=9780779742271\

Government of Alberta (2009b). *First Nation Consultation Plan—Lower Athabasca Region.* June 2009. Edmonton: Alberta Aboriginal Affairs and Northern Development, Land Use Secretariat.

Government of Alberta (2009c). *Terms of Reference for Developing the Lower Athabasca Regional Plan.* Alberta Aboriginal Affairs and Northern Development http://www.landuse.alberta.ca/RegionalPlans/LowerAthabasca/documents/TermsOfRefDevLowerAthabascaRegionalPlan-Jul2009.pdf

HLFN—Heart Lake First Nation (2008). *Heart Lake First Nation's Comments on Alberta's Draft Land-use Framework.* Submitted to the Government of Alberta, August 6, 2008.

LRRCN—Little Red River Cree Nation (2008). *Little Red River Cree Nation's Critique of the Draft Alberta Land Use Plan.* Submitted to the Government of Alberta, August 1, 2008.

Poschwatta-Yearsley, J. and A. Zelmer (2009). The *Alberta Land Stewardship Act*: Certainty or uncertainty? *Canadian Institute of Resource Law* 106: 1-10. www.cirl.ca.

Slattery, B. (2006). The metamorphosis of Aboriginal title. *Canadian Bar Review* 85: 255-286.

Statt, G. (2008). 'Consultation, cooperative management and the reconciliation of rights,' pp. 187-208 in D.C. Natcher, ed., *Seeing Beyond the Trees: The Social Dimensions of Aboriginal Forest Management*. ed. D.C. Natcher, 187-208. Concord, ON: Captus Press.

Water Matters Society of Alberta (2009). *Water Matters*. Lower Athabasca terms of reference explore new conservation areas and aggressive growth targets. September 14, 2009. http://www.water-matters.org/story/311

Webb, J.R. H. Sewepagaham and C. Sewepagaham (2009). 'Negotiating cultural sustainability: Deep consultation and the Little Red River Cree in the Wabasca-Mikkwa lowlands, Alberta,' pp. 107-126 in M.G. Stevenson and D.C. Natcher, eds. *Changing the culture of forestry in Canada: Building effective institutions for sustainable forest management*. Edmonton: CCI Press.

SECTION 2:
SETTING THE PLANNING STAGE
FOR CO-EXISTENCE

Chapter Three

Unfinished Business: The Intent of the Crown to Protect Treaty 8 Livelihood Interests (1922-1939)

Jimmie R. Webb

Introduction

This chapter examines federal Crown commitments to protect and safeguard 'livelihood' interests of Indians under the provisions of Treaty 8,[1] and post-Treaty Crown government policy initiatives to implement these commitments. Treaty 8 was the first numbered treaty to reflect a 'shared intent'[2] between the Crown and Indian peoples for the latter to exercise their "…usual vocations of hunting, trapping and fishing…"[3] within a large forest landscape that both parties believed would remain mostly unoccupied for the foreseeable future. This 'shared intent' was based on the insistence of the Indians that exploration, settlement and development within their territories not interfere with or diminish their way-of-life, and on the Crown's expectations that the Indians would remain economically self-reliant through the pursuit of their usual vocations of hunting, trapping and fishing, as these activities took place.

In *Delgamuukw v. British Columbia*[4] the Supreme Court determined that the honour of the Crown requires reconciliation of the pre-existence of Aboriginal societies with the sovereignty of the Crown. In *Haida Nation v. British Columbia*[5] the Supreme Court acknowledged that treaty-making gives

---

[1] Treaty No. 8, June 21, 1899, Queen's Printer, Ottawa, 1966.

[2] The Supreme Court of Canada in *R. v. Badger* [1996] 1 S.C.R. 771, provides guidelines for interpretation of the 'shared intent' of the Crown and Aboriginal peoples as the basis for Treaty relationships between the Crown and Treaty 8 peoples.

[3] Documentation of differences in meaning among and between the parties to these negotiations is relevant to interpretation of the 'shared intent' of these parties in relation to the Treaty phrase "their usual vocations of hunting, trapping and fishing." Analyzing a range of possible meanings grounded in the languages of the respective parties is a necessary step for selecting the best interpretation of the meaning of this phrase. The need for such an approach is a reflection of the fact that treaties are the product of a fusion between the Aboriginal legal systems and the common law of Canada, and the fact that treaty negotiations took place in several languages. These issues are discussed in the following chapter.

[4] *Delgamuukw v. British Columbia* [1997] 3 S.C.R. 1010.

[5] *Haida Nation v. British Columbia* [2004] 3 S.C.R. 511 para. 32.

rise to ongoing Crown obligations for the accommodation of First Nation interests:

> ...*the duty to consult and accommodate is part of a process of fair dealing and reconciliation that begins with the assertion of sovereignty and continues beyond formal claims resolution.* Reconciliation is not a final legal remedy in the usual sense... [It] flows from the Crown's duty of honorable dealing toward people and de facto control of land and resources that were formerly in the control of that people. (emphasis added)

Justice Binnie, in *Mikisew Cree Nation v. Canada*[6] interpreted Treaty 8 as affecting a surrender of Aboriginal title[7] and providing the Crown authority to 'take up lands.' However, he identified the need for reconciliation of ongoing tensions between the Crown's exercise of this right and its solemn commitments to protect and support the Indian's use of 'lands not taken up' for "...their usual vocations of hunting, trapping and fishing." As noted by Justice Binnie,[8] "negotiation of Treaty 8 was not expected to produce a future land use blueprint for the territory, and negotiations (1899) were no more than a first step in a long process of reconciliation."[9]

It follows that the honour of the Crown requires contemporary implementation of yet 'unsatisfied' Treaty 8 promises to protect and safeguard First Nation interests in lands and resources. The Crown's right to take up lands is subject to a duty to consult and, if appropriate, to accommodate Treaty-protected First Nation livelihood interests before reducing or otherwise degrading an area within which First Nation members continue to exercise their usual vocations of hunting, trapping and fishing. The large volume of documented claims alleging breach of Crown Treaty obligations, filed before the courts,[10] or presented to the federal Crown for negotiation,[11] demonstrate the

---

[6] *Mikisew Cree First Nation v. Canada* (Minister of Canadian Heritage) 2005 SCC 69, para. 56.

[7] There is ambiguity between the 'non-specific surrender' clause of the Treaty and the specific Crown agreement that Indians could use all lands not 'taken up' for conduct of their usual vocations, which becomes more salient in the context of differences in the understandings of the meaning evoked by translation of the English text of the Treaty into indigenous languages (*see* next chapter).

[8] *Mikisew* at para. 27.

[9] *Mikisew* at para. 56.

[10] One such 'statement of claim' was filed before the Court of Queen's Bench of Alberta, on behalf of the Little Red River Cree Nation, on 4 May 2000. This legal action is currently held in abeyance to allow for negotiation.

[11] In 2001, the Co-Chairs of the Indian Claims Commission (ICC) advised the House of Commons–Aboriginal Affairs Committee that some 585 specific claims had been filed nationally; 115 were being negotiated, and 61 were under review by the ICC. At that time, there was a backlog of 408 specific claims awaiting federal validation (ICC presentation to the House of Commons Committee, 29 May 2001, as cited by the National

salience of this tension. Reconciliation of Treaty 8 livelihood interests within Alberta is being seriously pursued under the federal Crown Specific Claims policy,[12] and the federal Crown Inherent Right to Self Government policies.[13] Treaty 8 First Nations are also attempting to use consultation with the Crown to affect interim accommodation of their asserted livelihood interests as negotiations take place.

Hunting, Trapping and Fishing, or Livelihood?

This chapter is about more than hunting, trapping and fishing; these were not the only historic natural resource harvesting activities undertaken by Treaty 8 Indians during the fur trade era, and fish, fur-bearers and wildlife were not the only natural resources used to support Indian livelihood activities prior to negotiation of Treaty 8. Indians gathered and used a broader range of natural resources to manufacture products and trade goods, and to support their provision of services associated with the fur-trade. This larger range of resource-based livelihood activities continued to be central to the identity of Treaty 8 Indians and a significant part of an Aboriginal 'mixed economy,' which involved the exchange of goods and services to sustain the fur trade and other European industries, both before and following the negotiation of Treaty 8 (Ray 1995; Standfer 2004; Tough 1996).[14] The term 'livelihood,' as used here, refers to First Nation rights of access to Treaty 8 lands 'not taken up' by the Crown and rights to harvest and use a broad range of natural resources to support historic and contemporary participation of Treaty 8 peoples in natural resource-based trade and commerce processes.[15] The term 'Treaty-Protected Aboriginal livelihood interest'[16] refers to that specific 'bundle'[17] of vocational interests reserved by

---

Aboriginal Forestry Association presentation to the United Nations Forum on Forests, April 2002).

[12] The Little Red River Cree Nation 'Hunting, Trapping, and Fishing' claim, was filed in October 2002, and is currently under review by the federal Department of Justice.

[13] The Treaty 8 First Nations of Alberta are engaged in Bilateral Treaty-based Governance and Treaty Implementation negotiations with Canada, under provisions of a 'Declaration of Intent,' signed on 22 June 1998. The provincial government attends these negotiations as an 'observer,' and has been requested to formally participate in order to facilitate negotiations about matters under provincial jurisdiction.

[14] Provision of this range of services and Indian livelihood activities related to provision of other manufactured goods from natural resources are often simply characterized as 'wage labour' for the Trading Posts, or for settlers. These commercial activities can be more accurately characterized as provision of contracted services, since most did not involve employee–employer relationships.

[15] Note: In *R. v. Horseman* [1990] 1 S.C.R. 901, Arthur Ray provided expert testimony that the Treaty 8 phrase "their usual vocations of hunting, trapping and fishing" also refers to Indian participation in the fur-trade.

[16] The definition of 'Treaty-protected Aboriginal Rights' refers to portions of Aboriginal title, which Indians reserved with the agreement of the Crown, under the terms of treaties. Use of lands and resources within

Treaty 8 Indians with agreement by the Crown. This chapter proposes that these vocational or 'livelihood interests' provide a *sui generis* basis for the conduct of Crown–First Nations negotiations focused on the need for equitable reallocation of natural resources[18] within Treaty 8 traditional territories, and the commercial development of this broader range of natural resources to create and sustain First Nation economies. See Chapter 4 for a lengthy discussion of the terms 'livelihood' and 'vocation.'

The Crown's Commitments to Protect the Livelihood Interests of Treaty 8 Indians

Prior to commencement of Treaty 8 negotiations, one of the Treaty 8 Commissioners, J.A.J. McKenna, wrote to the Minister of the Interior, Clifford Sifton[19] regarding the need for Crown commitments to protect the Indians' usual vocations of hunting, trapping and fishing from 'white' settlers. Excerpts from McKenna's letter demonstrate Crown awareness of a need for ongoing consultations with Treaty 8 peoples in relation to the protection of Indian

---

the Treaty 8 territory for conduct of their usual vocations was not a right granted by the Crown; Indians possessed this right under their natural law. The Treaty Commissioners agreed that the Indians could reserve it, and placed it under the protection of the Crown (*see* James [*Sakej*] Henderson-Youngblood 1997:75-78).

[17] The term 'bundle of rights' is used by Madam Justice L'Heureux-Dubé, in *R. v Van der Peet* [1996] 2 S.C.R. 507. She advised that a treaty may affirm the reservation of some aspects of Aboriginal title, and identified the need to examine the 'bundle of rights' reserved by the Indians, with agreement of the Crown, in order to determine whether this has resulted in the establishment of treaty-protected Aboriginal rights.

[18] The Supreme Court of Canada decision in *R. v Marshall* [1999] 3 S.C.R. 456 ruled that, while the subject of Micmac trading rights involved liberties enjoyed by all citizens, their conferral in a treaty afforded the Micmac peoples protection against Crown infringement. It follows that a treaty-protected livelihood interest is *sui generis.* Subsequent to this ruling, Crown governments undertook to negotiate equitable re-instatement of Micmac access to the maritime fishery, and to reconcile Micmac interests with other interests. I propose that Treaty 8 provides a similar *sui generis* right to livelihood use of natural resources by Treaty 8 peoples and requires similar Crown actions to effect reconciliation.

[19] Fumoleau's (1975:65) review of Treaty 8 negotiations reflects first-hand accounts of the Indians, Treaty Commissioners, Oblate missionaries, and other non-Aboriginal witnesses and participants party to Treaty negotiations. His review documents the active participation of Oblate priests and bishops in an on-going dialogue with the federal government about the need for protection of Indian livelihoods prior to, during and after negotiation of Treaties 8 and 11. He, and other historians, have documented the activist role of the Church in collecting and preserving accounts of first-hand witnesses to treaty negotiations, and the active correspondence between/among Bishop Breynat, James K. Cornwall, federal Ministers and senior officials within Indian Affairs and the Ministry of the Interior during the post-Treaty period.

livelihood interests following negotiation of the Treaty. There is a strong correspondence between McKenna's concerns for protection of Aboriginal livelihood interests as a means of inducing the Indians to enter into treaty relations, and James Cornwall's written affidavit regarding his recollections of Treaty 8 negotiations at Lesser Slave Lake and Peace River Crossing (as cited in Fumoleau 1975:74-75, format follows Fumoleau):

> 5. The Commissioners finally decided after going into the whole matter, that what the Indians suggested was only fair and right but that they had no authority to write it into the Treaty. They felt sure that the government on behalf of the Crown and the Great White Mother would include their request they made the following promises to the Indians:
>> a. Nothing would be allowed to interfere with their way of making a living as they were accustomed to and as their forefathers had done.
>> b. The old and the destitute would always be taken care of, their future existence would be carefully studied and provided for, and every effort would be made to improve their living conditions.
>> c. They were guaranteed protection in their way of living as hunters and trappers from white competition: they would not be prevented from hunting and fishing as they had always done, so as to enable them to earn their living and maintain their existence.
> 6. Much stress was laid on one point by the Indians as follows: They would not sign under any circumstances unless their right to hunt, trap and fish was guaranteed and it must be understood that these rights they would never surrender."[20]

Fumoleau's (1975) review of historic Treaty 8 negotiation processes led him to conclude that it was the Treaty Commissioners' promises that the Indians would be free to hunt, trap and fish for a living, and that their livelihood rights would be protected against the abuses of white hunters and trappers, that produced the Indians' consent to Treaty. There is little doubt that the Treaty 8 Commissioners intended to, and did, provide undertakings to protect Indian livelihood interests during Treaty 8 negotiations. They believed such undertakings were necessary to induce the Indians to agree to the Treaty, even though they do not appear in the official text of the Treaty 8, or in the Report of the Treaty Commission (Laird *et al.* 1899[1966]:47). Notwithstanding their

---

[20] Affidavit of James K. Cornwall, 1 November 1937. The Oblate Bishop Breynat also collected 37 signed affidavits from other first-hand witnesses to the Treaty negotiations in 1937. Cornwall, known as 'Peace River Jim' was a trader and entrepreneur, and an advocate of the need for protection of Indian interests.

status in some quarters as 'outside promises,'[21] there is a clear record that such undertakings were made by the Treaty Commissioners on behalf of the Crown. In *R. v. Horseman* (1990), Justice Wilson (in dissent), noted the importance of Treaty 8 livelihood promises to the establishment of Treaty 8 relations:

> Indeed, it seems to me to be of particular significance that the 8 commissioners, historians who have studied Treaty 8, and the Treaty 8 Indians of several different generations, unanimously affirm that the government of Canada's promise that hunting, fishing and trapping rights would be protected forever was the sine qua non for obtaining the Indians' agreement to enter into Treaty 8... . Hunting, fishing and trapping lay at the centre of their way of life. Provided that the source of their livelihood was protected, the Indians were prepared to allow the government of Canada to 'have title' to the land in the Treaty 8 area.

These 'outside promises' created Crown obligations to undertake post-treaty consultations with Treaty 8 Indians (Borrows and Rotman 1997:21), and to safeguard the treaty-protected Aboriginal livelihood interests reserved under the treaty.

The Natural Resources Transfer Agreement, 1930 (Alberta)

The NTRA and Robert Horseman

The enactment of the *Natural Resources Transfer Agreement* (NRTA) in 1930 made the livelihood interests of Treaty 8 Indians subject to provincial laws. Subsequently, the enforcement actions taken by Fish and Wildlife against Treaty Indians after enactment of the NRTA have resulted in Court rulings that restrict Indian hunting interests. This line of restrictive judicial reasoning includes the Supreme Court of Canada's decision in *R. v. Horseman*, which established a presumption at law that sec. 12 of the NRTA reflected a federal Crown intent to extinguish 'commercial interests' of Treaty 8 Indians within lands transferred to Alberta. This, as will be shown, was not Canada's intent at all.

There is a large body of previous judicial commentary[22] and academic criticism (Irwin 2000; Ray 1995; Tough 1996) regarding this presumption at law, the lack of evidence to support it, and the impact of the *Horseman* decision upon contemporary efforts of Treaty 8 First Nations to implement processes for

---

[21] 'Outside promises' is the term Price (1987:4) uses to refer to a range of commitments, undertakings and promises made by the Treaty commissioners during treaty negotiations. The Indians, whose custom and laws are grounded in oral history, consider them to be as much a part of the treaty as the English-language document, which they could not read, and which they had to trust was an accurate translation into the indigenous languages used for the conduct of negotiations.

[22] For example, the comments of dissenting Justices in *R. v. Horseman*, and those of dissenting Justices of the Alberta Court of Appeals, in *R v. Badger* [1996] 1 S.C.R. 771.

accommodation and reconciliation of so-called 'unproven'[23] Treaty 8 livelihood assertions and claims. This body of evidence is expanded and I propose that, if the Courts in the *Horseman* hearings had been provided the information outlined below as to the specific two-fold intent of the federal Crown and the applicability of additional sections of the NRTA (1930) for analysis of the matter then before the Courts, the presumption reached by the Supreme Court in the *Horseman* decision (1990) may have been very different. By comparison, more recent comments from the Trial Court in *Mikisew*, demonstrate that similar federal Crown arguments regarding the 'extinguishment' of Treaty 8 hunting, trapping and fishing interests incident to the creation and subsequent administration of Wood Buffalo National Park in 1922, are 'without substance and self-serving.'[24]

A Specific Crown Intent to Establish 'Special Reserves'

The historic record of federal-provincial negotiations regarding the establishment of 'special reserves' as a means of protecting Treaty 8 livelihood interests following the transfer of natural resources to the province of Alberta, clearly refutes the presumption at law that sec. 12 of the NRTA had the effect of extinguishing the commercial hunting interests of Treaty 8 First Nations. Archival records demonstrate that negotiations between Canada and Alberta for establishment of 'special reserves' to protect Indian livelihood interests within Treaty 8 territory commenced in 1923 and continued for another 16 years. These negotiations were but one aspect of ongoing Crown negotiations related to the NRTA (1930) and the establishment of a post-NRTA provincial governance regime over First Nation use of natural resources within provincial Crown lands.

Letters and memoranda used to support these federal-provincial negotiations (1922-1939)[25] provide *prima facie* support for the assertion that the federal Ministry of the Interior expressed a two-fold intent regarding Indian interests, incident to negotiation of the NRTA with the Alberta Government:

> 1) to protect the right of all Indians within the province, including Treaty 6, 7 and 8 Indian bands, and a number of known non-Treaty Indians, to hunt, trap and fish for their support and sustenance under para. 12 of the NRTA; and

---

[23] Note: These assertions are 'unproven' to the extent that the presumption at law (established by *Horseman*) can be relied on to deny the existence of these treaty-protected livelihood interests after enactment of the *Natural Resources Transfer Agreement* (1930).

[24] *Mikisew First Nation v. Canada*, Trial Division [2002] 1 C.N.L.R. 169

[25] Hunting, Trapping and Fishing Claim of the Little Red River Cree Nation, (Havlik–Metics Research Group 2002). This specific claim was prepared on behalf of the Little Red River Cree Nation, and submitted to the Department of Indian Affairs under provisions of the federal Specific Claims policy.

2) to protect and safeguard the livelihood interests of identified Treaty 8 Indian bands from 'white competition' within the northern portion of the province under general provisions for constitutional protection of existing trusts and interests within the NRTA.

There are three additional sections of the NRTA that must be considered when interpreting federal Crown intent related to the effect of the NRTA upon Treaty 8 First Nation livelihood interests:

Paragraph 1: this paragraph states that the transfer of lands to the Province is "subject to any *trusts* existing in respect thereof and to any interest other than that of the Crown in the same...."

Paragraph 2: this paragraph provides, in part, that the Province will carry out "every other arrangement whereby any person has become entitled to any interest therein as against the Crown, and further agrees not to affect or alter any term of any... arrangement by legislation or otherwise, except *either with the consent of all* the parties thereto other than Canada or insofar as any legislation may apply generally to all similar agreements relating to lands, mines or minerals in the Province or to interest therein, irrespective of who may be the parties thereto."

Paragraph 10: this paragraph provides that "... the province will from time to time, upon the request of the Superintendent General of Indian Affairs, set aside, out of the unoccupied Crown lands, hereby transferred to its administration, such further areas as the said Superintendent General may, in agreement with the appropriate Minister of the Province, *select as necessary to enable Canada to fulfil its obligations under the Treaties with the Indians of the province, and such areas shall thereafter be administered by Canada* ..." (emphasis added).

Historical documents recording post-treaty federal government efforts to identify and establish a number of 'special reserves' demonstrate a clear federal Crown intent to protect the livelihood interests of identified Treaty 8 Indian bands within Alberta, British Columbia and the Northwest Territories (Havlic-Metics Research Group 2002). These documents also show that the federal government's initiative to establish 'special reserves' for Treaty 8 First Nations within Alberta was part of a broader federal government policy for establishment of 'game preserves' within all Canadian provinces and territories. These documents record federal government actions (1922-1939) to identify specific areas within Alberta required for use by specific Treaty 8 Indian bands, and lengthy negotiations between the Dominion and the provincial government for the establishment of 'special reserves' on behalf of Treaty 8 Indian bands. These documents also reflect historic federal government awareness of obligations to consult with these Indians, and to obtain Indian agreement as to the provisions of the NRTA affecting Treaty 8 livelihood interests.

Seven such 'game preserves' were established under this federal policy for a number of Treaty 8 and Treaty 11 bands (1923). While return of lands by Alberta to Canada for fulfillment of treaty obligations was possible under para. 10 of the NRTA, no agreement was reached with Alberta for the return of provincial lands to the federal government or the establishment of 'special reserves' to protect Treaty 8 livelihood interests within provincial boundaries.

Following a review of the historic record below, it is argued that the abandonment of Crown negotiations regarding the establishment of 'special reserves' as a means to protect Treaty 8 livelihood interests within the province of Alberta (Government of Alberta 1938) created a fundamental breach of Crown Treaty obligations toward Treaty 8 peoples. This Crown obligation remains unsatisfied and implementation of a similar or alternative Crown regime for the ongoing accommodation of Treaty 8 livelihood interests constitutes *'unfinished business'* related to reconciliation of Aboriginal–Crown relations within the Treaty 8 area of Alberta.[26]

The Historical Record (1922–1939) Regarding Special Reserves in Alberta
The original Indian request for establishment of a 'special reserve' to protect Treaty 8 Indian livelihood interests within the province of Alberta is found in the 1922 *Report of The Treaty Annuity Payment for the Chipewyan Band.*[27] Formal discussions between Charles Stewart, federal Minister of the Interior, and George Hoadley, Minister of Alberta Agriculture (1923), led the provincial government to ask for identification of districts within Alberta where the federal government proposed to establish 'special reserves.'[28] Written documentation of federal Crown intent to establish 'special reserves' within the Edmonton, Saddle Lake and Lesser Slave Lake Indian Agencies is found in internal memoranda from Duncan Campbell Scott.[29] In these memoranda, Scott identifies eight groups of Indian Bands for which 'special reserves' were contemplated, and provides information about total population, the male population for each group and the locations and sizes of the eight proposed 'special reserves,' which varied from 50-75 square miles to 3,500-5,600 square miles. Scott's memoranda include a copy of the federal Order in Council for establishment of the seven 'game preserves' within the NWT, and provides an outline for the conduct of a

---

[26] As *R v. Badger* (1996) established the presumption at law that the Crown will honour its Treaty commitments.

[27] NAC RG10, Vol. 4049, File 361, 714. The record of Crown discussions and correspondence is documented in the Havlic-Metics Research Group (2002) report.

[28] NAC RG10, Vol.6732, File 420-2b: Letter from G. Hoadley to Hon. Charles Stewart, 6 March 1923; and letter to J. D. McLean, Secretary of the Department of Indian Affairs, 31 October 1923.

[29] NAC RG10, Vol. 6732, File 420 –2B, Deputy Superintendent General of Indian Affairs to Charles Stewart, 13 November 1923. Scott advised the Minister that the other Indian Agencies within Alberta were located within the more settled parts of the province where the Indians had already taken up farming.

proposed federal–provincial consultation process with identified Indian bands in Alberta, missionaries, fur trading posts and other interested parties, incident to establishment of 'special reserves,' within the province of Alberta.[30]

Ongoing discussions with Alberta in 1924 about the pending transfer of natural resources to Alberta, prompted the Deputy Minister of the Interior [31] to ask Minister Scott for an outline of federal Crown commitments to provide for the interests of Indians incident to the proposed transfer. Duncan Campbell Scott's response to this request clearly identifies the twofold intent of the Department of Indian Affairs to both protect the general right of all Indians to hunt for sustenance, and to establish hunting and fishing reserves as a means of protecting Treaty 8 Indian livelihood interests.[32] This memorandum also provides a clear indication by Scott that Indians would be subject to provincial laws:

> While the Indians shall be subject to the game laws of the Province, provision should be made for hunting and fishing reserves, and for exemptions in favour of Indians who are hunting and fishing purely for their own sustenance.

Another memorandum from Duncan Campbell Scott regarding the establishment of 'special reserves,' notes a demand made by the provincial Minister that Indians confine their trapping activities to the areas to be set aside for their exclusive use, and records the federal minister's observation:

> "...it is obvious that if the Indians are to confine their trapping activities to the areas set aside for their own exclusive use, they will in effect be waiving their Treaty right to trap anywhere in the province. It is assumed that any such waiver can only be made by the Indians themselves, and the attitude which they might take towards any such proposition has not been discussed with the Indian Department." [33]

This paragraph supports the observation, made by Justice Kerans in *R. v. Badger* (1993),[34] that the federal government viewed Treaty 8 Indian livelihood interests to be matters which reflect an 'arrangement requiring the informed consent' of the Indians for Crown actions to modify or extinguish such interests, incident to the transfer of natural resources to Alberta as required under para. 2 of the NRTA.

Provincial receipt of schedules providing metes and bounds (size and boundary) descriptions of the 'special reserves' are well documented.[35] J.D.

---

[30] Order in Council 1862, 22 September 1923.

[31] NAC RG10 Vol. 6820, File 492–4–2, Pt 1.

[32] NAC RG10 Vol. 6820, File 492–4–2, Pt 1 *"Memorandum far as the Indian Interest is concerned,"* 29 January 1925.

[33] NAC RG10, Vol. 6732, File 420–2B, 27 September 1926.

[34] Kerans J. A., in *R. v Badger*, Alberta Court of Appeal Docs, Edmonton Appeal 9203-0131-A, 9203-1032-A and 9203-1032-A, 1993.

[35] For example, letter from G. Hoadley, *Schedule of Proposed Hunting and Trapping Reserves...*, NAC RG10 Vol. 6732, File 420 – 2B, 5

McLean, Assistant Deputy Minister and Secretary for Indian Affairs, subsequently confirmed that these schedules identified all areas within Alberta required by the Dominion government of Canada for establishment of the proposed 'special reserves.'[36]

However, whether Indians would retain Treaty rights outside of the proposed 'special reserves' became an issue for the province, to which J.D. McLean provided the federal government's position:

> "In response to the question... I may say that I presume that in the event of the preserves being established, special regulations for their control would be enacted by the Lieutenant Governor in Council. In the Northwest Territories, where, as you are aware, similar preserves have been established, the Indians are not confined to them."[37]

This advisement confirms the understanding of the federal government that the livelihood interests of Indians within the 'special reserves' would be subject to provincial legislation, while indicating that the Dominion Government was not prepared to unilaterally extinguish Treaty 8 Indian livelihood interests outside 'special reserves.'

At a Conference of Federal and Provincial Game Officials in Ottawa in January 1928, the Dominion Government presented a policy statement,[38] which outlined their intent to establish 'game preserves' in a number of provinces and identified actions taken by Parliament to create seven such Indian 'game preserves' having a total land area of 518,230 square miles within the Northwest Territories.[39] This federal policy statement, which proposed that the creation of 'special reserves' was the only means for the Dominion Government to respond to the growing destitution of Indian peoples resulting from increasing white competition with Indian use of Crown lands, was supported by delegates to the conference via resolution and delivered to J.D. McLean on 28 January 1928.[40]

In 1928, Charles Stewart advised Parliament that the province of Quebec had set aside large exclusive Indian hunting, trapping and fishing areas within northern Quebec, and that he was undertaking discussions about the need for

---

April 1927. In another letter, Indian agent G. Card to J.D. McLean (NAC RG10, Vol. 6732, File 420-2B), identified two proposed 'special reserves' for Treaty 8 bands within Alberta under the administration of the Fort Smith Indian Agency. Note also that one of the 'special reserve' areas identified by Laird, on behalf of three Treaty 8 bands administered by the Lesser Slave Lake Indian Agency, is situated within the northeast portion of British Columbia.

[36] NAC RG10, Vol. 6732, File 420 –2B, 18 October 1927.

[37] NAC RG10, Vol. 6732, File 420 –2B, 17 November, 1927.

[38] Order in Council 1862, 22 September 1923.

[39] Note: Wood Buffalo Park, as it existed in 1923, was one of the 'game preserves' established within the NWT. This 'game preserve' includes federal Park lands situated north of the 59th parallel, within the province of Alberta.

[40] NAC Vol. 6731, File 420 –1 –2.

similar protective measures with representatives of the prairie provinces and Ontario.[41] On 14 February 1929, the Premier of Alberta wrote Stewart to advise that the province was:

> ...in sympathy with the idea that so far as reasonably possible our aborigines should be safeguarded, and we are not opposed to the principle of having *certain reserves* set aside for the exclusive use of hunting Indians. There are, however, certain questions to be worked out before any such policy could be put into effect... . I have reference to your letter to Mr. Hoadley, the Minister of Agriculture, with whose Department any such policy would have to be completed (emphasis added).[42]

These documents demonstrate the existence of a federal/provincial Crown Ministerial agreement-in-principle to protect and safeguard the livelihood interests of Treaty 8 Indian bands prior to enactment of the NRTA (1930). While the letter from the Premier of Alberta begs identification of "...*certain questions to be worked out...* ," research undertaken by the Indian Association of Alberta (n.d.) documents that after the NRTA was enacted, provincial government officials continued to profess agreement-in-principle with the federal government's proposal for establishment of 'special reserves,' subject to:

- a reduction in their size;
- provincial insistence that the Indians could only trap within these 'special reserves;'
- provincial insistence that establishment of the 'special reserves' would not restrict development of other natural resources; and
- the necessity for registering other trap line areas for whites and "half-breeds."[43]

Fiscal Responsibility for Alberta Métis and the Establishment of Special Reserves

The relationship between the reference by Alberta's Premier to 'our aborigines,' and the ultimate failure of the federal Crown to negotiate successfully for establishment of 'special reserves' on behalf of Treaty 8 Indian peoples within the province of Alberta, can be better understood by an examination of the historic federal–provincial discussions about the provision of government services to destitute Métis.[44] These government discussions were coloured by

---

[41] Charles Stewart, House of Commons, as reported in *Hansard*, 6 June 1928, pg. 3825.
[42] NAC RG10, Vol. 6731, File 420 –1.
[43] As cited in the 14 February 1929 letter from the Premier of Alberta, and Provincial Archives of Alberta; Premier's Papers file 0015, letter from B. Lawton to George Hoadley, 19 February 1929.
[44] *Constitution Act, sec.35.1* [1982] defines Aboriginal peoples as including Indians, Inuit and Métis peoples of Canada. In this chapter, the

the impact of a general economic collapse of provincial and federal revenues, the growing urgency to manage the aggressive expansion of the fur trade within northern Alberta and provincial revenue generated by fur taxes.

There is a clear record of the Treaty Commissioners' oral undertakings to "...protect the Indians, in the conduct of their usual vocations," to "...care for the elderly and the destitute..." and to provide for the education and health care of Indians.[45] No such Crown commitments regarding Aboriginal communal interests or the provision of government services were given to Métis individuals within these territories who elected to take scrip in return for a surrender of their Aboriginal title and interests (Madill 1986:66). After self-identification and acceptance of scrip from the Half-Breed Commission, these Métis men, their wives and children, and their descendents, were not considered to be Indians under the terms of article 91 (24) of the *Constitution Act* (1867) and the Dominion Government acknowledged no further obligations to protect their interests or to provide for their welfare. From the perspective of both the federal and provincial Crowns, these Métis individuals, through their acceptance of scrip, became citizens with the same individual rights and responsibilities as other Canadians. However, the poor physical and economic circumstances of the provincial Métis population, and the growing cost of providing provincial government services to the Métis during the period following establishment of the Province of Alberta (1905-1930), led Alberta to create a provincial government Royal Commission on the Half-breed Population (Government of Alberta 1936), in which Canada refused to take part (Martin 1989:250).[46] As this provincial Royal Commission process concluded, the Government of Alberta considered provincial establishment of a number of 'Half–breed Colonies' where destitute Métis individuals and their families could settle and be provided government assistance to undertake farming as a livelihood, with inclusion of enough forest lands within these Colonies for the Métis, who chose to settle within them, to pursue hunting, trapping and fishing as a means of subsistence (Saunders 1978:22).[47] At the same time, and in exchange for releasing provincial Crown land for the creation of special reserves to protect the livelihood interests of Treaty Indians, the province attempted to convince the Dominion Government to take the general Métis population of Alberta under federal 'wardship,' and assume fiscal responsibility for the health, education and social services to the Métis.[48]

---

terms 'Aboriginal' have a common definition.

[45] Report of the Commissioners for Treaty No.8, Winnipeg, Manitoba, 22 September 1899.

[46] H.V. Martin cites a provincial 'Memorandum,' filed by George Hoadley regarding federal refusal to participate in the Ewing Commission, page 256.

[47] The provincial government's initiative for establishment of Métis Colonies, and the provision of provincial government assistance to Métis who chose to settle within the Colonies, was not related to provincial Crown recognition of Métis Aboriginal claims, which were considered by the provincial government to have been extinguished through issuance of scrip.

[48] Alberta's initiative to convince Canada to make the general Métis

... the Province (of Alberta) is willing to co-operate fully with the Dominion in an effort to solve the problems, both of the Indian and the Half-breed population, and if the federal government would resume responsibility for the Half-breed as well as for the Indian, the Province would willingly make available to the dominion whatever lands might be required for the establishment of reserves.[49]

However, Canada rejected the province's request to assume fiscal responsibility for the provision of government services to the Métis as condition for provincial protection of the livelihoods interests for Treaty Indians:[50]

I read with much interest the report of the Royal Commission appointed in December, 1934. It would seem, however, the recommendations should be implemented by the responsible constitutional authority, which is the Government of the Province... . (Métis) legal status is not differentiated from that of the rest of the community. They accepted the scrip of their own choice and...it would be contrary to the policy of this administration generally to permit half–breeds to revert to wardship in any degree under Federal authority... . Under these circumstances, I regret to inform you that the Dominion Government cannot, as you suggest, accept any responsibility for them....

Following this exchange, Charles Camsell, the federal Deputy Minister of Mines and Resources,[51] proposed a strategy for working cooperatively with the Government of Alberta to establish a series of 'hunting and trapping areas' in support of both the Half–breed and Indian populations of northern Alberta. Federal-provincial discussions regarding the return of provincial Crown lands to federal administration for the establishment of exclusive 'hunting and trapping areas' for use by Métis and Indians were stalemated, however, by the Dominion Government's ongoing refusal to assume the provincial government's fiscal burden for provision of relief, hospitalization and education to the Métis of Alberta.

---

population of Alberta federal 'wards' reflects the Ewing Report recommendation that any Métis who chose not to settle within these proposed Colonies should have no claim to public assistance from Alberta.

[49] Indian Affairs, RG 10, Vol 6733, file 420-2-1, 10 December 1937, A.N. Tanner, Alberta Minister of Lands and Mines, to T.A. Crerar, Minister for Mines and Resources Canada,

[50] Indian Affairs, RG 10, Vol 6733, file 420-2-1, T.A. Crerar, Minister for Mines and Resources Canada, to N.E. Tanner, Minister, Alberta Lands and Mines, 10 January 1938; signed on 12/1/38.

[51] NAC RG10, Vol. 6733, File 420–2–1, 18 January 1938, 'Memorandum' from H.W. McGill, Director of the Indian Affairs Branch.

Thus, in 1938, the Alberta Legislature passed the *Métis Population Betterment Act*,[52] and established a number of Métis Colonies with a protected land base of 1.26 million acres for settlement of the destitute Métis population. Métis settlers were provided exclusive, provincially regulated, access to and use of other renewable natural resources within the Colonies, (e.g., fisheries, fur-bearers, game and timber), as well as commercial fishing rights (i.e., one-half of provincial quotas for commercial fisheries), within waters adjacent to the Colonies.

Shortly thereafter, the Province of Alberta proceeded to unilaterally develop and implement provincial trapline registration regulations, which prompted W.W. Cross,[53] the federal Deputy Minister of Mines and Resources, to request that the province consider issuing 'communal licenses' over large portions of northern trapping districts to Indian Bands, instead of implementing the proposed provincial scheme for issuing smaller trapping areas to individual trappers. However, the provincial Minister of Agriculture's[54] response to this request for establishment of communal Indian trapping areas reflects an emerging provincial belief that Indians should not have special rights and interests:

> ...the regulations established for Alberta would conflict with this plan, and as we are trying our best to show no partiality to any group of individuals we would at once be petitioned by other groups of Half–breeds and whites for similar privileges... . The Indian has the same right as the white person to a registered trapline so we do not see where any great difficulty will be experienced in obtaining registered lines.

This exchange of correspondence demonstrates that, while the Dominion Government was still undertaking to implement a policy for protection of the communal livelihood interests of Treaty 8 Indian bands, the provincial government would not recognize any difference in Indian and non-Indian interests associated with hunting, trapping and fishing. Subsequent to the passage of the *Métis Population Betterment Act* (1938), and implementation of the provincial trapline system (after 1939), the Dominion Government abandoned all efforts to establish 'special reserves' as a means of protecting the livelihoods of Treaty 8 Indians. In the end, no protected land base, comparable to the Métis Colonies, was established to protect and safeguard Treaty 8 Indian livelihood interests.

Notwithstanding the historic failure of the Dominion Government to implement the federal policy for establishment of 'special reserves' within the Province of Alberta, the archival record of these Dominion–provincial negotiations provides a well-documented historic federal Crown intent to identify and protect the livelihood interests of Treaty 8 Indians. Federal

---

[52] Eleven Métis Colonies were originally established. However, three were subsequently closed, leaving eight.

[53] NAC RG10, Vol. 6733, File 420–2–1, 5 December 1939.

[54] NAC RG10, Vol. 6733, file 420–2–1.

Ministerial actions in these regards, clearly refutes provincial Crown assertions that it was the federal Crown's intent to extinguish these livelihood interests incident to passage of the NRTA.[55]

'Unfinished Business,' Contemporary Negotiations and the Honour of the Crown

In *Haida*, the Supreme Court advised that sec. 35 (1) of the *Constitution Act*, provides a promise of rights recognition, which must be realized through a process of honourable negotiation.[56] The Treaty 8 First Nations of Alberta[57] are currently engaged in negotiations with the federal Crown for Treaty implementation and for the establishment of a Treaty-based First Nation governance regime. These bilateral negotiations include formal processes for undertaking joint research on the nature and scope of Treaty 8 livelihood interests, and Crown obligations to protect them, and for use of these findings under the auspices of an 'interim neutral facilitator,' or a Treaty Commission,[58] to develop mutual understandings as to the nature and scope of existing Treaty-protected Aboriginal livelihood interests.[59]

Crown commitments to protect and safeguard Treaty 8 livelihood interests, and the record of federal government actions to establish 'special reserves' within the Province of Alberta are part of the information provided to support these formal negotiation processes. The Government of Alberta is an 'observer' to the bilateral process and has received copies of this information. It is proposed that enactment of the NRTA, and establishment of a provincial Crown regime for discretionary control over treaty-protected Aboriginal livelihood interests without concluding the federal-provincial negotiations to protect and safeguard the livelihood interests of Treaty 8 Indians was/is a breach of Crown fiduciary obligations and that this outstanding obligation remains 'virtually unsatisfied.'[60] Following the rationale provided in *Haida*,[61] both the

---

[55] I question the actions of Alberta, which presented, but did not substantiate their argument as to the intent and motives of the federal Crown, which had opportunity, but failed, to exercise fiduciary diligence in acknowledging its historic intent to protect the Treaty 8 commercial interests then before the Courts. These questions impugn the honour of the Crown.

[56] *Haida Nation v British Columbia*, 2004 SCC 73, para.20.

[57] The Treaty 8 First Nations of Alberta is a provincial Treaty organization incorporated under the *Alberta Societies Act*, to coordinate the response of member First Nations to issues affecting Treaty rights and interests. The Treaty 8 Tribal Association of British Columbia has a similar mandate within that province.

[58] *Bilateral Process Framework Agreement Establishing the Treaty 8 First Nations of Alberta/Canada Treaty Discussions*, sec. 4.1.5 and 4.1.10-4.5.10, 23 April 2003.

[59] *Declaration of Intent between Canada and the Treaty 8 First Nations of* Alberta (1998). Canada and the Treaty 8 First Nations have requested that the provincial government join into the Treaty negotiations.

[60] Canada's admission that similar Treaty 11 provisions are 'virtually

federal and provincial Crown now have knowledge of the potential existence of asserted, but 'unproven' treaty-protected Aboriginal livelihood rights and interests, and are in possession of documents which demonstrate the probability of their existence.

Given the commitment of the federal government and the Treaty 8 First Nations of Alberta to undertake formal negotiations for treaty implementation, it is in the best interests of the provincial Crown to join these formal negotiations and seek reconciliation in order to uphold its honour, and to protect its own interests. The Treaty 8 First Nations of Alberta and the Government of Canada have requested that the Government of Alberta become an active party to ongoing negotiations between Canada and the Treaty 8 First Nations.

Given Canada's commitment to multi-culturism, to social and legal pluralism, and the prominence given Aboriginal rights within the constitutional fabric of our Nation, the need for Canada and Alberta to accept, and respond to, Treaty 8 peoples as a uniquely 'indigenous other' possessing *sui generis* resource-based livelihood rights and interests that are distinctly different in both origin and content from the rights and interests of non-aboriginal Canadians (*see* generally, Borrows and Rotman 1997), should be unquestioned. Few would argue with the proposition that mutual accommodation of Aboriginal and non-aboriginal natural resource interests in a manner which supports re-establishment of First Nation economic self-reliance and self-determination is a preferable future to a continued government reliance on a strategy of assimilation of impoverished Aboriginal peoples into the larger Canadian populace (Stevenson and Webb 2003). Negotiations between the Treaty 8 First Nations, the Government of Alberta and the Government of Canada, grounded in establishment of a mutually-shared understanding of the nature and scope of treaty-protected Aboriginal livelihood interests,[62] could provide a principled basis for reconciliation of these Treaty 8 interests as required by the *Canadian Constitution*.[63]

---

unsatisfied' provided the federal Crown rationale for negotiation with Treaty 11 First Nations under the provisions of the federal Comprehensive Claims policy. See, generally the *Tlicho Land Claims and Self-Government Agreement* (2003).

[61] *Haida Nation v British Columbia* [2004] 3 S.C.R. 511, para. 34.

[62] Henderson-Youngblood (1997:74) notes that a common linguistic understanding of the intent of the parties to a treaty is required for the contemporary interpretation of treaty rights and Crown obligations. Documentation of differences in meaning among and between the parties to the negotiations is relevant to determination of the 'shared intent' of these parties in relation to the treaty phrase "their usual vocations of hunting, trapping and fishing." The need for such an approach is a reflection of the fact that the historic treaties are the product of a fusion of the Aboriginal legal systems and the common law of Canada, and the facts that the negotiations took place in several languages.

[63] Principled negotiations incorporate and rely on use of interpretive principles and guidelines established by the Courts. Sec. 3.1.2 of the *Bilateral Framework Agreement* provides that treaty discussions shall be informed and guided by decisions of the Supreme Court of Canada.

Recognition, Accommodation & Reconciliation: Which Way Forward?

A large measure of what Justice Binnie characterizes as "inevitable tensions underlying implementation of Treaty 8,"[64] is grounded in the historic failure of the federal government to implement its' post-treaty policy regime for protection of Treaty 8 livelihood interests incident to the transfer of lands and resources to Alberta, and the long-standing indifference of the provincial government to address the livelihood concerns of Treaty 8 peoples. His opening comments, in the Mikisew decision were:

> The fundamental objective of the modern law of aboriginal and treaty rights is the reconciliation of aboriginal peoples and non-aboriginal peoples and their respective claims, interests and ambitions. The management of these relationships takes place in the shadow of a long history of grievances and misunderstanding. The multitude of smaller grievances created by the indifference of some government officials to aboriginal peoples concerns, and the lack of respect inherent in that indifference has been as destructive of the process of reconciliation as some of the larger and more explosive controversies.[65]

It follows that reconciliation of Treaty 8 First Nation claims arising from asserted, but 'unproven' Treaty 8 livelihood interests will be both a source of ongoing tension and a topic of ongoing importance for all Albertans. Collectively, we must struggle to manage these tensions, or collectively face the consequences of our failure to find a way to manage them. Tony Penikett (2006:271) advises that the consequences of such a failure could be the catalyst for development of our Balkans, our Beirut or our Belfast. These alternatives to treaty implementation are ours to debate and choose.

Treaty 8 First Nations in Alberta want to use treaty implementation negotiations and interim consultation processes to address matters of joint management over Treaty 8 lands and resources, resource revenue-sharing arrangements and First Nations access to resources for business and employment generation. As Canadians, and as Albertans, we no longer have the luxury of ignoring Treaty 8 First Nation requests for reconciliation. We do so at our peril. All Albertans should take this opportunity to insist that our federal and provincial Crown governments undertake to reconcile these matters through principled negotiations, and that both governments commit to undertake meaningful consultations and seek to develop interim accommodation of asserted Treaty 8 livelihood interests in recognition of the Supreme Courts reminder that "we are all here to stay."[66] While the archival record of Crown

---

[64] *Mikisew v. Canada* [2005] SCC 69, para. 33.
[65] *Mikisew v. Canada* [2005] SCC 69, para. 1.
[66] *Delgamuukw v British Columbia* [1997] 3 S.C.R. 1010.

negotiations related to establishment of 'special reserves' could provide a catalyst for recognition of the Crown's obligation to revisit their Treaty obligations and possibly act as the nexus for development of principled negotiations and meaningful consultations, these processes must be undertaken in the context of the contemporary circumstances within First Nation communities, Alberta and Canada. It remains to be seen whether we have the collective political will to enter into and complete such an important mission.

References Cited

Borrows, J. and L. Rotman (1997). The s*ui generis* nature of Aboriginal rights: Does it make a difference? *Alberta Law Review* 36(1): 21.

Fumoleau, R. (1975). *So Long as This Land Shall Last*. Toronto: McMillan and Stewart.

Government of Alberta (1936). Royal Commission on the Half-breed Population Report.

Government of Alberta (1938). *Métis Population Betterment Act*. Government of Alberta.

Havlik–Metics Research Group (2002). Hunting, trapping and fishing claim of the Little Red River Cree Nation. Unpublished manuscript on file.

Henderson-Youngblood, J. (1997). Interpreting s*ui generis* treaties. *Alberta Law Review* 36(1): 46-96.

Indian Association of Alberta (n.d.). Research papers: Implementation of Treaty Eight. Unpublished manuscript on file with Treaty 8 First Nations of Alberta.

Irwin, R. (2000). A clear intention to effect such a modification: The NRTA and treaty hunting and fishing rights. *Native Studies Review* 13(2): 47-89.

Laird, D., J.H. Ross, and J.A.J. McKenna (1899, 1966). *Report of commissioners for Treaty 8*. 22 September 1899. Winnipeg: Queen's Printer.

Madill, D. (1986). Treaty research report: Treaty Eight. Unpublished manuscript on file with the Treaties and Historical Research Centre. Indian and Northern Affairs.

Martin, H.W. (1989). 'Federal responsibility in the Metis settlements of Alberta,' pp. 273-296 in D.C. Hawkes, ed. *Aboriginal Peoples and Government Responsibility: Exploring Federal and Provincial Roles*. Carleton University Press.

Penikett, T. (2006). *Reconciliation First Nations Treaty Making in British Columbia*. Vancouver: Douglas & McIntyre.

Price, R. (1987). 'The spirit and terms of Treaty Eight,' in R. Price, ed., *The Spirit of the Alberta Indian Treaties*. Edmonton: Pia Pica Press.

Ray, A.J. (1995). Commentary of the economic history of the Treaty 8 area. *Native Studies Review* 10(2): 169-195.

Saunders, D. (1978). 'A legal analysis of the Ewing commission and the Métis colony system in Alberta,' in H.W. Daniels, ed., *The Forgotten People; Métis and Non-status Indian Land Claims*. Ottawa: Native Council of Canada.

Standfer, R.L. (2004). Breadth and depth of Canada's understanding of livelihood at time Treaty Eight was signed published. Unpublished paper on file with Treaty 8 First Nations of Alberta.

Stevenson, M.G. and J. Webb (2003). 'Just another stakeholder? First Nations and sustainable forest management in Canada's boreal forest,' pp. 65-112 in P.J. Burton, C. Messier, D.W. Smith, and W.L. Adamowicz, eds., *Towards Sustainable Management of the Boreal Forest*. Ottawa: NCR Press.

Tlicho Final Agreement (2003). *Land Claim and Self Government Agreement Among the Tlicho and the Government of the Northwest Territory and the Government of Canada*. 25 August 2003.

Tough, F. (1996). *As Their Natural Resources Fail: Native Peoples and the Economic History of Northern Manitoba, 1873-1930*. Vancouver: UBC Press.

Chapter Four

On Vocation and Livelihood: Interpretive Principles and Guidelines for Reconciliation of Treaty 8 Rights and Interests

Jimmie R. Webb

Introduction

While a number of contributors to this volume have for more than a decade been conducting research to support the development and implementation of effective Aboriginal institutions for sustainable forest management (Stevenson and Natcher 2009), First Nation peoples across Canada have been compelled to rely upon court action to protect and safeguard their Aboriginal and Treaty rights and interests in Canada's forests and waters. As a result of First Nations litigations, the Canadian courts have developed an impressive array of interpretive principles and guidelines related to the reconciliation of First Nations Aboriginal land and resource interests with respect to the sovereignty of the Crown. This chapter outlines how and why these interpretive principles can be used to reframe Aboriginal–Crown relations related to land and resource development within the Treaty 8 territory of northern Alberta.

This chapter builds upon the previous chapter that documented historic, post-treaty efforts of the Dominion government to safeguard and protect the 'usual vocations' of Treaty 8 peoples incident to the transfer of provincial lands and resources to Alberta under the provisions of the *Natural Resources Transfer Agreement* (NRTA 1930). It also constructs an interpretive framework for understanding what the respective parties to Treaty 8 may have meant by the terms 'livelihood' and 'vocations,' and uses these historic understandings to suggest that the Crown and Treaty 8 First Nations need to negotiate a mutual understanding as to the contemporary scope of Treaty-protected livelihood interests which must now be accommodated in the context of ongoing provincial Crown allocation of lands and resources to third party interests.

Within non-Treaty areas of Canada, the courts have developed and used a 'large, liberal and generous' interpretive regime to define and affirm the cultural and economic interests of Indians arising from 'Aboriginal title,' and certain other Aboriginal interests or rights which are not sufficient to constitute such title, but which are "…central to the identity and integral to the culture of the Indians."[1] Likewise, 'large, liberal and generous' interpretive guidelines have been developed by the courts for affirmation of the 'livelihood' or commercial

---

[1] *R. v Adams* [1996] 4 C.N.L.R. 1.

interests of Mikmaq Indians within Maritime provinces under the provisions of historic, pre-confederation "Treaties of Peace and Friendship."[2] More pertinent to this chapter, within the province of Alberta, the courts have provided similar guidance relevant to the need to develop a mutually shared understanding of what the English-language words 'vocation' and 'livelihood' meant to the Treaty Commissioners and the Indian leaders during negotiation of Treaty 8.[3] Justice Binnie has relied upon these guidelines in his analysis of matters before the Court in *Mikisew First Nation v. Canada*.[4] While he interpreted Treaty 8 as affecting a surrender of Aboriginal title[5]and providing the Crown authority to 'take up lands,' he identified the need for reconciliation of ongoing tensions between the Crown's exercise of this right and the Crown's solemn commitments to protect and support the Indian's use of 'lands not taken up' for "…their usual vocations of hunting, trapping and fishing." As the text of Treaty 8 states:

> And her Majesty the Queen HEREBY AGREES with the said Indians that they shall have right to pursue their usual vocations of hunting, trapping and fishing throughout the tract surrendered as heretofore described, subject to such regulations as may from time to time be made by the Government of the country, acting under the authority of Her Majesty, and saving and excepting such tracts as may be required or taken up from time to time for settlement, mining, lumbering, trading or other purposes.[6]

As noted by Justice Binnie, the negotiation of Treaty 8 was not expected to produce a future land use blueprint for the territory, and Treaty negotiations (1899) were no more than a first step in a long process of reconciliation.[7] It follows that the honour of the Crown requires contemporary implementation of yet 'unsatisfied' Treaty 8 promises to protect and safeguard First Nations 'livelihood' interests in lands and resources within lands not 'taken up.'

---

[2] *R. v Marshall* [1999] 3 S.C.R. 456; *R. v Bernard* [2003] 230 D.L.R. (4th) 57.

[3] *Delgamuukw v British Columbia* [1997] 3 S.C.R. 1010. This decision established the need for development of such a shared understanding as a basis for reconciliation, while *R. v Badger* [1996] 1 S.C, R. 771, applied this guideline to interpretation of Treaty 8.

[4] *Mikisew Cree First Nation v. Canada (Minister of Canadian Heritage)* [2005] SCC 69, (hereafter, *Mikisew*) para. 63.

[5] *Mikisew* para. 56. Note: I assert there is ambiguity between the 'non-specific surrender' clause of the Treaty, and the specific Crown agreement that Indians could use all lands 'not taken up' for conduct of their usual vocations. This ambiguity is exacerbated by clear evidence that the Crown had a well-defined legal understanding of 'Aboriginal title,' and the opportunity to be explicit regarding what they were asking the Indians to surrender, but chose not to do so. This ambiguity becomes more salient in the context of differences in the understandings of meaning evoked by translation of the English text of the treaty into indigenous languages.

[6] Treaty No. 8, June 21, 1899, Queen's Printer, Ottawa, 1966.

[7] *Mikisew* para. 27 and 56.

Moreover, according to Justice Binnie, the Crown's right to take up lands is subject to a duty to consult and, if appropriate, to accommodate treaty-protected First Nation livelihood interests before reducing or otherwise degrading an area within which First Nation members continue to exercise "...their usual vocations of hunting, trapping and fishing." All of this begs the questions of what this treaty phrase refers to, and what is meant by the words 'vocation' and 'livelihood.'

Treaty Discussions Relating to Livelihood: Are Our Meanings Shared?

As outlined in the previous chapter, research undertaken to support ongoing Bilateral treaty-based governance, and treaty implementation negotiations,[8] help to generate a growing awareness that hunting, trapping and fishing were not the only historic natural resource harvesting activities undertaken by Treaty 8 Indians during the fur trade era. In other words, fish, fur-bearers and game were not the only natural resources used to support Indian livelihood activities prior to negotiation of Treaty 8. Accordingly, in asking for implementation of yet unsatisfied livelihood claims, the Treaty 8 First Nations of Alberta assert that the term 'livelihood,' as used in these Bilateral negotiations, refers to First Nation rights:

- of access to Treaty 8 lands not taken up by the Crown;
- to harvest and use a broad range of natural resources to support their culture and way-of-life; and,
- to contemporary participation of Treaty 8 peoples in natural resource-based trade and commerce processes.[9]

For certainty, Treaty 8 First Nations are asserting a proprietary interest in lands and resources under provincial Crown jurisdiction.

Crown opinions about the interpretation of livelihood and the nature and scope of Crown Treaty commitments reflect "a shallow view of history,"[10] and

---

[8] The Treaty 8 First Nations of Alberta (T8FNs) are engaged in negotiations with Canada, under provisions of a 'Declaration of Intent,' signed on 22 June 1998. The provincial government attends these negotiations as an 'observer,' and has been requested to formally participate in order to facilitate negotiations about lands and resources currently under provincial jurisdiction.

[9] Note: In *R. v. Horseman* [1990] 1 S.C.R. 901, Arthur Ray provided expert testimony that the Treaty 8 phrase "their usual vocations of hunting, trapping and fishing" also refers to Indian participation in fur-trade era patterns of trade and commerce.

[10] Richard Howitt, during a lecture at the University of Alberta in 2004, used the term "a shallow view of history" to describe how 'our' language allows us to decide how to interpret reality as a reflection of what was important to us, and to ignore those aspects of historic reality which don't fit with our myths and worldviews.

are vestiges of the conventional wisdom that Treaty 8 is nothing more than an English-language document which can be interpreted in a straight forward manner under Canadian law. Tom Flanagan (2008:7) expressed this perspective by stating that, "The treaties mean what they say...."[11] Within the Crown's interpretive perspective, Indians have an ongoing right to hunt, trap and fish for subsistence upon Crown lands which have not been 'taken up' or put to some use which is visibly incompatible with such ongoing subsistence use. As outlined below, this perspective is reflected in the government of Alberta's policy on First Nations Consultation.[12] From this, it is apparent that the meanings which the respective parties bring to these negotiations are not shared. Moreover, as expressed by Flanagan (2008:7), "...Their reinterpretation... has the potential to be both expensive and mischievous for the economies of all provinces in which treaties have been signed."

Interpretative Principles and Reconciliation

A negotiated reconciliation of the asserted Treaty 8 livelihood claims with the sovereignty of the Crown will depend, in part, on the willingness of all parties to participate willingly in negotiation processes that develop and rely upon a 'large, liberal and generous'[13] interpretation of the meaning of the Treaty clause "...*their usual vocations of hunting, trapping and fishing...*,"[14] and the related concept of 'livelihood.' As demonstrated in the previous chapter, the negotiations leading to such reconciliation will also require a willingness to use a 'large, liberal and generous' interpretation of the intent of the federal government to protect and safeguard Treaty 8 livelihood interests incident to enactment of para. 1, 2 and 10 of the NRTA (1930).

Aboriginal scholars, such as James [*Sakej*] Henderson-Youngblood (1997:74) note that the treaties were negotiated among a number of peoples,

---

[11] Note: For the past three decades, Flanagan, a professor of political science at the University of Calgary and a member of the 'Calgary School'—a neo-conservative brain-trust—has argued against native land claims, and called for an end to Aboriginal rights. His book, *First Nations? Second Thoughts*, dismisses the continent's First Nations as merely its 'first immigrants,' and calls for outright assimilation of Aboriginal peoples into the Canadian polity. His arguments reflect the narrow world view and a Eurocentric understanding of history, political science and anthropological theory which Howitt terms "a shallow view of history."

[12] *Strengthening Relationships: The Government of Alberta's Aboriginal Policy Framework*, Government of Alberta (2003); *The Government of Alberta's First Nations Consultation Policy on Land Management and Resource Development* (Government of Alberta 2004).

[13] *Nowijecyk v R*, [1983] 1 S.C.R. 29.

[14] Arthur Ray (1995) has characterized this treaty phrase as providing Crown agreement for a First Nation's exclusive right to use Treaty 8 lands for pursuit of their usual vocations of hunting, trapping and fishing. He provided expert testimony on this matter in *R. v Horseman*, and the Supreme Court of Canada relied on his testimony to advise that Treaty 8 established a commercial interest on behalf of Treaty 8 Indians.

each of whom shared their own unique languages and linguistic understandings of treaty negotiations. Henderson-Youngblood (1997:74) notes that little attention has been paid to the need to establish a common linguistic understanding of the intent of the parties to treaty, as a condition of arriving at the original intent of the parties required for contemporary interpretation of Treaty rights and Crown obligations. He proposes (1997:81) that the interpretation of treaty rights must be undertaken in light of the dual linguistic contexts used for expression of the shared intent that led to the treaty. Development of a shared understanding of Treaty 8 livelihood commitments must be grounded in an appreciation of how 'vocations' and associated words such as 'livelihood' may have been translated into Cree or Dene dialects, and the meanings that such translations would evoke among Cree or Dene-language speakers. He advises that Aboriginal intent in relation to negotiation of treaty relationships are enfolded in their language and oral traditions and that these concepts are expressed by use of the English-language phrase 'the spirit and intent of the Treaties.' Further, Aboriginal scholars point out problems inherent to the translation of phrases and words from English into Dene or Cree dialects, and from Cree and Dene dialects into English, and advocate for addressing these problems as a precondition for the establishment of a common, shared, linguistic understanding of the range of meanings evoked by these phrases and words within each language.

One of the first steps in the search for a shared understanding of treaty commitments is the acknowledgement that many words, expressed within any language, have multiple meanings. To determine which meaning is intended in any expression uttered in any language is highly dependent on the context in which the words are used (Stevenson 2009). Given our focus on the English-language word 'vocation,' the following is an outline of what this word might have meant to English-speaking participants in historic Treaty 8 negotiations.

The Meaning of 'Vocation'

Treaty 7 was the first of the numbered treaties to include the phrase "...their usual vocations of hunting, trapping and fishing." The English-language word 'vocation' is most often used to denote one's livelihood, or occupation. The English-language word 'avocation,' which is found as parts of similar phrases in Treaties 3 through 6 (e.g., "...their traditional avocations of hunting, trapping and fishing") is generally thought of as a valued pastime or activity, which does not provide a basis for making a living, or constitute a livelihood. Use of the word 'avocations' in the text of earlier numbered treaties reflected the opinions of senior officials of Indian Affairs and the federal Crown Treaty Commissioners of the time that farming would be the future means of livelihood most likely available to Indians occupying these more southerly temperate regions. Further, the use of the word 'avocations' in these earlier treaties signalled the intent of the Crown to establish a post-treaty Indian reserve regime focused on forced abandonment of 'traditional avocations' commensurate with the development of agricultural livelihoods within a southern Canadian Indian population which was to be assimilated into a larger agrarian polity with settlement of the western prairies.[15]

The use of the word 'vocations' in the text of Treaties 8 through 11 is consistent with the historical opinions expressed by senior government officials, and federal Crown Treaty Commissioners, as to the ongoing ability of northern Indians within these Treaty areas to remain economically self-sufficient. This self-sufficiency was proposed to be contingent upon the Crown providing minimal levels of instrumental support of existing modes of livelihood; allowing these Indians to continue with traditional patterns of occupancy and use of unoccupied Crown lands (Price 1999); and protecting Indians from 'white competition' in pursuit of these usual vocations.[16]

This interpretive outline of the term 'vocation' within the English-language text of Treaty 8 is supported by the expert testimony of Arthur Ray in *R v Horseman* (1990) as to the meaning of the phrase "...their usual vocations of hunting, trapping and fishing...." Ray asserts that the entire phrase refers to Indian participation in a "complex of trade and commerce associated with the fur-trade," and his testimony provides a solid footing for demonstration of the fact that Treaty 8 resulted in an exclusive reservation of commercial interests associated with conduct of their usual modes of livelihood within Treaty 8 territory by the Indians (*see* previous chapter).

As a final aid to interpretation, while the English-language word 'vocation' is generally understood to refer to a means of making a living, use of 'vocation' within certain contexts is understood to refer to a spiritual imperative, unique to some livelihoods, to act toward others in a particular manner—what is known as a 'calling.' Within this larger concept of 'vocation as calling,' English-language speakers recognize that the word 'vocation' sometimes refers to more than a job. Some vocations entail a prescriptive range of moral, cultural and legal obligations toward others associated with particular livelihood practices, such as the law or the priesthood. As it turns out, this notion of 'vocation as calling' is important when one undertakes to determine how 'vocation' might have been translated into Cree, and understood by Cree speakers.

---

[15] In the words of the Honourable Minister, Duncan Campbell Scott: "Our object is to continue until there is not a single Indian in Canada that has not been absorbed into the body politic, and there is no Indian question, and no Indian Department" (*cited in* Borrows and Rotman 1997:27).

[16] As cited in Fumoleau (1975:65), Treaty 8 Commissioner J.A.J. McKenna wrote to to the Minister of the Interior, Clifford Sifton, on 17 April 1898, regarding the need for Crown commitments to protect the Indians' usual vocations of hunting, trapping and fishing from 'white' settlers. Excerpts from McKenna's letter demonstrate Crown awareness of a need for ongoing consultations with Treaty 8 peoples in relation to the protection of Indian livelihood interests following negotiation of the treaty.

Cree Understandings of 'Livelihood' and Why They are Important to Reconciliation

Historically, Crown governments have paid little or no attention to how the English-language words 'vocation' and 'livelihood' were translated into the several Indigenous languages used for oral explanations of the English-language text of Treaty 8 during treaty negotiations. Neither has much attention been accorded any roughly equivalent Indigenous-language words and what meanings they might have had for Indians who relied on these Indigenous translations as a basis for agreement to the treaty.

The contemporary Courts use the phrase 'natural understanding'[17] to refer to the meanings attributed to translations of the English text of Treaty documents by Aboriginal participants of treaty negotiations. In undertaking to establish a 'natural understanding' as to how the English-language word 'livelihood' could have been expressed within the Cree language,[18] and what cultural meanings would have been conveyed through translation into a corresponding Cree-language word, the Saskatchewan Office of the Treaty Commissioner (1998:40), worked with the Federation of Saskatchewan Indian Colleges to develop a broad, culturally-specific understanding of the relationship between the English-language word 'livelihood' and the Cree word, *pimâcihowan*.

First Nations elders and leaders identify *pimâcihowan* (pima-atchee-hoo-win)[19] as a central concept in the negotiation of Treaty 8. Its English translation corresponds roughly, in part, to 'livelihood,' or earning a living, but includes broader cultural meanings related to 'walking the way of life,' including all of the holistic practices and beliefs associated with Cree identity and culture, the relationships of the Cree peoples to the lands, the resources of the land, and the relationships of the Cree peoples to the 'others' with whom the Cree agreed to share use of their lands and its resources.[20] Cardinal and Hildebrant (2001:34) summarize the larger meanings evoked through use of this Cree word:

> When Treaty Elders use the word pimacihowan, they are describing a holistic concept that includes a spiritual as well as a physical dimension. It is an integral component of traditional First Nation's doctrines, laws. principles, values, and teachings regarding the sources of life, the responsibilities associated with them, including those elements seen as necessary for enhancing the

---

[17] The term 'natural understanding' is used by the courts to refer to the meanings attributed to the Aboriginal participants of treaty negotiations. See, for example, *R v Badger* [1996] 1 S.C.R. 771 para. 53; *Nowagijick v the Queen* [1983] 1 S.C.R. 29, para. 36.

[18] My review of these concepts relies on work by W. Wheeler and S. Krasowski (2003).

[19] Note: The Report of the Treaty Commissioner spells this Cree word as *pimacihowan*. Harold Cardinal used phonetic spelling of *pimacihowan* to indicate to English speaking readers how this Cree word sounds.

[20] *Statement of Treaty Issues: Treaties as a Bridge to the Future*, Office of the Treaty Commissioner, Saskatoon, 1998.

spiritual components of life and those associated with making a living.

These natural understandings of *pimacihowan*, or 'livelihood as way of life,' appear to be similar in meaning to the English-language concept of 'vocation as calling,' and provide for a deeper understanding as to Indian perspectives about the comprehensive scope of the 'bundle of rights'[21] associated with 'livelihood,' which Indian leaders undertook to preserve under Treaty 8 (1899). These natural understandings allow for deeper exploration of how the terms 'their usual vocations of hunting, trapping and fishing' and 'livelihood' would come to be interpreted, and what they would mean, when informed by the Natural Law of Treaty 8 peoples (Cardinal and Hildebrant 2001:43).[22] According to Harold Cardinal (1997:52, 127), while the treaty relationship, expressed in Cree as *Whit-Aski-Toowin*,[23] was an agreement to share use of Treaty 8 lands with others, those lands identified as 'not taken up' by the Crown under Treaty 8 were understood, within the natural understanding of the Cree, to be lands which the Indians saved for themselves. This natural understanding is expressed in Cree as *Skun-Ga-Na*.[24]

Given these preliminary understandings of what each of the Treaty parties may have meant in relation to use of the words 'livelihood' and 'vocation,' the second step in searching for a common, shared understanding is to acknowledge that within the Canadian system of legal pluralism, these terms have an 'intersocietal nature,'[25] i.e., their legal meaning does not rely solely on Canadian

---

[21] The term 'bundle of rights' is used by Madam Justice L'Heureux-Dube in *R. v Van der Peet* [1996] 2 S.C.R. 507, where she advises a treaty may affirm the reservation of Aboriginal title, and the need to examine the 'bundle of rights' reserved by the Indians, with agreement of the Crown, in order to determine whether this has resulted in establishment of treaty-protected Aboriginal rights.

[22] Borrows and Rotman (1997:12-13) cite *R. v Sparrow* [1990] 1 S.C.R. 1075, note 4 at 411, to advise that Aboriginal conceptions of the meanings of rights can and must be considered in judicial formulation of Aboriginal rights.

[23] Harold James Cardinal (1997:52, 127) advises that *Whit-Aski-Toowin* is an agreement to co-exist or live in peace and harmony with another nation on the land.

[24] Cardinal (1997:66-78) also advises that lands set aside as Indian Reserves, and those lands where Indians could continue to exercize their usual vocations of hunting, trapping and fishing, (i.e., 'lands not taken up') continued, within their Natural Understanding, to be *Skun-Ga-Na* or lands which they reserved for themselves, excepting for those portions which may be taken up by the Crown, from time-to-time for purposes of settlement.

[25] Brian Slattery (1992:120-121) proposes that the doctrine of Aboriginal land rights is an autonomous body of law created to reconcile Aboriginal tenure laws with English or French tenure laws, and uses the term 'intersocietal' to emphasize a intermingling of Aboriginal law and common law concepts which recognize Aboriginal difference while building strong ties of cooperation and unity between Aboriginal and non-

law. The term 'intersocietal' emphasizes that any complete understanding must reflect an intermingling of Aboriginal law and common law concepts which recognize Aboriginal differences while building strong ties of cooperation and unity between Aboriginal and non-Aboriginal peoples. From this, it is clear that any 'large, liberal and generous interpretation' of these terms must consider the range of possible meanings evoked by use of both the English-language word and the Cree-language translations during Treaty negotiations. Moreover, as the emerging doctrine on Treaty interpretation laid out by the Supreme Court of Canada specifies, in searching for the shared understanding which fits best the joint intent of the Treaty parties, ambiguities must be resolved in favour of the Indians.[26]

Implications for Principled and Ethical Reconciliation

The utility of adopting these interpretive principles and guidelines for analysis of Treaty 8 livelihood interests should be apparent from clear statements of the courts that Treaties 8 through 11 were negotiated in the context of a Federal Crown intent to protect and safeguard the livelihood interests of Treaty Indians in the lands and resources to be opened for development by the Crown.[27] It follows that, because of the presumption that the Crown intends to honour commitments made in negotiation of treaties with Indians, such principles and guidelines should be applied to analyses of Crown decisions affecting Treaty 8 Indian livelihood interests.

The objective to be reached through a principled approach to negotiation is the creation of a process for ongoing reconciliation of treaty-protected Aboriginal livelihood interests with the Provincial government's existing natural resource management regimes.[28] Reconciliation will require that the parties first reach agreement about the need for recognition and equitable reinstatement of First Nation natural resource interests within provincial Crown lands. Such recognition will then allow for the modification of existing provincial resource management regimes and the reallocation of natural resources within Treaty 8 territories in order to ensure equitable First Nation access to and benefit from natural resources on provincial Crown land. Equitable access to these natural resources will, in turn, support sustainable improvements in the lives of Treaty 8 peoples, ecological sustainability, economic equity and cultural survival (Howitt 2001:10). As stated by Borrows and Rotman (1997:28):

> Aboriginal communities…are seeking a patriation of their spiritual, cultural and institutional homelands… . They do not want to dismantle Canada. They simply want to redesign it to more closely address their needs and have it reflect their position in Canadian society.

Aboriginal peoples. This use is noted in Borrows and Rotman (1997:37-38).

[26] *Nowegijcyk v. R* [1983] 1 S.C.R. 29.

[27] *R. v Badger* [1996] 1 S.C, R. 771.

[28] *Delgamuukw v British Columbia* [1997] 3 S.C.R. 1010.

Principled negotiation provides our best hope for establishment of these processes for the reconciliation of the *sui generis* livelihood interests of Treaty 8 Indians with the historic dependence of settler populations on access to and economic use of the forest.[29]

Treaty 8 First Nations in Alberta want to use treaty implementation negotiations to discuss joint-management over Treaty 8 lands and resources, ongoing First Nations access to resources to support their way-of-life and for business and employment generation, and Crown–First Nation resource revenue-sharing arrangements. While these negotiations are taking place, Treaty 8 First Nations want to use interim consultation processes to address two legitimate expectations regarding Crown management of Treaty 8 lands, cultural sustainability and equitable benefit:[30]

- *Cultural sustainability* refers to the obligation of the Crown, when taking up land or allocating resources for development, to manage Treaty 8 lands not taken up so as to provide an environment which will sustain the culture and way-of-life of Treaty 8 peoples. This expectation presupposes that the Crowns right to take up lands is fettered by an obligation to leave sufficient Treaty 8 lands unallocated and to maintain an environment within these lands capable of sustaining the culture of Treaty 8 peoples.

- *Equitable benefit* refers to the obligation of the Crown, when taking up land or allocating resources for development, to ensure that the processes used for allocation, and the actual patterns of allocation produced by these processes are equitable to Treaty 8 peoples and settler populations.

These expectations have prompted the Treaty 8 Chiefs of Alberta to recommend that Alberta use the current land use planning process to discuss and reach agreement on the need for allocation of 30% of Treaty 8 forest lands to Treaty 8 First Nations in a new form of provincial forest tenure—a Forest Conservation Management Agreement—which would enable these First Nations to address their aspirations for cultural sustainability and also use these provincial tenures to participate in an emerging conservation offsets market to help revitalize First Nation economies.

---

[29] The need for such reconciliation has been addressed by the Supreme Court of Canada, in *Delgamuukw v British Columbia* [1997] 3 S.C.R. 1010. The Court advised that both the processes used for resource allocation, and the patterns of allocation resulting from the process, must be equitable.

[30] This request was one of a number of recommendations contained in the Treaty 8 First Nations presentation to Hon. Ted Morton, December, 2008, Recommendations for Alberta Land Use Framework.

Constructive Notice and Meaningful Consultation

Most resource conflicts in the early post-treaty period (1905-1938) centered around direct competition between whites and Indians for access to furs and trapping areas. While access to lands and resources is still being contested, current Crown-First Nation resource tensions have focused on the negative socio-cultural impacts of, and economic opportunities associated with, extensive forestry, oil/gas, and mineable oil sands operations.[31] The scope and scale of forestry and oil/gas development initiatives are both a threat to the way of life of Treaty 8 Indians and an opportunity for economic revitalization of Treaty 8 First Nation communities. Achieving an acceptable balance between ongoing natural resource development and accommodation of the livelihood interests of Treaty 8 peoples is a matter of national importance, and one that affects the honour of the Crown, and is dependent upon meaningful consultation.

These circumstances have prompted Treaty 8 First Nations to request that the federal and provincial governments undertake 'meaningful consultations' focused on the need for interim accommodation of asserted, but unproven, treaty-protected Aboriginal livelihood interests during ongoing treaty implementation negotiations. In *Haida Nation v. British Columbia*, the Supreme Court has outlined why such consultations and interim accommodation are required:[32]

> Where a strong prima facie case exists for the claim, and the consequences of the government's proposed decision may adversely affect it in a significant way, addressing the Aboriginal concerns may require taking steps to avoid irreparable harm or to minimize the effects of infringement, pending final resolution of the underlying claim. Accommodation is achieved through consultation.

The federal government is currently undertaking discussions with Treaty 8 First Nations about how court decisions in *Haida* and *Mikisew* have modified

---

[31] Northern Alberta contains 90% of Alberta's forest lands. Commercial timber resources within northern Alberta have been fully allocated to support regional milling operations. This northern forest area produces 42% of Alberta's conventional oil, 37% of its natural gas and produces 100% of heavy oil from the minable tar sands deposits. There are estimated to be some 12,000 new oil and gas dispositions awarded each year.

[32] *Haida Nation v British Columbia*, [2004] S.C.R. 73, para. 47. At para. 18: "Where the Crown has assumed discretionary control over specific Aboriginal interests, the honour of the Crown gives rise to a discretionary duty … ." At para. 27: "it may continue to manage the resource in question pending claims resolution. But, depending on the circumstances, the honour of the Crown may require it to consult with and reasonably accommodate Aboriginal interests pending resolution of the claim."

Crown obligations to consult and how the establishment of a federal government consultation regime might satisfy these new obligations.

While the Government of Alberta developed policy statements to guide government and industry consultations with Aboriginal peoples,[33] until recently, it has taken every opportunity to deny that it has consultation obligations toward First Nation peoples arising from the directions provided by the Supreme Court of Canada in the *Haida* decision. Alberta can be described as a province where the government has a "shallow view of history," and appears unwilling to recognize the legitimacy of Aboriginal aspirations of self-determination, or to negotiate development of equitable principles for co-existence with First Nations. The provincial government views Treaty 8 livelihood claims as a threat to the province's territorial integrity (Abele and Graham 1989:150-151), and as demonstrated below, the Government of Alberta continues to resist First Nations requests for consultation about asserted, but unproven, proprietary livelihood rights.

In a series of recent provincial interventions in *Mikisew*,[34] Alberta asserted that the terms of Treaty 8 gave the Crown an unrestricted right to 'take up land' for a variety of purposes,[35] and that Treaty 8 imposed no obligation on the Crown to consult with Indians when doing so.[36] The Federal Court of Appeal accepted this argument,[37] but Madam Justice Sharlow,[38] in dissent, rejected Alberta's argument as unsound and grounded in a 'literal interpretation' of the Treaty, as opposed to a 'large, liberal and generous' approach specified by the Supreme Court. She advised that for consultation to be 'meaningful' there must be evidence that First Nation rights are 'on the table,' that these rights are understood, and that the Crown is undertaking good faith efforts to address

---

[33] *Strengthening Relationships: The Government of Alberta's Aboriginal Policy Framework* (Government of Alberta 2003); *The Government of Alberta's First Nations Consultation Policy on Land Management and Resource Development* (Government of Alberta 2004).

[34] *Mikisew First Nation v Canada*, Factum filed by the Attorney General for Alberta.

[35] The clause in the text of Treaty 8, "...*and saving and excepting such tracts as may be required or taken up from time to time for settlement, mining, lumbering, trading or other purposes*" is relied upon by the Provincial government to support this assertion.

[36] *Mikisew First Nation v Canada*, Factum filed by the Attorney General for Alberta.

[37] *Mikisew First Nation v Canada*, Federal Court of Appeal Decision [2004] 236 D.L.R. (4th) 648. The majority of the Court ruled that under the specific circumstances before the Court, the decision of the Minister of Heritage Canada to approve construction of a winter road within Wood Buffalo National Park without consulting more fully with the Mikisew Cree Nation, was justified, since the road would affect a very small corridor within an otherwise undisturbed 44,000 square kilometer protected forest landscape.

[38] *Mikisew First Nation v Canada*, Federal Court of Appeal Decision, [2004] 236 D.L.R. (4th) 648.

them.[39] In the Supreme Court decision on *Mikisew*, Justice Binnie provided the following support for Sharlow's comments. He proposed that:

> ... Treaty 8 provides a framework within which to manage the continuing changes in land use already foreseen in 1899 and expected, even now, to continue well into the future. In that context, consultation is the key to achievement of the overall objective of the modern law of Treaty and Aboriginal rights, namely reconciliation. [40]

Notwithstanding the growing body of evidence (*see* previous chapter) that the Supreme Court's reasoning in *Horseman* was flawed, the Government of Alberta continues to rely on narrow, technical interpretations of Treaty 8 traditional use rights, and the Supreme Court's presumption that sec. 12 of the NRTA (1930) extinguished Treaty-protected commercial interests, in order to deny any obligation to consult about what Alberta considers 'unproven' livelihood interests.

In 2003, the North Peace Tribal Council,[41] acting on behalf of its member First Nations, provided a 'Public Notice' to the federal and provincial governments, and to third-party corporations known to be undertaking or contemplating resource development activities within the 'special reserve' areas previously identified on behalf of these bands by the federal government in 1927 (*see* previous chapter).[42] The Tribal Council requested consultations regarding certain beneficial or equity interests within these lands protected under the provisions of Treaty 8,[43] and under para. 2 of the NRTA. Stan Rutwind,[44] then leader of the Aboriginal Law Team, Alberta Justice, responded to this Public Notice and request for consultation, on behalf of the provincial government:

> Please be advised that Her Majesty the Queen in right of Alberta (Alberta) does not recognize the priorities asserted in the Public Notice. Furthermore, Alberta does not recognize a guarantee of protection from white competition to the lands set out in the Public Notice.

---

[39] The British Columbia Court of Appeal, in *Halfway River v British Columbia* [1997] 4 C.N.L.R.45, provided a rationale similar to the one set out in Madame Justice Sharlow's dissent. Treaty rights must be accommodated incident to consideration of resource management decisions within the province of British Columbia.

[40] *Mikisew Cree First Nation v. Canada* [2005] SCC 69, para. 63.

[41] Public Notice, North Peace Tribal Council. Note: A Tribal Council is a First Nation organization, incorporated under the auspices of the *Alberta Societies Act*, to represent the interests of member First Nations.

[42] NAC RG 10, Vol. 6732, File 420–2B.

[43] Havlik-Metics Research Group (2002). The research undertaken identified seven proposed 'special reserve' areas on behalf of 18 Treaty 8 First Nations.

[44] Letter from S.H. Rutwind, Alberta Justice, to Jim Webb, North Peace Tribal Council, dated 5 July 2004.

In further correspondence, Pearl Calahasen,[45] then provincial Minister of Aboriginal Affairs and Northern Development, advised the North Peace Tribal Council:

> For what it's worth, my own understanding of the current legal position of this government is that the interests forwarded in the public notice do not exist. The position of this government is based on a different interpretation of historic and archival information available.

The provincial government's *First Nations Consultation Policy on Land Management and Resource Development* (Government of Alberta 2004) includes the following definition of matters on which Alberta will consult:

> Rights and Traditional uses include (or 'sic') existing constitutionally protected rights to hunt, trap and fish and other uses of public lands such as burial grounds, gathering sites and historic or ceremonial locations*, and does not refer to proprietary interests in the land.* (emphasis added)

The Treaty 8 First Nations of Alberta are currently undertaking policy discussions with the Government of Alberta about the need for the provincial Crown to acknowledge its obligation to consult with Treaty 8 First Nations under circumstances where:[46]

- the First Nations assert the existence of a treaty-protected Aboriginal livelihood interest;
- there is evidence sufficient to establish a strong *prima facie* case for the probable existence of this livelihood interest; and
- the First Nations are seriously pursuing the resolution of livelihood claims through formal bilateral negotiations under the observation of the province.

Without prejudice to the appropriateness of Crown or First Nation understandings about the nature of First Nation livelihood interests affirmed and protected under Treaty 8, it is apparent that the provincial Crown's refusal to acknowledge any obligation to consult about asserted, but 'unproven,' treaty-protected interests, and their reliance upon narrow, technical interpretations of the Treaty make meaningful consultation, as mandated by the courts, virtually impossible. The Supreme Court, in *Haida*,[47] has advised that it is no longer acceptable for the Crown to avoid consultation by denying the existence of

---

[45] Letter from Hon. Pearl Calahasen, to Chief Johnsen Sewepagaham, North Peace Tribal Council, dated 28 September 2004.
[46] Note: These guidelines are set out in the Supreme Court decision of *Haida Nation v. British Columbia*, [2004] S.C.R. 73.
[47] *Haida Nation v British Columbia*, [2004] S.C.R.

rights in circumstances where the right or interest has not been proven in the eyes of the court:

> In circumstances where a right or interest is unproven, the scope of the Crown's duty is proportionate to a preliminary assessment of the strength of the case supporting the existence of the interest, and the seriousness of potentially adverse effect upon the right being asserted.

The provincial Crown's position that the livelihood interests established by Treaty 8 are of no contemporary force or effect in relation to provincial government allocation of natural resources appears to be irreconcilable with these judicial guidelines.[48] As Richard Howitt (2004) observes:

> While the law is changing, old privileges die slowly as the legal landscapes which were familiar and previously served the interests of the Crown, come to be reinterpreted in a strangely unfamiliar manner.

How then will Alberta, and Albertans, respond to the recent ruling by the Supreme Court in *Mikisew*? According to Flanagan (2009), *Mikisew* may have major implications for resource development within Treaty 8 territory. He proposes that "… If First Nations have a right to be consulted on any future developments that could affect hunting, fishing, and trapping, they have an ill-defined but still real property right on Crown land in this vast area."

Hopefully, in a post-*Mikisew* reality, the Government of Alberta will be willing to actively participate in ongoing Treaty 8 bilateral negotiation processes, to undertake meaningful consultations, and to consider interim accommodation of Treaty 8 livelihood interests in order to protect the honour of the Crown and its interests in the Treaty 8 settlement area.

References Cited

Abele, F. and K. Graham (1989). 'High politics is not enough: Policies and programs for Aboriginal peoples in Alberta and Ontario,' pp. 141-171 in D.C. Hawkes, ed., *Aboriginal Peoples and Government Responsibility*. Ottawa: Carlton University Press.

Borrows, J. and L Rotman (1997). The *sui generis* nature of Aboriginal rights: Does it make a difference? *Alberta Law Review* 36(1):9-45.

Cardinal, H.J. (1997). *Treaty 8 Right to Livelihood*. LLM thesis, Harvard Law School.

Cardinal, H.J. and W. Hildebrant (2001). *Treaty Elders of Saskatchewan*. Calgary: University of Calgary Press.

Flanagan, T. (2008). *First Nations? Second Thoughts*. 2nd ed. Montreal: McGill-Queen's University Press.

---

[48] See, for example, the Factum of the Intervener in *Haida v British Columbia*, Attorney General of Alberta, Stan Rutwind, Alberta Department of Justice, 18 December 2003.

Flanagan, T. (2009). Resource industries and security issues in northern Alberta. ©Canadian Defence & Foreign Affairs Institute. www.cdfai.org. June 2009.

Fumoleau, R. (1975). *So Long as this Land Shall Last*. Toronto: McMillan and Stewart.

Government of Alberta (2003). *Strengthening Relationships: The Government of Alberta's Aboriginal Policy Framework.*

Government of Alberta (2004). *The Government of Alberta's First Nations Consultation Policy on Land Management and Resource Development.* Edmonton: Aboriginal Affairs and Northern Development. August 31, 2004.

Henderson-Youngblood, J.(S). (1997). Interpreting *sui generis* treaties. *Alberta Law Review* 36(1): 46-96.

Howitt, R. (2001). *Rethinking Resource Management: Justice, Sustainability and Indigenous Peoples.* New York: Routledge.

Howitt, R. (2004). Sustainable forest management network. Lecture at the University of Alberta, Edmonton.

Office of the Treaty Commissioner (1998). *Statement of Treaty Issues: Treaties as a Bridge to the Future.* Saskatoon.

Price, R. (1999). *The Spirit of the Alberta Indian Treaties.* 3rd ed. Edmonton: University of Alberta Press.

Ray, A.J. (1995). Commentary of the economic history of the Treaty 8 area. *Native Studies Review* 10(2): 169-195.

Slattery, B. (1992). 'The legal basis of Aboriginal title,' pp. 113-132 in F. Cassidy, ed., *Aboriginal Title in British Columbia: Delgamuukw v. the Queen.* Lantzville, BC: Oolican Books.

Stevenson, M.G. (2009). 'Negotiating research relationships with Aboriginal communities: Ethical considerations and principles,' pp. 197-210 in M.G. Stevenson and D.C. Natcher, eds., *Changing the Culture of Forestry in Canada: Building Effective Institutions for Aboriginal Engagement in Sustainable Forest Management.* Edmonton: CCI Press.

Stevenson, M.G. and D.C. Natcher, eds. (2009). *Changing the Culture of Forestry in Canada: Building Effective Institutions for Aboriginal Engagement in Sustainable Forest Management.* Edmonton: CCI Press.

Wheeler, W. and S. Krasowski (2003). Treaty 8 livelihood discussion Paper. Treaty 8 First Nations of Alberta. On file with T8 FNs of Alberta.

Chapter Five

Aboriginal Rights and Aboriginal Forest Tenures: Toward 'Reconciliation' in British Columbia

Monique Passelac-Ross

Introduction

British Columbia is the Canadian province with the greatest number of First Nations-initiated legal cases challenging traditional forest management by the provincial government. Most of the jurisprudence on the Crown's duty to consult and accommodate originates from BC. The First Nations launching these court challenges seek not only to be adequately consulted and accommodated by government in forest decision-making, but also seek access to forest tenures and a share in revenues from forest development. Beyond that, First Nations also seek to regain control over land and resource development within their traditional territories. BC First Nations continue to assert their Aboriginal rights and Aboriginal title to lands that the government generally allocates to forest and other resource companies.

The research described in this chapter was part of a larger Sustainable Forest Management Network funded project that explored the features and feasibility of Aboriginal forest tenures (i.e., forest tenures that significantly address and accommodate Aboriginal needs, rights and interests) in British Columbia. Led by Dr. David Natcher, and derived from community-based research, this project sought to outline some of the key attributes of Aboriginal forest tenures (Swaak 2008; Swaak *et al.* 2009; Weber 2008; Weber *et al.* 2009), and to explore the legal framework upon which they might be built. In this chapter, I ask:

- What is the legal context in BC for the development of Aboriginal forest tenures?
- How has the provincial government responded to the courts' admonitions to negotiate rather than litigate differences with First Nations?
- What needs to occur legally for the development and implementation of culturally appropriate Aboriginal forest tenures?

To illustrate the analysis, the Carrier Sekani Tribal Council (CSTC) is used as a case study. The CSTC has for many years sought to achieve recognition and protection of its First Nations members' title and rights, and has struggled to obtain tenure rights over the forest resources located within the traditional territories of its member communities.

This chapter provides background information on the negotiation of treaties in the BC context. The analysis then moves to a brief review of the jurisprudence on Aboriginal rights, title and governance. Implications of the case law for the management of forest lands and resources are drawn, followed by a review of the steps taken by the BC government to address recent legal developments. The last section discusses some issues and options in relation to the development of Aboriginal forest tenures in BC.

Treaties in the BC Context

By contrast with other Canadian provinces, and in particular the Prairie provinces, very few historical treaties were signed between First Nations and the (British) Crown in British Columbia. Until very recently, the only existing historical treaties were the Douglas treaties, signed on Vancouver Island between 1850 and 1854, and Treaty 8, signed in 1899 in northeast BC.[1] In the rest of the province, the provincial government persisted in denying the existence of Aboriginal rights and title, and refused to negotiate treaties with the First Nations throughout most of the 20th century.

In 1973, the landmark Supreme Court decision in the *Calder* case established that Aboriginal title existed as a right within the common law.[2] Prior to this decision, many questioned whether there was any surviving Aboriginal title in Canada. Mr. Calder, on behalf of the Nisga'a people, sought a declaration that they had Aboriginal title to lands occupied by their ancestors since time immemorial, and that this title had never been lawfully extinguished. The Supreme Court affirmed the existence of Aboriginal title as a legal right.

The *Calder* decision led to a fundamental change in federal policies. The federal government declared its willingness to begin negotiating Aboriginal title claims.[3] In 1981, the federal government published a native claim policy and established a Comprehensive Claims Office to negotiate comprehensive settlements of Aboriginal title (Government of Canada 1981).[4] British Columbia refused to participate in the land claims process until 1991, when the tri-partite

---

[1] Treaty 8, the largest of the so-called 'numbered treaties,' extends from the northwestern corner of Saskatchewan, through the northern half of Alberta, to the northeastern corner of British Columbia, and its northern limits reach as far as the south shore of Great Slave Lake in the Northwest Territories. Several bands in northeast BC did not sign the treaty in 1899. The Slaves of Fort Nelson only adhered to Treaty 8 in 1910. Other bands adhered a few years later: e.g., the Saulteau First Nation and West Moberly First Nation joined Treaty 8 in 1914. See Treaty 8 Tribal Association, Chronology at:
www.treaty8.bc.ca/about/chronology.table.html
[2] *Calder v. A.G.B.C.*, [1973] S.C.R. 313.
[3] *Statement by Honourable Jean Chrétien, Minister of Indian and Northern Affairs, Regarding Indian and Inuit Claims*, August 8, 1973.
[4] This was followed by a Specific Claims Policy (Government of Canada 1982).

BC Claims Task Force prepared a report on the land claims situation in B.C. and made recommendations for a treaty negotiations process. The provincial government agreed to enter into a treaty process with First Nations and, along with First Nations and Canada, established the B.C. Treaty Commission in 1992.

Since 1992, the provincial and federal governments have been negotiating treaties with First Nations at various 'tables' around the province.[5] To date (October 2009), only two land claims have been settled: the *Tsawwassen First Nation Final Agreement*, ratified in June 2008, and the *Maa-nulth First Nations Final Agreement*, ratified by the provincial legislature in November 2007 and expecting ratification by Parliament. The *Nisga'a Final Agreement*, the first modern-day treaty in B.C., came into effect in 2000.[6] However, the agreement was not settled under the BC Treaty Commission process, but under the Federal Comprehensive Claims process. As of October 2009, seven BC First Nations are in the final stage of treaty negotiations (Stage 5) after having signed agreements-in-principle.

Progress on the negotiation of treaties has been very slow. In late 2006, a group of 40 First Nations in British Columbia formed the Unity Protocol to jointly engage the federal and provincial governments in the negotiation of a common position on six areas that were creating an impasse in treaty negotiations. The six key issues identified include: certainty, constitutional status of treaty lands, governance, co-management throughout traditional territories, fiscal relations and taxation, and fisheries. In January 2007, five more First Nations signed the *Unity Protocol Agreement*.[7] The intention is to negotiate principles or options that all parties can rely upon when negotiating individual treaties.

The efforts of the Unity Protocol group were successful, and in March 2008, the BC Treaty Commission announced the formation of a common table that "will identify the issues and obstacles to progress and seek options that may be applied at individual treaty tables to expedite treaty negotiations," and credited the First Nations Summit for the establishment of that table.[8] The six issues selected for consideration are identical to the ones identified by the Unity Protocol group. Discussions at the common table, which involved more than 60 First Nations and senior officials for the governments of Canada and BC, ended in July 2008. The BC Treaty Commission reports that "discussions at the common table ended on a positive note because there is agreement among First Nations and senior government officials on a number of areas where progress

---

[5] The process involves six stages, from Stage 1—Filing a Statement of Intent to Negotiate a Treaty, to Stage 6—Implementing the Treaty. The BC Treaty Commission Annual Report 2008 states that there are 60 First Nations participating in the treaty process. Because some First Nations negotiate at a common table, there are 49 sets of negotiations. See www.bctreaty.net.

[6] *Nisga'a Final Agreement*, S.C. 2000, c. C-7; available at www.aaf.gov.bc.ca/aaf/treaty/nisgaa/docs/nisga_agreement.html.

[7] News Release, First Nations Unity Protocol Agreement, 18 January 2007.

[8] BC Treaty Commission, March 2008 Update.

may be possible."[9] The Commission also reports that the provincial government and certain First Nations are negotiating 'incremental treaty agreements (ITA),' in which elements of the treaty package are implemented by BC and the First Nation in advance of the full agreement.[10]

Carrier Sekani Tribal Council (CSTC)

The Carrier Sekani Tribal Council (CSTC), which represents eight First Nations and has a combined population of over 10,000 members, is located in north-central British Columbia. The traditional territory of its members encompasses 78,700 km^2 (7.87 million hectares). The CSTC submitted its comprehensive land claim to the federal government in 1982, and the federal government accepted the claim of a territory and a traditional system of governance in 1983. In 1993, the CSTC submitted its Statement of Intent to the BC Treaty Commission on behalf of its member bands, and has been in the treaty negotiations process since 1994.

Over the years, the negotiating team has attempted to negotiate interim protection measures and interim economic measures in forestry (efforts to obtain forest licences were made in 1999 and in 2002). In 2006, after 12 years of negotiations, the Tribal Council had reached an agreement-in-principle (AIP), i.e., Stage 4 of the treaty process. Frustrated by the slow pace of progress and the significant impediments to the resolution of their claim, the CSTC announced in 2007 that it was withdrawing from the treaty process. Chief David Luggi of the CSTC stated that the stumbling block to settling treaties was the extinguishment of Aboriginal rights and the need to give up most of their traditional territories.[11] Treaty negotiations are currently inactive. However, Chief Leonard Thomas of the Nak'azdli First Nation, one of the CSTC member First Nations, has stated that it will be up to individual communities to decide if they want to pull out of the treaty process. The CSTC has stated its intention to negotiate one-on-one deals with private resource companies, especially forest companies, e.g., Canfor Corp (Brethour 2007).[12]

---

[9] BC Treaty Commission, News Release, "Mood Upbeat as Common Table Talks Conclude," 24 July 2008.

[10] *BC Treaty Commission Annual Report 2008, supra* note 6 at p. 2. The *Tla-o-qui-aht* have already signed an ITA, and the Haisla Nation is about to sign one by the end of 2008.

[11] In a Letter addressed to Jim Prentice, then Minister of Indian Affairs & Northern development, on 30 May 2007, Chief Luggi identified several key areas of concern for the CSTC, including the federal and provincial representatives' mandates vs. positions, certainty, lands and co-management, interim measures, governance, compensation and the BCTC loan funding, concluding that "our people decided it is time to walk away from this process as there is no possibility of achieving a just reconciliation with the government's current framework." See www.cstc.bc.ca/cstc/24/treaty+negotiations.

[12] Patrick Brethour, 'Band to withdraw from treaty talks,' *Globe and Mail*, 31 March 2007.

Aboriginal Rights, Aboriginal Title and Issues of Governance: The Jurisprudence

The Existence of Aboriginal Rights and Title in BC

The Supreme Court of Canada decision in the 1973 *Calder* case recognized Aboriginal title as a legal right, although it split equally on the question of the legal foundation of this title and on whether title had been lawfully extinguished in British Columbia. In the *Guérin* decision, the Court reasserted that the Indians' "interest in their lands is a pre-existing legal right not created by *Royal Proclamation*, by s. 88(1) of the *Indian Act*, or by any other executive order or legislative provision."[13] In *Van der Peet*, the Court explained that "Aboriginal rights arise from the prior occupation of land, but they also arise from the prior social organization and distinctive cultures of Aboriginal peoples on that land."[14] In *Delgamuukw*, Justice Lamer summed up the developments in legal doctrine as follows:

> It had originally been thought that the source of Aboriginal title in Canada was the *Royal Proclamation, 1763*: see *St Catherine's Milling*. However, it is now clear that although recognized by the *Proclamation*, it arises from prior occupation of Canada by Aboriginal peoples. That prior occupation, however, is relevant in two different ways, both of which illustrate the *sui generis* nature of Aboriginal title. The first is the physical fact of occupation which derives from the common law principle that occupation is proof of possession in law [...]. [...] a second source for Aboriginal title – the relationship between common law and pre-existing systems of Aboriginal law.[15]

With respect to Aboriginal rights specifically, the case that established their continued existence in British Columbia is the *Sparrow* decision.[16] Whereas the lower court had held, following *Calder*, that Aboriginal rights no longer existed in the province, the Court of Appeal, and ultimately the Supreme Court, recognized an Aboriginal right to fish for food.

In *Delgamuukw*, the Supreme Court confirmed that the province lacked the constitutional authority to extinguish Aboriginal rights, including Aboriginal title, since it joined Canada in 1871.[17] Further, provincial laws of general application cannot extinguish Aboriginal rights. As a result of the recognition

---

[13] *Guérin v. R.* [1984] 6 W.W.R. 481 at para. 497.

[14] *R. Van der Peet* [1996] 2 S.C.R. 507 at para. 74.

[15] *Delgamuukw v. B.C.* [1997] 3 S.C.R., at para. 114.

[16] *R. v. Sparrow* [1990] 1 S.C.R. 1075.

[17] *Delgamuukw, supra* note 16 at para. 173: "Since 1871, the exclusive power to legislate in relation to 'Indians, and Lands reserved for the Indians' has been vested with the federal government by virtue of s. 91(24) of the *Constitution Act, 1867*. That head of jurisdiction, in my opinion, encompasses within it the exclusive power to extinguish Aboriginal rights, including Aboriginal title."

and affirmation of Aboriginal rights in s. 35(1) of the *Constitution Act* (1982), Aboriginal rights and title can only be extinguished by consent of the Aboriginal peoples.

It is noteworthy that, even though asserted Aboriginal rights and Aboriginal title have not yet been established in court or recognized under the terms of a treaty, they still enjoy a measure of legal protection. The Supreme Court in *Haida* stated that when the government has knowledge of the *potential existence* of an Aboriginal right or title and contemplates conduct that may adversely affect that right, it is under a duty to consult with Aboriginal peoples and to accommodate their interests. This point is developed below.

What are Aboriginal Rights and Aboriginal Title?

Aboriginal rights are said to be *sui generis* or unique, in that they do not fit into categories of English or French law. In *Delgamuukw*, Justice Lamer stated that Aboriginal rights stem from the reconciliation of prior Indigenous occupation with the Crown's assertion of sovereignty. Reconciliation is achieved by building a bridge between Indigenous and non-Indigenous cultures, which gives rise to common law rights that are truly *sui generis.*[18] Aboriginal rights fall along a spectrum with respect to their degree of connection with the land. At one end of the spectrum are Aboriginal rights which confer the right to engage in site-specific activities on a tract of land to which Aboriginal peoples may not have title. At the other end is Aboriginal title, which confers the right to the land itself.[19]

Aboriginal rights are more than just rights to exercise site-specific activities (e.g., harvesting resources) on the land. The Supreme Court 1996 trilogy on Aboriginal rights—*Van der Peet, Gladstone* and *NTC Smokehouse*[20]— defined Aboriginal rights as practices, customs or traditions that were 'integral to the distinctive culture' of the Aboriginal group in question prior to contact with European societies. The practice, the Court said, must be more than "an aspect of the Aboriginal society, it must be "one of the things which made the society what it was."[21] Aboriginal rights are "collective rights which contribute to the cultural and physical survival of Aboriginal peoples."[22] They include the right to use the land to hunt, fish and gather materials for sustenance, social, spiritual and ceremonial purposes. They also include governance rights (e.g., the right for Aboriginal peoples to adopt their own customary rules of access to and allocation of land and resources, and the right to control activities on the land so as to preserve natural resources). Aboriginal rights can evolve in a modern form: for example, harvesting timber to construct a bungalow is a modern evolution of the right to harvest timber to construct a wigwam (*Sundown*).

Following the 1996 Supreme Court trilogy, Aboriginal rights to hunt, fish and gather had been interpreted as being resource-specific rights. For instance,

---

[18] *Delgamuukw, supra* note 16 at paras. 81-82.

[19] *Ibid.* at para. 138.

[20] *R. Van der Peet, supra* note 15; *R. v. Gladstone*, [1996] 2 S.C.R. 723; *R. v. NTC Smokehouse*, [1996] 2 S.C.R. 672.

[21] *Van der Peet, ibid.* at para. 55.

[22] *Ibid.* at para. 74.

the Aboriginal right to fish was not a right to fish any species of fish for food, but a right to fish salmon (*Sparrow*) or herring spawn on kelp (*Gladstone*). Moreover, the Supreme Court decisions had been understood as supporting the proposition that a practice undertaken merely for survival purposes could not be considered 'integral to the distinctive culture' of an Aboriginal people. Ten years later, in the *Sappier* and *Gray* cases,[23] the Court clarified its position. Both cases involve the unauthorized cutting of trees for the construction of furniture and housing. The Court stated that Aboriginal rights are not to be defined in terms of a specific resource "because to do so would be to treat it akin to a common law property right."[24] Justice Bastarache characterized the right to harvest timber in a broader way as "a right to harvest wood for domestic uses as a member of the Aboriginal community."[25] The Court rejected the notion that practices associated with mere survival could not constitute Aboriginal rights:

> I can therefore find no jurisprudential authority to support the proposition that a practice undertaken merely for survival purposes cannot be considered integral to the distinctive culture of an [A]boriginal people. Rather, I find that the jurisprudence weighs in favour of protecting the traditional means of survival of an [A]boriginal community.[26]

The majority position was that the right had no commercial dimension, excluding even barter from its scope.[27] Further, the Court emphasized that the right is communal: "the right is not one to be exercised by any member of the Aboriginal community independently of the Aboriginal society it is meant to preserve. It is a right that assists the society in maintaining its distinctive character."[28]

In the recent *Tsilhqot'in* case in BC's Supreme Court, Justice Vickers takes a more generous approach to Aboriginal rights, holding that hunting, trapping, fishing and trading rights practiced by the Tsilhqot'in in direct reliance on their landbase are an integral element of their traditional way of life and pattern of survival.[29] The trading right so acknowledged is limited to earning a

---

[23] *R. v. Sappier; R. v. Gray*, [2006] 2 S.C.R. 686.

[24] *Ibid.* at para. 21.

[25] *Ibid.* at para. 24.

[26] *Ibid.* at para. 37.

[27] *Ibid.* at para. 25: "the harvested wood cannot be sold, traded or bartered to produce assets or raise money." Note that Justice Binnie (at para. 74), writing concurring reasons, disagreed with this aspect of the majority decision, stating: "barter (and its modern equivalent, sale) within the reserve or other local Aboriginal community would reflect a more efficient use of human resources than requiring all members of the reserve or other local Aboriginal community to which the right pertains to do everything for themselves."

[28] *Ibid.* at para. 26.

[29] *Tsilhqot'in Nation v. British Columbia*, [2008] 1 C.N.L.R. 112. This is a lower court decision, and the majority of the lengthy decision is a non-binding opinion offering Justice Vickers' views as to what he would

'moderate livelihood,' a term defined by the Supreme Court in the *Marshall* case as meeting day-to-day needs supplemented by a few amenities, but not wealth accumulation.[30]

Aboriginal rights encompass Aboriginal title, which "is a sub-category of Aboriginal rights which deals solely with claims of rights to land."[31] In the *Delgamuukw* case, the Supreme Court defined Aboriginal title as a *sui generis* interest in land. It confers a right to exclusive use and possession of land, including forests, minerals, oil and gas, for a variety of purposes that need not be aspects of tradition, customs and practices which are integral to the distinctive Aboriginal cultures.[32] It includes the right to choose to what uses the land may be put, with the proviso that these uses cannot destroy the ability of the land to sustain the kinds of activity which made it Aboriginal title land. Title lands, said the Court, must be used in a way which is consistent with the nature of Aboriginal Peoples' attachment to the land.[33] Brian Slattery (2006:279) sums up Justice Lamer's identification of the three distinctive features of Aboriginal title as follows: 1) it cannot be alienated other than to the federal government; 2) it stems from possession of land prior to the assertion of British sovereignty; and 3) it is a collective right held communally and governed by communal decisions. Further, Aboriginal title has an 'inescapable economic component'; it entitles Aboriginal peoples to reap economic benefits from land and resource use (e.g., forest harvesting).[34]

Aboriginal title includes the right to manage, control or govern the land and its resources. It gives Aboriginal peoples the right to decide whether or not to harvest timber and other non-timber resources, how much, where, under which conditions, etc. The implication is that Aboriginal peoples have the authority to make laws to protect their lands and resources, and to act as stewards of a territory. Kent McNeil (2006:282) notes:

> [Aboriginal title] … cannot be conceptualized as a mere property right. Instead, it is in the nature of a territorial right that has both proprietary and jurisdictional elements.

---

have found, had the case been presented slightly differently. This included recognition that half of the claimed area was Aboriginal title land, and that provincial legislation purporting to regulate that land would be of no effect. Despite the fact that the major part of the decision is non-binding, its significance cannot be overstated. Justice Vickers made important and binding rulings with respect to Aboriginal title and hunting and trapping rights.

[30] *R. v. Marshall*, [1999] 3 S.C.R. 456 at para. 59.

[31] *Van der Peet*, *supra* note 15 at para. 74.

[32] *Delgamuukw*, *supra* note 16 at paras. 111, 117 and 124.

[33] *Ibid.* at para. 1088: "Implicit in the recognition of historic patterns of occupation is a recognition of the importance of the continuity of the relationship of an Aboriginal community to its land over time (inherent limit)." Also at para. 129: "The land has an inherent and unique value in itself, which is enjoyed by the community with Aboriginal title. The community cannot put the land to uses which would destroy that value."

[34] *Delgamuukw*, *supra* note 16 at para. 169.

… in situations where a title-holding community already had laws
governing such things as land tenure and use, as Lamer C.J
envisaged in *Delgamuukw*, one would expect those laws to retain
their validity and continue to govern those matters until the
community decided, through the exercise of its decision-making
authority, to change them.

The Co-existence of Aboriginal and non-Aboriginal Societies
The recognition of Aboriginal jurisdiction over lands and resources raises the
issue of the provincial powers to legislate with respect to Aboriginal title and
Aboriginal rights. In *Delgamuukw*, the Supreme Court had affirmed that there
was a core of jurisdiction that was off-limits to provincial legislatures. This is
referred to as the doctrine of interjurisdictional immunity.[35] Even though they
may be valid as laws of general application, provincial laws that affect matters
that are within the jurisdiction of the federal government do not apply to
Aboriginal title lands or to Aboriginal rights in relation to land.[36]

The *Tsilhqot'in* decision applies the findings of the Supreme Court in
Delgamuukw. Justice Vickers confirms that Aboriginal title is the source of
Aboriginal jurisdiction:

> To have any significance for Aboriginal people, Aboriginal title
> must bring with it the collective right to plan for the use and
> enjoyment of that land for generations to come. Prior to European
> colonization, the lands and forests of Tsilhqot'in traditional
> territory supplied Tsilhqot'in people with sustenance and
> protection from the elements, as well as a moderate livelihood.
> Tsilhqot'in people were able to make all land use decisions with
> respect to that territory. The imposition of the provincial forestry
> management scheme removes the ability of Tsilhqot'in people to
> control the use to which the land is put.[37]

Any provincial legislation which interferes with the exercise of this
Aboriginal jurisdiction—such as the provisions of BC.'s *Forest Act*, which
purports to authorize the management, acquisition, removal and sale of timber
on Crown lands—is not applicable on Aboriginal title lands. Provincial forest
legislation that allows timber cutting on title lands affects the very core of
Aboriginal title by interfering with Aboriginal jurisdiction.

---

[35] The doctrine holds that the federal government has exclusive
jurisdiction over 'Indians and lands reserved for the Indians,' as provided
under s. 91(24) of the Constitution. 'Lands reserved' has been interpreted
as including Aboriginal title lands as well as Aboriginal rights in relation
to land (*Delgamuukw, supra* note 16 at para. 178.

[36] *Delgamuukw, supra* note 16 at 175-176. As noted by some legal
scholars, the division of powers analysis is more protective of Aboriginal
rights than an analysis based on s. 35 of the *Constitution Act, 1982*,
because it escapes the justificatory approach that characterizes a s. 35
analysis (*see* e.g., Bankes 1998:320).

[37] *Tsilhqot'in, supra* note 30 at para. 1066.

The right to exclusive use and possession is fundamental to Aboriginal title. Aboriginal title confers a right to the land itself, and the right to determine how it will be used. Legislation that authorizes the granting of rights to harvest timber from these lands to third parties strikes at the very core of Aboriginal title. These legislative enactments are beyond the constitutional reach of the Province. They fall within the exclusive domain of Parliament under s. 91(24).[38]

While Justice Vickers views provincial forest legislation as inapplicable to Aboriginal title lands, this is not the case with respect to *Aboriginal rights*. In his view, provincial forest legislation is constitutionally applicable to lands over which Aboriginal people have Aboriginal rights, because it does not go to the core of Tsilhqot'in Aboriginal rights. However, inasmuch as the legislative scheme interferes with Aboriginal rights, it will have to be justified by the province, as discussed in the following section.[39]

With respect to *treaty rights*, however, the Supreme Court has found in the *Morris* case that the doctrine of interjurisdictional immunity also protects treaty rights from the application of provincial laws that infringe significantly upon such rights.[40] The Court found that provincial legislative provisions imposing a categorical ban on night hunting and hunting with illumination in the whole province interfered significantly with the rights to hunt guaranteed to the Tsarlip by treaty. These provisions were "overbroad, inconsistent with the common intention of the parties to the treaties, and completely eliminated a chosen method of exercising their treaty right."[41] They were inapplicable to the Tsarlip[42] and could not be incorporated under s. 88 of the *Indian Act*. Consequently, the court acquitted the accused.

Justification of Rights Infringements

The courts have asserted that Aboriginal and treaty rights, even though they are protected under s. 35(1) of the *Constitution Act, 1982*, are not absolute. Government (federal and provincial) may infringe these rights, but only if it justifies the infringement. The infringement and justification analysis was developed in the *Sparrow* case in 1990 and applied to Aboriginal title in *Delgamuukw* in 1997.[43] The justification analysis is in two parts. First, the

---

[38] *Ibid.* at para. 1048.

[39] *Ibid.* at para. 1042 and 1045

[40] *R. v. Morris*, [2006] 25 S.C.R. 915. Two members of the Tsarlip Indian Band of the Saanich Nation were convicted of hunting at night with a light, contrary to the provisions of BC's *Wildlife Act*. As a defence, they invoked their treaty right to hunt under the 1852 North Saanich Treaty, one of the Douglas Treaties.

[41] *Ibid.* at para. 58.

[42] *Ibid.* at para. 55: "The purpose of the *Sparrow/Badger* analysis is to determine whether an infringement by a government acting within its constitutionally mandated powers can be justified. This justification analysis does not alter the division of powers, which is dealt with in s. 88."

[43] *Delgamuukw*, *supra* note 16 at para. 160. See however the discussion in section 3.2 above with respect to the doctrine of interjurisdictional immunity, which eliminates the need for a justification analysis.

Aboriginal people must show that the legislation or decision in question interferes with their Aboriginal rights. The onus then shifts to government to justify such infringement by demonstrating that: 1) it is acting to achieve a compelling and substantial objective (including economic development); and 2) its actions uphold the honour of the Crown (including a duty to consult and accommodate, and potentially to compensate).

Sparrow had defined compelling and substantial objectives that may justify infringement of Aboriginal rights as those that either conserve and manage resources or ensure public safety. Post *Sparrow*, in a series of decisions the Supreme Court expanded the meaning of 'compelling and substantial' to include objectives such as the pursuit of economic and regional fairness and the interests of non-Aboriginal fishers.[44] This interpretation, endorsed in *Delgamuukw*, has been criticized by Justice McLachlin as overly permissive and inconsistent with earlier authorities, as it extends the meaning "to any goal which can be justified for the good of the community as a whole, Aboriginal and non-aboriginal."[45] The Court also 'watered down' the second branch of the justification test" (McNeil 1997:37). Justice McLachlin pointed out that:

> To reallocate the benefit of the right from aboriginals to non-aboriginals would be to diminish the substance of the rights that section 35(1) of the *Constitution Act* (1982) guarantees to the Aboriginal people. This no court can do.[46]

The Crown's duty to consult and accommodate forms part of the justification analysis, although the courts have held that they are based on the concept of 'the honour of the Crown' rather than on the need to justify rights infringements. As discussed below, judicial decisions have established that governments are under an obligation to consult and accommodate Aboriginal peoples whose rights may be infringed as a result of resource development, even when those rights are simply 'asserted' and have not yet been established. The scope and content of that duty vary with the strength of the claim and the seriousness of the impact of government's actions or decisions on the right or title claimed. In the *Taku River* case, the Supreme Court found that lengthy consultations with the Taku River Tlingit First Nation in the environmental assessment process had been "sufficient to uphold the Province's honour and meet the requirements of the duty."[47]

Consultation with Aboriginal peoples and accommodation of their rights may help to mitigate the negative impacts of resource development on Aboriginal communities or to identify the least intrusive means of achieving government objectives. But as currently implemented, these Crown obligations often do not suffice to fulfill the requirements of Aboriginal rights' recognition and protection under s. 35 of the *Constitution Act*.

---

[44] *Gladstone, supra* note 21 at para. 75.

[45] *Van der Peet, supra* note 15 at para. 304.

[46] *Van der Peet, supra* note 15 at para. 315.

[47] *Taku River Tlingit First Nation v. British Columbia (Project Assessment Director)*, [2004] 3 S.C.R. 511, at para. 2.

Reconciliation

The Supreme Court has repeatedly emphasized the importance of the concept of reconciliation.[48] In *Mikisew*, Justice Binnie stated that the reconciliation of Aboriginal peoples and non-Aboriginal peoples and their respective claims, ambitions and interests is "the fundamental objective of the modern law of Aboriginal and treaty rights."[49] Brian Slattery (2006) suggests that "reconciliation must strike a balance between the need to remedy past injustices and the need to accommodate contemporary interests. This can only happen if the nature and scope of historical Aboriginal rights and title are fully recognized. Slattery (2006:282-83) suggests that, in order to foster a new legal order for Aboriginal rights, the twin set of Principles of Recognition and Reconciliation must be acknowledged:

> ...The successful settlement of Aboriginal claims must involve the full and unstinting recognition of the historical reality of Aboriginal title, the true scope and effects of Indigenous dispossession, and the continuing links between an Indigenous people and its traditional lands. ...However, by the same token, the recognition of historical title, while a necessary precondition for modern reconciliation, is not in itself a sufficient basis for reconciliation, which must take into account a range of other factors. ...To suggest that historical Aboriginal title gives rise to modern rights that automatically trump third party and public interests constitutes an attempt to remedy one grave injustice by committing another.

The fundamental principle of the honour of the Crown developed in recent cases mandates the Crown to negotiate with Aboriginal peoples to identify their rights, to consult with them when its activities affect these rights and, where appropriate, to accommodate Aboriginal rights by adjusting its activities.[50]

In the recent *Tsilhqot'in* case, Justice Vickers acknowledges that "the time to reach an honourable resolution and reconciliation is with us today."[51] However, he also recognizes that the courts are "ill equipped to effect a reconciliation of competing interests."[52] Quoting extensively from Brian Slattery's above-mentioned article, he insists that the governments, both federal and provincial, must "discharge their constitutional obligations" [53] and negotiate Aboriginal title claims in an honourable way.

---

[48] *Ibid.*; *Delgamuukw, supra* note 16; *Haida Nation v. British Columbia (Minister of Forests)*, [2004] 3 S.C.R. 511.

[49] *Mikisew Cree First Nation v. Canada (Minister of Canadian Heritage)*, [2005] 3 S.C.R. 388, at para. 1.

[50] *Haida, supra* note 52 at para.25.

[51] *Tsilhqot'in, supra* note 30 at para. 1338.

[52] *Ibid.* at para. 1357.

[53] *Ibid.* at para.1368: see Slattery, *supra* note 35.

Implications for the Management of Forest Lands and Resources

As noted earlier, a large number of BC First Nations are asserting Aboriginal rights, including title to traditional lands, and are seeking to protect these rights from the potential negative impacts of land and resource development through treaty negotiations. In other areas of the province, treaties are already in place. These are either historical treaties such as Treaty 8 in northeastern BC, or modern treaties such as the Nisga'a Final Agreement, the Tsawwassen First Nation Treaty and the Maa-nulth First Nation Treaty. The treaty rights of these First Nations benefit from constitutional protection and must be acknowledged and protected by the provincial government in the process of resource development.

The courts put the onus on government, while negotiating comprehensive claims, to respect the asserted but still unproven Aboriginal interests. This implies at a minimum an obligation to consult and attempt to accommodate the First Nations whose rights and interests may be affected by government actions and decisions. In the *Haida* and *Taku River* decisions in 2004, the Supreme Court clearly established that *asserted* Aboriginal rights and title trigger the Crown's duty to consult and potentially accommodate, stating:

> The Crown, acting honourably, cannot cavalierly run roughshod over Aboriginal interests where claims affecting these interests are being seriously pursued in the process of treaty negotiations and proof. It must respect these potential, but yet unproven, interests. …To unilaterally exploit a claimed resource during the process of proving and resolving the Aboriginal claim to that resource, may be to deprive the Aboriginal claimants of some or all of the benefit of the resource. That is not honourable.[54]

The Supreme Court stated that "the duty (to consult) arises when the Crown has knowledge, real or constructive, of the potential existence of the Aboriginal right or title and contemplates conduct that might adversely affect it."[55] The consultation process may reveal a duty to accommodate, leading the Crown to make changes to its proposed action based on information obtained through consultation, and to take steps to avoid irreparable harm or to minimize the infringement.[56] However, the Chief Justice was careful to state that pending final proof of a claim, Aboriginal groups do not have a veto over what can be done with the land, and that "what is required is a process of balancing interests, give and take."[57]

The Supreme Court has laid out the implications of the obligation to consult and accommodate by specifying the timing and the type of consultation that the government needs to undertake with potentially affected Aboriginal

---

[54] *Haida, supra* note 52 at para. 27.
[55] *Ibid.* at para. 35.
[56] *Ibid.* at paras. 46-47.
[57] *Ibid.* at para. 48.

peoples. With respect to forestry developments, it has found that consultation at the operational level is not meaningful. In *Haida*, the province had argued that it would consult with the Haida prior to authorizing cutting permits or other operational plans. The Court stated that in order to be meaningful, consultation had to occur at the strategic level, at the stage of granting or renewing Tree Farm Licences (TFLs). It reasoned that the allocation of a TFL triggers preparation of a management plan by the tenure holder, including inventories and a timber supply analysis on which the determination of an Annual Allowable Cut (AAC) is based. Decisions at this strategic planning stage may have potentially serious impacts on Aboriginal rights and title.[58]

Similarly in the *Huu-Ay-Aht* case, BC's Supreme Court found that the Crown had an obligation to consult and accommodate the First Nation on its Forest and Range Agreement (FRA) program.[59] The Crown had argued that it was premature to hold any consultation with a First Nation on a policy strategy such as the FRA program, which did not imply any specific conduct that might affect Aboriginal interests. For the Crown, the general continuing of forest operations were "insufficient and too broad to trigger the duty."[60] The court disagreed and found that merely contemplating an infringement, coupled with the knowledge of a claim, gave rise to the duty to consult, regardless of whether or not there was specific conduct that might adversely affect Aboriginal interests.[61]

In the recent *Wii'litswx* case involving the judicial review of a decision by the Regional Director of Forests to approve the replacement of six forest licences (FLs) in the Gitanyow traditional territory, BC's Supreme Court found that the director unreasonably minimized the potential impact of the FLs replacements on the Gitanyow's Aboriginal interests, and as a result underestimated the extent and scope of his duty to consult and accommodate.[62] Further, the court found that the Crowns' reliance on future operational decisions to minimize the potential negative effects of logging on the Gitanyow's interests was not reasonable in terms of accommodation:

> While consultation at the operational level is desirable, I am not satisfied that reliance on future discretionary decisions, over which Mr. Warner has no control, can be viewed as reasonable accommodation for the decision to replace the FLs. That decision was the first step, and the only strategic step, in the process that would ultimately permit logging on Gitanyow's traditional territory.[63]

---

[58] *Ibid.* at para. 76.
[59] *Huu-Ay-Aht First Nation v. British Columbia (Minister of Forests)*, [2005] 3 C.N.L.R. 74.
[60] *Ibid.* at para. 109.
[61] *Ibid.* at para. 112.
[62] *Wii'litswx v. British Columbia (Minister of Forests)*, [2008] B.C.J. No. 1599, at paras. 166-167.
[63] *Ibid.* at para. 186.

These cases all emphasize the need for government to hold meaningful consultations with Aboriginal peoples on a broad range of strategic planning as well as operational decisions in forest planning and management, and to make serious attempts to accommodate their rights. The enactment of legislation, programs and policies, as well as the granting or replacement of forest tenures, are all strategic decisions that mandate consultation and accommodation. BC's courts are willing to scrutinize a variety of government decisions that may infringe Aboriginal rights, even if these rights are only asserted, and to assess the adequacy of consultation and accommodation measures offered to affected Aboriginal peoples.

But even though consultation with Aboriginal peoples and accommodation of their rights, especially at the strategic level, may achieve some level of protection of Aboriginal interests, they are no substitute for the actual control and management by Aboriginal peoples of their lands and resources. As stated earlier in this chapter, Aboriginal title includes the right for Aboriginal peoples to make decisions with respect to their lands and resources, under their own laws and customs. The implication for forest management, as stated in the *Tsilhqot'in* case, is that the province cannot unilaterally grant resource extraction rights, such as forest tenures, on title lands. In Justice Vickers' opinion, "provincial management of timber and the acquisition, removal and sale of timber by third parties under the provisions of the *Forest Act* …do not apply to Aboriginal title lands under the doctrine of interjurisdictional immunity."[64]

As noted by Kent McNeil (2008), if the *Forest Act* provisions authorizing the removal and sale of timber to third parties do not apply to Aboriginal title lands, the implication is that provincial allocations of timber harvesting rights on Aboriginal title lands are not valid. The same would hold true for any provincial grant of interests, such as leaseholds and mineral rights that would infringe the Aboriginal title holders' "right to exclusive use and possession of land, including the use of natural resources."[65]

By contrast, provincial forestry legislation is applicable to Aboriginal rights short of title, and the province may infringe these rights, provided that the infringement is justified. In *Tsilhqot'in*, Justice Vickers examined the provincial forestry regime and found that forest harvesting activities are a *prima facie* infringement on the Tsilhqot'in hunting and trapping rights.[66] He then found that the province could not and did not justify the infringement of these rights. In his view, to justify the infringement of Tsilhqot'in Aboriginal rights, the province would have had to acknowledge the priority of these rights, and to collect sufficient information to allow a proper assessment of the impact of forestry activities on these rights. Consultation did occur with the Tsilhqot'in, but this consultation did not acknowledge their Aboriginal rights and therefore could not justify the infringement of these rights.[67]

---

[64] *Tsilhqot'in*, *supra* note 30 at para. 1045.
[65] *Ibid.* at 13.
[66] *Tsilhqot'in*, *supra* note 30 at para. 1288. Justice Vickers' findings with respect to forest legislation are discussed further below.
[67] *Ibid.* at para. 1294.

These judicial developments have compelled the BC government to review its resource policies and legislation to account for the rights of BC's First Nations in the resource development process. What steps has the provincial government taken to consult with First Nations and accommodate their rights as it allows resource development? Specifically, has the government amended its legislative provisions governing the allocation and management of forest lands to acknowledge and protect Aboriginal and treaty rights? This is the topic of the next section.

BC Government's Response to Legal Developments

Policy Developments

For a period of 30 years the legal position of the province as argued in all cases from *Calder* to *Delgamuukw* was that Aboriginal title had been extinguished. In the wake of the *Delgamuukw* decision, the government had to reverse its position. The decision to enter into a treaty process with the First Nations, beginning in 1992, was the first major response of the provincial government to the judicial decisions discussed above. While engaged in the treaty negotiation process, the government has nevertheless been reluctant to "acknowledge the full impact of s. 35(1)."[68] In its 2008 Annual Report (Government of BC 2008:6), the BC Treaty Commission comments on the persistent lack of government recognition of Aboriginal rights, stating:

> [...] the BC government has spent considerable time defending itself in court as First Nations seek the court's assistance in resolving disputes primarily over land and resources. Its hard-line stance in court has made its position on Aboriginal issues somewhat confusing.

Indeed, as noted by Louise Mandell (2008), even following *Delgamuukw*, the government's policy of denial continued. The province took the position that, pending proof of Aboriginal title, it retained the right to dispose of lands and resources as it pleased. Before they could be recognized and reconciled, Aboriginal rights and Aboriginal title first had to be proven in court or established by treaty. When the Supreme Court once again rejected the provincial position in the *Haida* and *Taku River* cases, the province developed a new theory to justify its lack of recognition of Aboriginal title. The 'small spots' theory, based on the province's interpretation of the *Bernard* and *Marshall* cases, holds that Aboriginal title only exists in relation to small sites that were physically and intensively occupied, such as village sites and cultivated fields. In *Tsilhqot'in*, the BC Supreme Court rejected this 'postage stamp' approach to title, stating:

> What is clear to me is that the impoverished view of Aboriginal title advanced by Canada and British Columbia, characterized by the plaintiff as a 'postage stamp' approach to title, cannot be

---

[68] *Tsilhqot'in, supra* note 30 at para. 1340.

allowed to pervade and inhibit genuine negotiation. A tract of land is not just a hunting blind or a favourite fishing hole. Individual sites such as hunting blinds and fishing holes are part of the land that has provided 'cultural security and continuity' to Tsilhqot'in People for better than two centuries.[69]

The province now admits that certain Aboriginal rights exist over larger tracts of Crown land, but these are very limited rights. In *Tsilhqot'in*, the province admitted that the nation had Aboriginal hunting and trapping rights over the claim area, but did not recognize their right to trade. The province did not recognize the economic and livelihood components of the Tsilhqot'in peoples' Aboriginal rights.

This entrenched legal position colours the negotiation mandates of government officials and negatively affects current efforts to address accommodation and reconciliation. Despite the legal posturing, however, BC has entered into various initiatives with First Nations. These include consultation policies and protocols, interim measures agreements with First Nations, tenure allocation, and First Nation involvement in land/resource planning and management. Some of these initiatives are discussed below.

Starting in 1995 after the BC Court of Appeal decision in *Delgamuukw*, BC developed various Aboriginal consultation policies. The 1995 Crown Land Activities and Aboriginal Rights Policy Framework (Government of BC 1995a) spelled out essential principles of consultation intended "to avoid any infringement of known Aboriginal rights during the conduct of [the Crown's] business." Agencies of the provincial Crown were to consult with Aboriginal groups to determine the existence of Aboriginal rights, the implications of resource management activities on these rights, and whether the granting of a Crown grant or tenure could co-exist with the Aboriginal rights. The Policy was amended in 1997 to reflect other court decisions, most importantly the decision of the Supreme Court in *Delgamuukw*. Consultation Guidelines released in 1998 described how decision-makers were to carry out their responsibilities for the allocation, management and development of Crown lands and resources. The most recent amendment of the consultation policy in 2002 follows the *Haida* and *Taku River* decisions, and stresses the need for decision-makers to consider asserted but unproven Aboriginal rights and title (Government of BC 2002). It is noteworthy that the Aboriginal consultation policies do not deal with treaty rights.

The Ministry of Forests (MOF) developed its own consultation policy in 1995 based on that of the province (Government of BC 1995a,b). This policy was replaced in 1999, and then again in 2003 to conform to the most recent provincial policy statement on consultation (Government of BC 1999, 2003). These policies are meant to be consistent with the province-wide consultation policy.

Even though they are a step forward in building relations with Aboriginal communities, the province's Aboriginal consultation policies are lacking in many respects. To begin with, they were developed and implemented

---

[69] *Tsilhqot'in, supra* note 30 at para. 1376.

unilaterally by the province. Second, as underlined by Tara Marsden (2005:74), the focus of these policies is on justifying the infringement of Aboriginal rights and title, rather than on accommodating them: "The attempted justification of an infringement almost always precedes the avoidance of infringement through accommodation of First Nations. Accommodation is usually the last effort in the consultation process." As noted by the BC Forest Practices Board (2007:16-17) in its investigation of the approval of forest harvesting in the traditional territory of the Tolquah Nation, the Ministry of Forests and the district manager in that case viewed the consultation process as an information-sharing exercise. While the First Nation was invited to identify its interests in the area and how these interests would be affected by the proposed harvesting operations, neither the MOF nor the district manager were willing to engage in a discussion of alternatives or possible substantive modifications of the proposal. Further, the issue of First Nations' consent to resource development, which is central to the discussion on consultation and accommodation, is opposed by the provincial government which maintains that it would undermine its jurisdiction over lands and resources. However, as discussed earlier, First Nations' consent to any resource development on their lands is a fundamental component of their right to determine how Aboriginal title lands will be used.

The most significant provincial policy document to address the relationship between government and First Nations is undoubtedly *The New Relationship* developed by senior provincial government officials and leaders from the First Nations Leadership Council in the Spring of 2005.[70] It represents a turning point in the BC government's position. Its statement of vision reads as follows:

> We are all here to stay. We agree to a new government-to-government relationship based on respect, recognition and accommodation of Aboriginal title and rights. Our shared vision includes respect for our respective laws and responsibilities. Through this new relationship, we commit to reconciliation of Aboriginal and Crown titles and jurisdictions.[71]

The document recognizes a number of First Nations' goals, including the need to achieve self-determination by being allowed to exercise "their jurisdiction over the use of land and resources through their own structures," "to ensure that lands and resources are managed according to First Nations laws, knowledge and values," and that First Nations are given "the primary responsibility of preserving healthy lands, resources and ecosystems for present and future generations." The document outlines several action plans, notably:

- to develop new institutions or structures to negotiate Government-to-Government agreements for shared decision-

---

[70] The First Nations Leadership Council is made up of the First Nations Summit, the Union of BC Indian Chiefs and the BC Assembly of First Nations–BC Region.

[71] See First Nations Summit: www.fns.bc.ca/info/newrelationship.htm.

making regarding land use planning, management, tenuring and resource revenue and benefit sharing;

- to identify institutional, legislative and policy changes to implement this vision and these action items;
- to identify and develop new mechanisms on a priority basis for land and resource protection, including interim agreements.

The first legislative support for this policy, the *New Relationship Trust Act*, was passed in March 2006. It provides for a $100 million fund to assist First Nations build capacity to participate in land and resource management and take advantage of economic, cultural and social opportunities. The question remains whether the new processes and structures for shared decision-making promised by the *New Relationship*, and the legislative changes that are needed to implement the action plans, will be developed.

Legislative Reforms: Are Aboriginal Rights/Title to Forest Lands and Resources Recognized in Forest Legislation?

Forsyth and Hoberg (2005) describe how the BC Liberal Party proceeded to change the provincial forest policies for First Nations, starting with its first term of office in 2001. They highlight the preeminent role of government/forest industry discussions in the formulation of these policies and the lack of involvement of First Nations' governments. These new policies were implemented through major legislative reforms in which the First Nations played no part.

In order to address Aboriginal communities' desire for a share in the economic benefits of forest development on their traditional territories, the provincial government amended the *Forest Act* in 2002 to allow the direct award of small to medium scale forest tenures to First Nations, provided that the First Nation enter into a treaty-related, economic or interim measures agreement with the province.[72] In 2003, the province announced the *Forest Revitalization Plan*, aimed at restoring the vitality of British Columbia's forest industry and fostering 'a new era of reconciliation with First Nations.' Part of the Plan was to redistribute forest tenures. The government stated it would 'take back' 20% of the volume allocated to major licensees, with half of this volume to be reallocated to First Nations and small community forest licences.

The government then enacted a series of significant legislative reforms to implement the Plan. These major forest policy changes were completed over the objections of BC's First Nations and without meaningful consultation with them (Forsyth and Hoberg 2005). The Title and Rights Alliance was formed in 2003 to oppose these sweeping changes.[73] The First Nations Summit passed a

---

[72] Bill 41, *Forest (First Nations Development) Amendment Act 2002*.

[73] According to its website, the Alliance resulted from two historic unity conferences that brought together First Nations from across British Columbia to discuss the provincial Crown's sweeping amendments to the *Forest Act* and the legal framework for forest practices, land use planning and land designations without meaningful consultation and

resolution asking the Minister to postpone the proposed legislative changes in order to consult meaningfully with First Nations. Opponents of the legislative amendments point, for example, to the repeal of provisions of the *Forest Act* that required ministerial consent to tenure transfers. As a consequence of these changes, the government can claim that it does not need to consult or accommodate Aboriginal peoples when forest tenures change hands.[74]

A cornerstone of the new BC forest policy was the increase in the number of interim measures agreements with First Nations. The government used the direct award provisions of the 2002 *Forest (First Nations Development) Amendment Act* to negotiate agreements with First Nations including both Direct Award Agreements and Forest and Range Agreements (FRAs) (Forsyth and Hoberg 2005).

The Forest and Range Agreements (FRAs) offer First Nations economic benefits in the form of revenue-sharing and direct tenure awards (a timber volume under a 5-year non-replaceable forest licence). In exchange for these economic benefits, which are pre-set by the MOF, the First Nations must not only agree "to not engage in direct action to exercise their rights and title, but must also agree to the MOF-led consultation process and are bound to participate in it" (Marsden 2005:68). Many First Nation communities have signed FRAs. Nevertheless, there is a high degree of dissatisfaction with government's approach in negotiating these agreements. As noted by Browne and Robertson (2004, 2006), the Gitanyow First Nation was the first to initiate a judicial review of a consultation process under the FRA program in 2004. The Huu-Ay-Aht First Nation (HFN) followed suit in 2005.[75] In both cases, the court found that the government had not adequately fulfilled its obligation to consult and accommodate by offering a FRA to the First Nations. In a strongly worded decision in *Huu-Ay-Aht*, Justice Dillon found that "the FRA policy does not meet the Crown's constitutional obligation to consult the HFN" and that "the Crown failed to follow its own process for consultation as set out in the Provincial Policy for Consultation with First Nations and the Ministry Policy."[76] Following persistent criticisms and court challenges of the FRAs by First Nations, in 2006 the province proposed a modified template, the Interim Agreements on Forest and Range Opportunities (FROs), with input from First Nation leaders. This revised agreement corrects some of the problems of the FRAs, but it still does not meet the expectations of the Union of British Columbia Indian Chiefs (2006).

In the *Tsilhqot'in* case, the court examined both the overall objective and certain specific provisions of forest legislation in order to assess whether the legislation accounts for Aboriginal rights. Justice Vickers stated that the *Forest Act* is silent with respect to Aboriginal title and rights.[77] Consequently, in

---

accommodation of indigenous peoples (Marsden 2005:84). See also www.titleandrightsalliance.org.

[74] These changes were enacted through Bill 29, the *Forest (Revitalization) Amendment Act 2003* (Forsyth and Hoberg 2005:20).

[75] *Huu-Ay-Aht*, *supra* note 64.

[76] *Ibid*, at paras. 104 and 123.

[77] *Tsilhqot'in*, *supra* note 30 at para. 1125.

determining the AAC, one of the most significant decisions in the forest management process, the Chief Forester does not take the potential existence of these rights into account. Indeed, the Chief Forester testified that he did not and believed he could not adjust his AAC determination on the basis of a claim for Aboriginal rights and title.[78]

> [...] the Chief Forester considered it to be his statutory duty to fully incorporate the Claim Area into the timber harvesting land base and to ignore the potential for Tsilhqot'in Aboriginal title.[79]

Elsewhere in the judgment, Justice Vickers examined the purpose section of the *Ministry of Forests and Range Act* and concluded that "there is no doubt that the Ministry seeks to maximize the economic return from provincial forests."[80] In his view, the timber-focused objective of forest legislation is an impediment to the accommodation of Aboriginal rights:

> A legislative scheme that manages solely for timber, with all other values as a constraint on that objective, faces a formidable challenge when called upon to balance Aboriginal rights with the economic interests of the larger society.[81]

In the *Wii'litswx* case,[82] Justice Neilson discussed at length the processes that were developed by the Ministry in consultation with the Gitanyow to enable their participation in key decisions with respect to land use planning and forest harvesting on their traditional territory. These efforts culminated in the signing of the Gitanyow Forestry Agreement (GFA) between the Gitanyow Hereditary Chiefs and the Minister of Forests and Range, on August 3, 2006 (Government of BC 2006).

The stated purposes of the GFA are to implement the order of the BC Supreme Court decision in the 2002 *Gitxsan* decision[83] and "provide a period of stability to forest and range resource development within the traditional territory." The five-year agreement recognizes the existence of the Gitanyow Huwilp (Houses) and their territories and outlines processes for planning and managing logging in the territories. The agreement includes the following commitments:

- revenue-sharing ($375,000/year);
- establishment of a Joint Resources Council (JRC) for cooperative planning, consultation and implementation of the agreement ($275,000);

---

[78] *Ibid.* at para. 1128.
[79] *Ibid.* at para. 1127.
[80] *Ibid.* at para. 1286.
[81] *Ibid.* at para. 1099.
[82] *Wii'litswx, supra* note 66.
[83] *Gitxsan First Nation v. British Columbia* (Minister of Forests), (2002) BCSC 1701, 10 B.C.L.R. (4h) 126.

- an invitation to apply for a forest tenure for 86,000 cubic meters/year;
- provincial commitment of $1 million on reforestation, silviculture work and watershed restoration;
- development of a joint sustainable resource management plan for the Gitanyow territory and deferral of logging in watersheds; and,
- consultation on the timber supply review leading to determination of the AAC within Gitanyow traditional territory.

Shortly after the signature of the GFA and the launch of the Joint Resources Council (JRC) and other initiatives, the District Manager approved the replacement of seven Forest Licences (FL), despite the Gitanyow's outstanding concerns over that decision. In his final assessment of the facts of the *Wii'litswx* case, Justice Nielson found that the GFA provides a useful framework for future consultations about FL replacements, but read as a whole, it did not provide accommodation to Gitanyow for the replacement of the licences.[84] Despite extensive consultation between the parties, in the end the Crown did not modify its position to accommodate Gitanyow's interests.[85]

The province is using land-use provisions under its *Land Act* to implement an ecosystem-based forest management scheme that should be more protective of First Nations' forest uses, including watersheds. As part of its Land and Resource Management Planning (LRMP) process, in February 2006 the government announced land use decisions for the Central Coast area and North Coast area. These decisions preserved a total protected area of 1.9 million hectares out of a total combined planning area of 6.4 million hectares. In July 2007, the provincial Minister of Agriculture and Lands signed a ministerial order establishing land use objectives for the south-central coast. These legal objectives are established pursuant to the *Land Use Objectives Regulation* under the *Land Act*.[86] The South Central Coast Order establishes land use objectives for the purposes of the *Forest and Range Practices Act* for specific landscape units identified in the order.[87] These objectives are for the protection of First Nations' traditional forest resources, First Nations' traditional heritage features, culturally significant trees such as monumental cedars, as well as the protection of aquatic habitats and biodiversity. They apply to forest companies or licensees as they prepare their forest development plans. By developing land use objectives that are protective of Aboriginal forest uses, the provincial government appears to be moving towards a more sustainable forestry regime.

---

[84] *Supra* note 67 at para. 184-185.

[85] *Ibid.* at para. 194.

[86] The *Land Amendment Act, 2003*, S.B.C. 2003, c. 74, amends the *Land Act* by adding Part 7.1: Land Designation and Establishment of Objectives; the *Land Use Objectives Regulation* was made in December 2005: O.I.C. 865/2005.

[87] South Central Coast Order, 27/31 July 2007.

Issues and Options for the Development of Aboriginal Forest Tenures

The above discussion of legal and policy developments shows that some progress towards 'reconciliation' is being achieved in British Columbia, although on-the-ground implementation of government initiatives is often deficient. Browne and Robertson (2004:11-12) suggest that in order to support the reconciliation of Aboriginal rights and title with the province's interest in ongoing development of the forestry economy, the following steps need to be taken:

- meaningful First Nation involvement in high level decision-making (e.g., the AAC determination);
- joint land-use planning;
- joint cumulative impact management;
- revenue-sharing and a fair share of timber resources or an option for a revenue-sharing only agreement.

As noted earlier, Aboriginal title includes the right to choose to what uses lands and resources will be put, and the methods of conservation, use and development of these resources. In forestry terms, this means that on title lands, an Aboriginal community as 'tenure' holder should be entitled to decide which and how much of the forest resources can be harvested (be they timber or other forest products), the rate and location of harvest, who is allowed to harvest, which methods of harvesting are to be used, etc. As stated earlier in this chapter, on lands where there is a strong probability of Aboriginal title, provincial forest legislation likely does not apply and the government is not entitled to grant rights to harvest timber unilaterally.[88] On lands where First Nations have Aboriginal rights short of Aboriginal title—rights which have a cultural as well as an economic dimension—an Aboriginal tenure holder should at a minimum be entitled to participate meaningfully in the decision-making process so as to ensure that the First Nations' rights are not infringed or extinguished. Ultimately, tenure is about governance and stewardship as much as about revenue generation.

For the majority of First Nations in BC who are in the process of negotiating their land claims, and whose Aboriginal title is not yet established, the objective is to protect as much of their traditional lands as possible as an interim measure. These lands are crucial to their sustenance, their culture and their identity. In many areas, traditional territories have already been extensively developed for forestry or other resource developments, and First Nations are faced with the need to preserve remaining critical areas and to restore damaged lands. The government's reliance on the FRA program to fulfill its obligations to consult and accommodate First Nations has had mixed results. The FRAs and FROs are temporary, and in many ways inadequate, measures. The form of forest tenure that the agreements offer is in most cases a short term (5 years)

---

[88] *Tsilhqot'in, supra* note 30.

non-renewable forest licence with limited volume. This is not the kind of forest tenure that enables First Nations communities to take on a stewardship role or carry out significant economic development in their forests.

Sarah Weber (2008; Weber *et al.* 2009) has used a community-based approach to identify how a particular First Nation community, the Stellat'en First Nation in central British Columbia, values the forest, what its goals and objectives are for forest tenures and governance, and how these goals may translate into alternative forest tenure and governance models.[89] The community has identified three overarching goals and objectives which are strongly interconnected:

- protect Stellat'en traditional territory for future generations,
- protect Stellat'en culture, and
- support Stellat'en economic self-determination.

For the Stellat'en First Nation, protection of its traditional territory for future generations is a critical overarching goal. Achievement of this goal requires a much higher level of decision-making authority in forest management than the community currently has under its FRA (the Stellat'en signed and accepted a FRA in 2005), and the community wants to have a leadership role in the stewardship of forestlands. Weber (2008:68) states that "the ability of Stellat'en to find their right 'balance point' would be facilitated by reforms to the western concept of tenure, to make it more compatible with the Aboriginal worldview—e.g., tenure as contingent proprietorship with an emphasis on good management and reciprocal social obligations." Her research emphasizes the importance of devolving authority from the provincial government to the Stellat'en, and that the "Stellat'en should play a leadership role in strategic and operational land-use planning, the setting of policy objectives, forest management standards and harvest levels (2008:69)." They should also be involved in monitoring and enforcement, in the allocation of harvesting rights, and should share revenues and profits with government and companies. This sharing of authority calls for models of co-management or co-governance.

In 1996, the Royal Commission on Aboriginal Peoples (RCAP) recognized the importance of co-governance or joint management arrangements as interim measures pending the negotiation of treaties.[90] The Commission reviewed various models of co-management and suggested ways to improve these models. The Commission also warned against the dangers of not giving proper consideration to Aboriginal values and admonished governments to show greater flexibility in their forest policies and practices (Government of Canada 1996:636).

---

[89] The Stellat'en First Nation is one of the eight First Nations members of the Carrier Sekani Tribal Council.

[90] In its final report, the RCAP recommended that "co-management arrangements serve as interim measures until the conclusion of treaty negotiations with the Aboriginal party concerned." See Report of the Royal Commission on Aboriginal Peoples (Government of Canada 1996:679).

Forsyth and Bull (2006) suggest that governance mechanisms are key to ensuring that Aboriginal rights are effectively incorporated into sustainable forest management. These authors analyze two examples of innovative forest management arrangements by the Nuu-chah-nulth First Nations in BC and the Innu Nation in Central Labrador. They conclude that these groups were successful in accommodating their rights and values in forest management because they were "involved in innovative and broadly-based co-management agreements with their provincial counterparts" (Forsyth and Bull 2006:20). In their view, granting forest tenures to First Nations in the absence of such co-management agreements would likely decrease the chances of success.

A co-governance model implies that First Nations use their own models of governance and stewardship, and their own Aboriginal laws. First Nations in BC are asserting their right to govern themselves in accordance with their traditional systems. As discussed earlier, the Gitanyow Forestry Agreement recognizes the Gitanyow Huwilp (houses) and their territories for the purposes of the agreement. For their part, the Carrier Sekani people highly value their traditional governance structures, the Keyoh and Bahlats system.

> The Keyoh is the system of land ownership and management which delineates use and access by clan membership. The clans of the Carrier Sekani are matrilineal entities that are maintained through exogamy (i.e., marriages allowed only with members of other clans). Each clan has a distinct Keyoh or traditional territory that it owns and controls. ...The Bahlats is the central institution through which the Keyoh are managed, owned and protected (CSTC 2006, 14).

As stated by Weber (2008:19), while the Stellat'en community (one of the CSTC member nations) elects a Chief and Council under the *Indian Act*, "traditional governance structures such as the Bahlats (potlatch) continue to be strongly valued."

The Keyoh system is being revived within some of the First Nations members of the CSTC. Thus, the Maiyoo Keyoh of the Central Carrier Nation asserts Aboriginal title to its family territory and has given notice to the federal and provincial governments of its assertion of title along with a map of its territory.[91] In February 2008, members of the Maiyoo Keyoh Society set up a protest camp on a logging road in order to prevent access by a logging contractor to an area that Canadian Forest Products (Canfor) was planning to clearcut. Neighbouring Keyoh titleholders joined the roadblock to show their support.

From an Aboriginal perspective, the ability to control multiple land and resources uses on Aboriginal title lands is central to the protection of their culture. It is the cumulative impact of resource development that often proves more damaging to the land than a single forestry development. This means that Aboriginal communities must first and foremost be involved in strategic level land use decisions over their territory. One of the ways in which this may be

---

[91] See www.maiyookeyoh.ca/Notice_of_Aboriginal_Title.jpg.

achieved is through a process of land use planning or 'ecosystem stewardship planning.' Resource development decisions and the allocation of resource tenures must be subordinate to strategic land use plans and broad land use decisions for a region.

The development of land and resource policies and plans has become a priority for many Aboriginal communities in BC. The CSTC has built capacity for Land Use Planning since 1994 and has developed a Land Use Planning Vision as a starting point for developing a Territorial Stewardship Plan at a regional scale.[92] In June 2009, the CSTC initiated the Carrier Sekani Conservation Strategy Project which will enable its peoples to develop a strategic approach on how they want to conserve and protect areas of their territories. [93] Several of its member First Nations, either at the Keyoh and Clan level, or at the Band Council level, are developing their own land use vision for their territory. The Carrier and Sekani are working towards developing new governance structures based on traditional governance systems that will work towards harmonizing these land use visions.[94]

Conclusion

The co-governance model of Aboriginal forest tenure necessitates legislative changes, not only in forest legislation, but also in other resource development statutes and environmental legislation, including the EA process under which projects are authorized and tenured. The *New Relationship* can only be given effect by such legislative amendments, which indeed are among the action plans outlined in that policy. In its 2008 Annual Report, the BC Treaty Commission states that the First Nations Leadership Council and the provincial government are discussing rights, title, governance and consultation in a proposed 'recognition and reconciliation act.' This would be "overarching legislation to supersede the myriad outdated statutes that do not reflect current understandings of Aboriginal rights and title" (BC 2008). It is only through such profound legislative change that the 'reconciliation' of Aboriginal rights and title with the province's interests and those of the broader society will be achieved.

References Cited

Bankes, N. (1998). Delgamuukw, division of powers and provincial land and resource laws: Some implications for provincial resource rights. *University of British Columbia Law Review* 32(2): 317-351.

BC Forest Practices Board (2007). *First Nations Interests and the Approval of Forest Harvesting near Ucluelet, BC.* Complaint Investigation 050708, FPB/IRC/131. December 2007.

Brethour, P. (2007). Band to withdraw from treaty talks. *Globe and Mail,* March 31, 2007.

---

[92] CSTC website www.cstc.bc.ca under Natural Resources—Planning —CSTC Land Use Vision.

[93] CSTC website www.cstc.bc.ca under Natural Resources— Conservation Project.

[94] *Ibid.*

Browne, M. and K. Robertson (2004). 'Title and rights alliance background paper: Forest and range agreements,' prepared for the *Title and Rights Alliance Conference: Moving Forward in Unity.* Saanich, BC. 19 May 2004.

Browne, M. and K. Robertson (2006). 'Aboriginal issues in the forestry sector,' in *The New Realities of Doing Business: Indigenous Peoples and Natural Resources.* Canadian Bar Association Conference. Calgary. March 10-11.

CSTC—Carrier Sekani Tribal Council (2006). *Carrier Sekani Tribal Council Aboriginal interests and use study on the Enbridge gateway pipeline.* May 2006.

Forsyth, J. and G. Bull (2006). Innovations in Aboriginal forest management: Lessons for developing an Aboriginal tenure. Paper on file with the authors. June 2006.

Forsyth, J. and G. Hoberg (forthcoming). 'In search of certainty: The 'new era' approach to forest policy for First Nations in British Columbia,' Chapter 2 of Jay Forsyth Master's Thesis (August 2005).

Government of British Columbia (1995a). *Crown Land Activities and Aboriginal Rights Policy Framework.* Victoria, BC: Ministry of Aboriginal Affairs, January 1995.

Government of British Columbia (1995b) *Protection of Aboriginal Rights Policy.* Victoria, BC: Ministry of Aboriginal Affairs, March 1995.

Government of British Columbia (1999). *Aboriginal Rights and Title Policy* and *Consultation Guidelines.* Victoria, BC: Ministry of Aboriginal Affairs.

Government of British Columbia (2002). *Provincial Policy for Consultations with First Nations.* Victoria, BC. : Ministry of Aboriginal Affairs, October 2002. www.gov.bc.ca/cpp/docs/ConsultationPolicyFN.pdf.

Government of British Columbia (2003). *Aboriginal Rights and Title Policy* and *Ministry of Forests Consultation Guidelines.* Victoria, BC.: Ministry of Aboriginal Affairs.

Government of British Columbia (2006). *Agreement Recognizes Gitanyow, Helps Boost Northwest.* Ministry of Forests and Range News Release, 3 August 2006.

Government of British Columbia (2008). *BC Treaty Commission Annual Report 2008.* Victoria, BC.

Government of Canada (1981). *In All Fairness: A Native Claims Policy—Comprehensive Claims.* Ottawa: Queen's Printer.

Government of Canada (1982). *Outstanding Business: A Native Claims Policy—Specific Claims.* Ottawa: Queen's Printer.

Government of Canada (1996). *Report of the Royal Commission on Aboriginal Peoples.* Vol. 2, Part 2, Recommendation 2.4.78(b). Minister of Supply and Services. Ottawa.

Mandell, L. (2008). 'The implications of Aboriginal title,' in *Tsilhqot'in First Nation v. British Columbia: The Immediate Impact and Next Steps.* Pacific Business and Law Institute Conference, Vancouver, March 4-5, 2008.

Marsden, T. (2005). *From the Land to the Supreme Court, and Back Again: Defining Meaningful Consultation with First Nations in Northern British Columbia.* Master's thesis, University of Northern British Columbia.

McNeil, K. (1997). How can infringements of the constitutional rights of Aboriginal peoples be justified? *Constitutional Forum* 8(2): 3-39 at 37.

McNeil, K. (2006). Aboriginal title and the Supreme Court: What's happening? *Saskatchewan Law Review* 69: 181-308.

McNeil, K. (2008). 'The significance of Tsilhqot'in Nation v. British Columbia,' in *Tsilhqot'in First Nation v. British Columbia: The Immediate Impact and Next Steps.* Pacific Law and Business Institute Conference, Vancouver. March 4-5, 2008.

Slattery, B. (2006). The metamorphosis of Aboriginal title. *Canadian Bar Review* 85: 255-286.

Swaak, N.D. (2008). *Forest Tenures and their Implications for Exercising Aboriginal and Treaty Rights on the Kaska Traditional Territory.* Master's thesis, University of Toronto.

Swaak, N., S. Kant and D.C. Natcher (2009). 'Aboriginal forest tenure attributes for the Kwadacha traditional territory,' pp. 127-142 in M.G. Stevenson and D.C. Natcher, eds., *Changing the Culture of Forestry in Canada: Building Effective Institutions for Aboriginal Engagement in Sustainable Forest Management.* Edmonton: CCI Press and Sustainable Forest Management Network.

Union of British Columbia Indian Chiefs (2006). Resolution no. 2006-06 Re: *Interim Agreement on Forest and Range Opportunities.* Vancouver. 30-31 January 2006.

Weber, S.E. (2008). *Aboriginal Forest Tenure and Governance in British Columbia: Exploring Alternatives from a Stellat'en First Nation Community Perspective.* Master's thesis, University of British Columbia.

Weber, S., R. Trosper and T. Maness (2009). Evaluation of two forest governance models based on Stellat'en First Nation goals,' pp. 143-162 in M.G. Stevenson and D.C. Natcher, eds., *Changing the Culture of Forestry in Canada: Building Effective Institutions for Aboriginal Engagement in Sustainable Forest Management.* Edmonton: CCI Press and Sustainable Forest Management Network.

Chapter Six

Designing a New Governance Structure: Analysis of a Stellat'en Model for Implementing Forest Management Devolution in British Columbia

Eddison Lee-Johnson and Ronald Trosper

Introduction

The Stellaquo, or Stellaten people,[1] are currently engaged in exploring ways of designing a new governance structure that combines both Aboriginal traditional styles of governance with elements of contemporary Canadian governance systems. Stellaquo's proposal considers land use planning as a governance process that is interconnected with major aspects of Aboriginal community well-being, including social, traditional, economic and cultural contexts. A product of data collected from a year's research with the Stellat'en First Nations and program implementation by the Stellaquo, this chapter discusses the Stellaquo's pre-contact forms of governance, and looks at their current governance structure and its relationship with the federal and provincial governments. It examines how Stellaquo's past relationships with settler society dissolved their traditional forms of governance, which negatively affected their identity, sovereignty and rights over forestland and resources.

The Stellaquo have proposed a tripartite or co-governance status with Canada and British Columbia in all aspects relating to the management of forestland and resources within their territory. This is a call for the devolution of authority over the management of land, including forestland and other resources to a local Aboriginal community. This model of self-governance varies from that of other First Nations in that it calls for the institutionalization of internal government structures, such as a legislature, and the wielding of higher authority than is presently exercised by First Nations governments in Canada (Natcher and Davies 2007:271-277). This is known as a 'bottom-up' approach, with devolved authority to First Nations that emphasizes interdependence, collaboration, and policy learning among governments and First Nations communities.

---

[1] The terms Stellaquo, Stellat'en and Stellat'en First Nation are used interchangeably to represent the First Nation communities and peoples discussed in this chapter.

In British Columbia, land use planning represents an increasingly important foundation for government-to-government engagement; land use plans need to be approved by both Aboriginal and provincial governments prior to implementation. Based on research with the Stellaquo First Nation, this chapter describes a proposed institutional framework for structuring a government-to-government relationship focused on lands and resources in a way that integrates Stellaquo's traditional governance systems with contemporary institutions of the provincial government. Unlike most other studies that advance governance models suitable for First Nations in the absence of defining their structures, this research offers a detailed description of the structural roles and relationships of different arms of their proposed government. Removed from the external influence of the Province, the Stellaquo governance model provides an alternative means to that by which 'Indians' are assimilated into a contemporary Eurocentric governing culture. The Stellaquo governance model is driven by a nation-building process aspiring to transform their currently imposed band council system of government to a devolved tripartite governance model that involves: 1) the recognition and implementation of Aboriginal rights and sovereignty, 2) the institutionalization of Stellat'en self-governance, and 3) a co-governance/ tripartite government–government relationship.

Various scholars have promoted the reinstatement of Aboriginal systems of self-government and have proposed a synthesis between traditional governmental structures and current Canadian governmental instruments (Peeling 2004; Newhouse and Belanger 2001; Tennant 1985; Mills 1994). Concurrently, they believe state governments should devolve forestland and resource authority to First Nations. Yet others advise caution, suggesting that anyone involved in transforming power relations should attempt to retain the 'assumed Canadian culture'[2] of integration and multiculturalism (Stephen and Taylor 2000; RCAP 1996; Riodan n.d.). Their claim is that First Nations must not be impervious to change, but rather should respond to new situations and challenges in a manner that allows people to retain a connection to their past. Essentially, any new designs for governance should not pose insurmountable obstacles to parallel systems of Canadian governance. Though the Stellaquo model is not a blueprint for all First Nations, lessons gleaned from this research should add to the growing set of case studies, and can provide directions for the implementation of self-governance and further research.

---

[2] We assume Canadian culture does exist, but is not adequately described as a common set of values or beliefs; but see Saul (2008) for a recent and popular opinion of how Aboriginal values underlay Canadian cultural identity. The cultural qualities assumed are broadly thematic, such as multiculturalism, peacefulness, friendliness, which in themselves can fit any western industrialized democracy, depending on the extent to which particular values are shared.

Background Study

The Stellaquo describe themselves, and have been recognized by others, as a nation that has always had a structured government and laws that governed their clan relationships (Brown 2002). This socio-political system was disrupted when they were dispossessed of their lands and became subject to a new form of government. In assuming jurisdiction over forestland and resources, federal and provincial governments took possession of lands which many authorities have acknowledged belong to these first inhabitants of Canada (Alfred and Corntassel 2005; McKee 2000:4-10). As a result, Aboriginal peoples have been systemically denied their sovereignty, treated as wards of the state, subjected to assimilation attempts and stripped of their traditional forms of governance (Menzies and Anthony 2005; Alfred and Corntassel 2005; Cole 2002; Chartrand 2002; Nietschmann 1995; Dickason 1992; Moss and Gardner-O'Toole 1991). Typical of European colonialism's contribution to the cultural devastation and demoralization of Indigenous People worldwide (Alfred and Corntassel 2005; Fisher 1976, 1977), these processes were carried out through the establishment of reserves, tenure regimes and the controlling instrument of the *Indian Act* (Alfred and Corntassel 2005; Nietschmann 1995). In this regard, Canada's Aboriginal peoples have much in common with colonized Africans, including suppressive elements that subjected them to abject poverty and sub-standard living conditions (Mayers 2006). On the Stellaquo reserve and other reserves across the country, unemployment rates are astonishingly high, average household incomes are well below the poverty level, dependence on welfare and other transfer payments is extensive, and indicators of ill health and poverty are unacceptably high (Jorgensen 2007; Tarnopolsky and Pentney 1985:1-11). The United Nations report of 2006 states that if Canada's Human Development Index (HDI) was to be measured against other countries using First Nations welfare and human rights issues (e.g., unemployment, health services, poor housing facilities, income, access to land and resources) as indicators, Canada would rank 174th out of 194 countries in the world (UNDP 2006). This is particularly so because the legacy of colonialism is still evident in Canada's Aboriginal communities, an effect that their colonized counterparts in Africa, with similarly low HDIs, are also experiencing (UNDP 2006).

Traditional Governance before Contact

On a global level, Aboriginal peoples' activism has expanded to the socio-political realm in its call for the reinstatement of traditional Indigenous land rights, governance and political self sufficiency[3] (Glück *et al.* 2007; Omlowsk and Menzies 2004; Brown 2002; Government of Canada n.d.). Many First Nations are pressing for devolution of lands and resources management that will enable them to integrate their values, interests and practices in a form of self-governance that is internally designed, institutionalized and managed by

---

[3] Political self-sufficiency means, at its most basic levels, the 'ability to set goals and to act on them without seeking permission from others'— something that Canada has consistently denied Aboriginal nations.

individual nations or a council of nations (Glück *et al.* 2007; Peeling 2004). This new trend has provoked much debate within the field of political economy. Conservative thinkers argue that the political and economic transformations of the twenty-first century indicate a shift in First Nations' demands for governance that mimics their traditional authority structure before contact. They contend that separate governing structures, based on cultural and traditional differences, which reflect First Nations interests and identity will radically undermine good relations with non-aboriginal peoples (Flanagan 2000).

Methods

Consistent with participatory social research methods, the Stellat'en First Nation took ownership of the research process in this study by incorporating its members' values into the research design. There was extensive collaborative engagement of community representatives at each stage of the research process. Two workshops, one for youth and one for adults, were organized in August of 2006 and May 2007. A final workshop was conducted in October 2007, in fulfillment of an agreement with the Stellat'en Research Council (SRC) for the researcher to provide the community with research findings. The purpose of the final workshop was to assess which portions of the research the community considered confidential or proprietary. Workshops were conducted as focus groups, with facilitators asking a series of questions and participants brainstorming the answers in small groups. There were also plenary sessions in which the general participants could share ideas from their group discussions and provide consensus on the answers. The Stellaquo established a research council comprising 12 members, each of whom represented a category specific to clan, gender, age, traditional leadership experience, and those with knowledge about forestry management, pre-contact governance, and Stellat'en history, traditions and cultural values. The SRC's mandate was to review and comment on the overall research, and the related research (Weber *et al.* 2009).[4] We jointly conducted workshops and most of the interviews.

Sixteen interviews were conducted with people from the four categories of respondents: Stellat'en community members, Registered Professional Foresters, Ministry of Forest and Range personnel, and non-First Nations. While this research might have benefited greatly from having representatives from the logging industry, members of the logging industries based in Stellat'en territory, namely Canfor and West Fraser, were unfortunately unavailable.

---

[4] Sarah Weber is a colleague who was studying Aboriginal tenure systems, alternate tenure systems that will incorporate the values of the Stellat'en First Nation. Weber's research and Lee-Johnson's complemented one another's in that, while Sarah studied alternative tenures, Eddison researched the governance structure needed to execute these tenures in Stellat'en traditional territory and British Columbia.

Findings and Results

The Stellat'en Current Government Structure: Internal Governance

The current governance structure of the Stellat'en is manifested by two structural tendencies which can best be described as traditional-hereditary governance (THG) and Band Council governance (BCG). The former is represented by the elders and the Keyah Whutduchan hereditary chiefs, while the latter is represented by the Band chief and councilors.[5] Each has different forms of operation and authority within the community, and collectively exhibit the following features:

1. There is a weak coordinating relationship between the two different entities.
2. There is evidence of disrespect and mistrust of the BCG among THG advocates because it is a 'white man's' model of governance that selects leaders without traditional names, and without a majority of the people supporting them.
3. The BCG functions more as an administrative unit than a government with true legislative powers.
4. THG advocates are prominent members of the community, who wield respect not only because they possess traditional names and titles, but also because of their knowledge of the traditions and culture, and contributions to the informal governance of the Stellat'en community.
5. THG commands very limited authority in executing current governance functions as traditional forms of government are prohibited by the *Indian Act*, i.e., and thus has no governance mandate.
5. There is a growing need to incorporate every unit of government in the new governance structure.

As the THG is more respected by the community, a series of conflicts arising between the two sets of authorities defines their relationship. While BCG finds its legitimacy in the *Indian Act*, the authority of the THG is grounded in pre-contact customary laws, traditions, culture and Aboriginal identity, which limit the influence of the elected officials of the BCG in the community. Most BCG officials do not carry traditional names, which are symbols of authority, respect and reverence in the community. This, in turn, tends to restrict their authority, particularly in dealing with community members with traditional names or titles. Without the latter, BCG officials often encounter criticism from named THG members, who question the legitimacy and source of authority of the band council.

---

[5] Band Councils are the *Indian Act*-mandated governments in First Nations communities. The Band Council is elected according to procedures laid out in the *Indian Act*. The councilors are elected by eligible band members and serve two year terms.

As an imposed Euro-Canadian system of governance, the BGG has come under the scrutiny of scholars such as Newhouse and Belanger (2001) and Moss and Gardner-O'Toole (1991:1-9). In both Stellaquo's land use research and planning consultation, it was the elders who were best able to express pertinent elements of the Stellat'en vision of land use, including land claims, territorial boundaries, ecological management and traditional knowledge, which is embedded in their oral history. The community had confidence in the THG's knowledge about the old ways of governance and land management. According to the Stellaquo's pre-contact political design, the THG is considered the legitimate authority to discuss these issues, which leaves the BCG handicapped in the performance of its external relations roles.

The Stellaquo's Governance Relationship with the Federal and Provincial Governments

The Stellat'en describe the current governance relationship with Canada and British Columbia as a subordinate relationship rather than a government-to-government regime. They indicated that they are being controlled as a nation by BC and Canada, as illustrated in Figure 1. As a nation, they consider themselves to have been subsumed into the Euro-Canadian polity, which hinders them from functioning as a nation with distinctive values, customs and traditional forms of governance (Brown 2002). They have also expressed frustration over the limited options and benefits that the treaty process has thus far provided First Nations, which only furthers the ambitions of the state to keep them governed (Jensen and Brooks 1991:67-79).

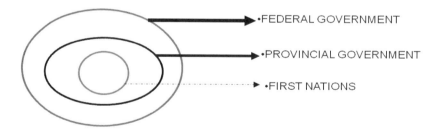

Figure 1. *Present governance structure and relationships.*

The Stellaquos' Proposed Governance Structure

The Stellat'en First Nation's proposed model of self-governance is therefore designed on the premise that the present governance structure is not only deficient, but inappropriate (Ottmann 2005). This is supported by workshop and interview responses, informal discussions with community members and feedback from the Research Council, which reveal weaknesses and internal frictions not only between elected band officers and clan leaders, but also in their relationship with the state.

The New Government Structure
After several discussions with the community and the Research Council a draft of their new government structure was constructed, as shown in Figure 2. The Stellaquo-proposed self-governance structure has six distinct components or arms of government:

1. the Community People/Citizens
2. the Co-governance board
3. the Clan Leadership
4. the Chief and Band Councilors (administrators)
5. the Legislature and Judicial Arm
6. the Youth Council

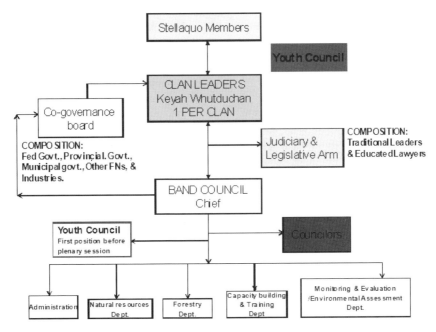

Figure 2. *Stellaquo Proposed self-governance structure.*

The Community/Citizens
Stellaquo citizens are demanding a change in the present form of governance so that their participation in government affairs becomes more recognized, vibrant and inclusive. Stellaquo members wish to input their concerns in all areas of government, particularly in the management of forestland and resources. They want and expect the administration, clan leaders and judicial arms of the government to hold consultations and report back to the community or citizenry on a regular basis. This periodic reporting is a strategy that is expected to make the government accountable and transparent, while also providing a forum in which community members could discuss and contribute views to help shape

land use planning processes in governmental operations (Reyntjens and Wilson 2004; Siar *et al.* 2006; Scheldler 1999).

The Co-governance or Tripartite Board

The term co-governance represents a new approach to the roles played by stakeholders in governance. In their proposal, the Stellaquo are suggesting tripartite governance with the province and the federal governments, whereby they contribute to the formulation and implementation of policies and management of forestland and resources in their traditional territory. Figure 3, represents the Stellat'en proposal for a shared or tripartite co-governance arrangement in decision-making and policy.

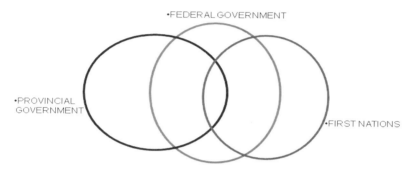

Figure 3. *Stellaquo-proposed tripartite/co-governance relationship.*

The current treaty process allocates a limited percentage of the total lands claimed by a First Nation for their direct control, while the remainder is co-managed by the province, federal and Aboriginal governments. The Stellaquo are proposing a co-governance instead of a co-management institutional design for the remaining portion of their territory. According to the Stellaquo, the Co-governance Board will serve as an entry point for development projects and programs. For example, a company wanting to undertake resource extraction in Stellat'en territory will present their proposal and have it thoroughly discussed by the Co-governance Board. Together, the Stellaquo, the federal and provincial governments, and industry representatives will establish the rules, expectations of revenue sharing and contracts through the Co-governance Board. The Stellaquo are re-claiming their jurisdiction and sovereign status through this governance arrangement ensuring federal, province and other stakeholders comply with rules, policies and set agreements (Jorgensen 2007; Lee-Johnson 2006; McNeil 2006; Ottmann 2005:4; Stephen and Taylor 2000; Rafoss 2005:9-13). In simple terms, the rules are jointly set by the board, but are regulated by the Stellat'en.

The Stellaquo people have anticipated some issues that the co-government or tripartite board is expected to address. For example, a co-government or tripartite governance model should respect the rights and responsibilities of all stakeholders. No longer content to be the recipients of policies and laws, the Stellaquo First Nations expect to assume an equal role in decision-making. Respect, equity and the reconciliation of rights should form the bases of this relationship and shift it from one of assimilation to one of co-existence, recognition and accommodation of all stakeholders' jurisdictions, values and cultures. In accepting this approach, the federal and the provincial governments must ensure that Stellaquo peoples are equal partners in negotiating changes in how rules are set in order to shape mutually satisfactory outcomes. Fitting this to the proposed governance structure shown in Figure 2, co-governance will require the emergence of a modern co-jurisdictional state, whereby the federal, provincial and municipal levels of government will be expected to dialogue with First Nations on land use planning and resource management, while at the same time acting as facilitators in the development of Stellaquo governance institutions. The major question that remains is: Are the provincial and federal governments prepared to relinquish this power?

A co-government or tripartite governance arrangement will involve re-defining historical relationships and developing new decision-making structures in which the Stellaquo will have an equal say in land management decisions (Glück *et al.* 2007; Edmunds and Wollenberg 2004). Within this context, the possibility of achieving cooperation and collaboration will also be high if new and equitable relationships are formed. The development of successful relationships among parties in a co-government or tripartite governing agreement will depend on the ability to effectively integrate goals and objectives based on cooperation, collaboration and mutual respect.

Composition and Functions of the Proposed Co-governance Board

There are two sets of memberships in the proposed model:1) permanent positions for representatives from the three levels of government, and 2) non-permanent positions for those representing other stakeholders in Stellaquo traditional territory:

1. Tripartite government:
- The Stellaquo First Nation, represented by Clan leaders, the Band Chief and Councilors, and some members of the legislature.
- Federal Government representatives, chosen by the Federal Government.
- Provincial government representatives, chosen by the Provincial Government.

2. Multi-stakeholders:
- Industries: Their representation hinges on the activities or interests they share regarding resources in Stellaquo territory in the form of projects within a given time frame.
- Other First Nations in the Carrier Sekani Tribal Council, particularly those adjacent to Stellat'en territory, with whom the Stellaquo share overlapping territorial interests.

- Non-First Nations, including those living in Fraser Lake, who are expected to be within Stellaquo jurisdiction after land claims are settled.

The Stellaquo will hold veto power over all extraction and zoning of forestland and resources within their traditional territory. With this in place, the Stellaquo will be able to negotiate with governments in new ways appropriate to re-defining power relationships and developing new decision-making structures in which the Stellaquo have an equal say and/or vote in land use planning and management decisions. The Board will evaluate projects or programs submitted by industries and potential investors regarding resources within their territory.

An acknowledgement of the board mandate to carry out shared authority and facilitate reconciliation of differences over previous land allocations, including proper consultation before changing current legislation that affects Stellaquo sovereignty, is necessary. The Board will also assist in jointly implementing equitable sharing of revenue from resource extraction, income and sales taxes, and in facilitating institutional and individual capacity building programs within the Stellat'en community.

Clan Leadership

Clan leaders represent the Stellaquo government in the tripartite relationship with the federal and provincial governments, forming a part of the Co-governance Board that makes decisions and policies relating to Stellaquo forestland and resources (*see* Fig. 2). Clan leaders, as the head of the political structure of the government, are responsible for planning, endorsing decisions and serving as primary advisors to the Stellaquo government. The Stellat'en are proposing that Clan leaders engage in the governance of the nation in a number of interrelated ways that integrate their traditional roles, responsibilities and obligations into the current and emerging structure of socio-political relations between First Nations and the Canadian state. Clan leaders must be sensitive to their relations with other First Nations and be able to lead the Stellaquo in building new relationships on the Co-governance Board, through negotiations with multiple stakeholders, namely, industry and all levels of government (municipal, regional, provincial and federal).

Functions of and Selection Processes of Clan Leaders

The three proposed functions of Clan Leaders are:

1. Represent Stellaquo in the tripartite relationship with the federal and provincial governments.
2. Pursue legislative and policy changes that devolve jurisdictional authority from the federal and provincial Crowns to First Nations, and
3. Identify capacity building initiatives that support self-sufficiency and promote a productive labour force.

Membership in this arm of the government is by appointment by elders of the community, with the main criterion based on matrilineal heritage (Peeling

2004; McKee 2000). Each Clan leader succeeds to an office within his or her matrilineal clan by taking a hereditary name that carries with it specific obligations and authorities (Peeling 2004; McKee 2000; Mills 1994). Hereditary chiefs succeed to office through the authority of the elders, starting with nominations and rigorous clan selection processes. Nominees are observed over a period of time by elders, who, through lengthy and rigorous scrutiny, identify successors. Brown (2000:21-67) and Aasen *et al.* (2006:6-22) describe this system of appointing leadership as hereditary and democratic, as each clan is fully represented, and clan selections are undertaken through rigorous procedures, including dialogue, negotiations and consensus. Possessing a traditional name bestowed on the basis of behaviour as well as birth right is an essential requirement for appointment equivalent to one's place of birth.

Clan appointment is strictly for individuals with matrilineal links to the Stellaquo community. It is recognized that within the Stellaquo community there are some Keyah Whutduchan members who moved to Stellaquo territory, and form part of the nation either through marriage or residence. While the Stellat'en are firm about who qualifies for Clan leadership, they do not consider marriage, residency or other social reasons as requirements for these positions. However, after a lengthy debate, participants in the governance workshop agreed to include individuals coming from other First Nations communities[6] with traditional names, who have been officially adopted by a Stellat'en clan.

Appointment is also based on solid social qualities such as the level of personal integrity and earned respect within the community.[7] Integrity is a high standard for Clan leaders, who are expected to be mentors to other community members. As a result, Clan leaders should be mindful of their actions and words, and be very temperate in their personality. A Clan leader who fails to meet these conditions can be deposed if his character is considered non-conforming to the cultural and traditional standards of Stellaquo leadership. One elder stressed the importance of integrity, as named members or clan leaders are expected to "walk the walk and talk the talk." Overall, Clan leaders are expected to live a balanced healthy life, be honest, sharing, caring, and command authority within the community.

Being a 'visionary' is a major quality proposed for members of this arm of government. Workshop participants stressed the need to have Clan leaders who are determined to initiate projects that will advance the interests of the community and improve the general welfare of the Stellat'en people. Whether there are appropriate criteria and indicators in place to measure this characteristic is an area that requires further research. With attributes as described above, Clan leaders are expected to be role models for the community.

---

[6] Participants in the governance workshop held in May 2007 indicated that some of members from other communities/First Nation bands have joined their community as a result of marriage, or other social reasons, and most have also been officially adopted as members of Stellat'en clans.

[7] Integrity implies that Clan leaders must not have been convicted for financial misappropriation, do not use alcohol or drugs and have earned the respect of community members.

Duration or Term of Clan Leaders

A Clan leader's tenure can go on indefinitely. Succession only occurs when the individual is replaced, chooses to step down, or is deceased. The Stellaquo's vision for self-governance indicates that a leader who does not meet the social expectations discussed above is liable to be replaced by the elders of the community. During their tenure of office, Clan leaders are expected to mentor youth who will assume their roles when deceased. This approach is used to ensure the survival and continuity of the government. The integration of the two systems of government—traditional and contemporary—in the Stellaquo proposal, will require Clan leaders to adopt a 'strong political commitment' to reforming, building and sustaining effective coordination among the different institutional structures, which will promote good governance (Siar *et al.* 2006:60-70).

Chief and Band Councilors

The Chief and Band Councilors comprise the technical and administrative arms of the government. Unlike the present *Indian Act* model, the Stellaquo model of governance would remove them from most political functions and policy-making roles, positioning them as managers of programs and staff members within each program (*see* Fig. 4). Thus, the current administrative roles of band councilors would be retained. However, they would have to relinquish their current role of making and enforcing laws, a responsibility proposed for Clan leaders and the legislative and judicial arms of the new government. Similar to public servants, e.g., deputy ministers, their role would now be more technical in nature and limited to managing programs that promote the welfare of the community.

Duration or Term of a Band Council Chief

Before 1999, the term for an elected Chief was three years and Councilors two years. Currently, the Band's Chief and Councillors positions have a two-year term. Workshop participants and the SRC have indicated the need to reduce the term to one year for the Chief's position and two years for Councilors, with an opportunity for re-election. The rationale for thedecrease in their tenure is as a result of the community's experience with an elected chief who deserted his position, because the two-year term was too long. Stellaquo's observation that many Councilors are currently serving reluctantly, with most looking forward to the end of their terms, is another indicator that the current term is too long. With a shorter term, they anticipate that officers will offer more energetic service, which would prepare the way for re-election.

Participants in the governance workshop identified the following functions as appropriate to the Chief and Band Council positions:

1) *Fulfill the need for grants.* They are expected to mobilize financial resources to meet the needs of the community:
2) *Implement programs.* All programs shall be implemented by the administrative arm of this government. Members of each of the individual offices or departments would have the

responsibility of creating programs specific to the community's needs. Workshop participants indicated that the full responsibility for the implementation of social programs be transferred from Indian Affairs to the Band Council.

3) *Negotiate agreements in consultation with Clan leaders.* This role positions Chief and Band Councilors as major participants in dealing with other stakeholders, such as the province, industries, other First Nations, and non-First Nations on the co-governance board.

4) *Hire and Layoff staff members.* Staffing in each department shall be the duty of the Band Council in consultation with the Clan leadership and the legislature. This paves the way for equal opportunity in the selection of staff and fosters collaboration and supportive relations among these three arms of government.

Legislature and Judicial Arms

The Stellaquo are seeking to institute a legislature that will facilitate the establishment of a constitution tailored for them. This division of the government will play a pivotal role in prescribing the tripartite and co-governance relationship the Stellat'en people seek with the state and other stakeholders. They are also advancing a judicial unit and the formation of a court system that will uniquely integrate Stellat'en and Canadian laws and enable them to function on a daily basis. The main function of the proposed judicial system is the maintenance of law and order through an innovative legal system that is independent, fair, responsive, and consistent with the First Nation's culture and traditional value systems (Quinn 2006; Palys n.d.).

The Stellaquo emphasize that the purpose of the call to integrate their customary laws into the legislative and judicial arms of their proposed government is to formalize these two legal mechanisms with proceedings regularized and carried out according to codified rules that combine Stellaquo laws and Canadian laws (Quinn 2006). Their legal system is expected to have checks and balances and, in cases in which no Stellaquo law exists for a particular crime, the court may apply federal and provincial laws. The courts would be conducted in both English and Carrier language, and interpreters will be used when necessary.

Functions of the Legislative and Judicial Arms

The legislative and judicial arms shall:

1. Facilitate the establishment of a constitution tailored to Stellaquo traditional values.
2. Play a pivotal role in prescribing the tripartite and co-governance relationship Stellat'en will have with the state and other stakeholders, and be responsible to consult with Clan leaders.
3. Establish a court system that uniquely integrates Stellat'en and Canadian laws.

4. Effectively maintain law and order through an innovative legal system that is independent, fair, responsive and consistent with the Nation's culture.

5. Decide the requirements and processes of Stellaquo band membership (status).

6. Establish an accountability code, which will govern the accountability of Stellat'en employees (including the Band Chief and Councilors).

7. Approve periodic budgets that will be presented by the Chief and Band Councilors, which is the administrative arm of the government.[8]

Composition and Requirements for Appointment of the Legislature and Judiciary

According to Stellat'en standards, bar membership will require formal training in both Stellaquo traditional law and Canadian law, as a condition to practice in the courts. Thus, the nation is proposing that formally educated lawyers be versed in both Canadian law and Stellaquo customary law. These individuals would work alongside traditional leaders, who are experts in traditional knowledge and justice, to participate in the legislature and court system. The traditional leaders would be there to ensure that practices are in conformity with customary Stellaquo values and systems of law.

The Youth Council

A Youth Council in the new government structure was called for by youth during the governance workshop in May 2006. At a plenary session comprising youth and adults, younger participants proposed specific functions for the Youth Council, which they felt should be comparable to those of the Band Council and similarly strategically positioned. However, some adult participants modified the position of the Youth Council, giving it less authority than the youth had originally proposed.

Eighty percent of the adult participants were strongly opposed to the Youth Council, because it contradicted Stellaquo tradition. In Stellaquo culture, youth are expected to be silent, passive and receive instructions from elders through apprenticeship for future roles (Thom 2001:1-7). Stellaquo elders asserted that, traditionally, youth developed leadership roles through mentorship, which consisted of the elders sharing their knowledge and nurturing leadership qualities in aspiring youths. When elders found a candidate with the right combination of qualities, they would train and mentor them, in preparation for the leadership role that the young person would one day assume. The elders also insisted that provisions will be made in the self-government structure to train youth in the Stellaquo leadership model, rather than adopting a design from outside Stellat'en culture, which had found favour among the youth.

Table 1 describes the different views held by the two categories of participants—those for and those against a Youth Council—that would execute

---

[8] May 2007 Governance workshop participants suggested this function as a measure of the government's accountability to its people.

roles equivalent to those of the Band Councilors. The discussion referenced in Table 1 reveals that young people generally feel powerless and excluded from Aboriginal traditional political process (Molley *et al.* 2002). The May 2007 Governance workshop underscored the profound issues that need to be resolved within many First Nations in order to effectively design appropriate government structures that would give Aboriginal peoples a voice in land use planning and management.

Table 1. Reasons supporting/rejecting motions re: Youth Council (with functions equal to Band Councilors).

Motion #	80% of Participants Against	20% of Participants in Favour
1	Youth Council with powers identical to the Band Council would create enormous financial strains on the government. Salaries for Youth Council representatives would further stretch the Band's already thin budget.	The Youth Council representatives should be paid since they will carry similar administrative functions as the Band Councilors.
2	Youth would become uncontrollable and disrespectful to their elders.	Youth are the future generations of Stellaquo who will assume political positions presently proposed by Stellaquo. They should be active participants in the proposed new government right from the beginning.
3	There is already a youth representative in the community, which is sufficient.	The current youth representative is dysfunctional and does not fully represent the wishes of youth. A Youth Council will be able to identify programs that would particularly suit the needs of Stellaquo youth.
4	The Canadian Charter of Rights does not make allowance for youth representation in the Provincial and Federal governments; why should it be the case for the Stellaquo?	Youth Council will help facilitate a more constructive youth voice, which will advance a productive participation in the new government.
5	The Youth Council should not possess authority analogous to that of Band Councilors because it contradicts Stellaquo tradition. According to Stellaquo tradition, youth are expected to be silent and take instructions from elders: *...children shouldn't have jurisdiction over parents in Stellat'en law.*	The effective participation of representatives of the proposed Youth Council can increase young people's levels of interest in local government. It would also increase young people's confidence and self-esteem and perceptions of their own capabilities. *In the olden days the elders and youth were one. To leave out the youth will be to ignore the future*

Source: Stellaquo governance workshop, May 2007.

Steps for Implementing Devolution of Forestland and Resources Management

Building Blocks for a Co-Governance or Tripartite Relationship

Figure 4 shows three steps advanced by the Stellaquo for affecting the devolution of governance of forestlands and resources to the nation. These steps, which are discussed in detail in *The Governance of Forestland and Resources in British Columbia: Case study of Stellat'en First Nation* (Lee-Johnson 2008), are:

> Step A. Recognizing and implementing Aboriginal rights, sovereignty and jurisdiction.
> Step B. Institutionalizing Stellat'en self-governance.
> Step C. Developing a Co-governance/Tripartite Government-to-Government relationship

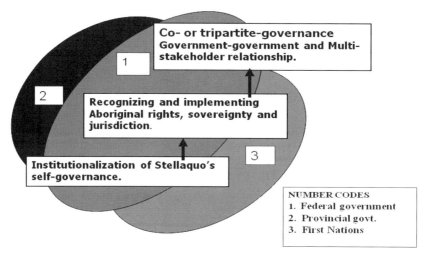

Figure 4. *Steps for implementing devolution of forestland and resources: Building blocks for a co- governance or tripartite relationship.*

Evaluating- the Stellaquo Governance Model

Strengths

The various arms of proposed government (clan leadership, Co-governance Board, the chief and councilors, legislature and judiciary, the community and Youth Council) demonstrate a division of labour in the execution of duties. The structure also reflects the determination of the Stellaquo to integrate their hereditary forms of government with contemporary systems. This integration also demonstrates their desire to create a mutual working relationship with the state, through the institutionalization of co-operative roles within the Co-governance Board. Also, Stellaquo people have clearly described the level of authority they aspire to, which requires a shift from state domination to a self-actualized system of governance ready to forge its own initiatives for social, cultural and economic development. The veto authority proposed by Stellaquo in deciding how land use and resources are to be managed is expected to support them in making decisions as they participate in the design and execution of policies. The inclusion of industries and non-First Nations in the Co-governance Board portrays their willingness to accommodate the rights and interests of different stakeholders which is the basis for any power-sharing relationship.

Weaknesses of the Stellaquo Governance Model
The Stellaquo government model requires further discussion, deliberation and refinement in areas such as the level of authority granted the Youth Council. Moreover, thorough discussion and reflection is needed to distinguish and define the roles of the clan leaders vis-à-vis the Chief and Band Council. This is important, as it will prevent redundancy in the execution of duties and avoid clashes between personnel in the two arms of government. This model of governance also does not make provision for an operational authority that would enforce laws and policies set by the new government. The Stellaquo will have to discuss the merits of this authority, which would depend on their decision about nomenclature, as well as identifying specific functions with which the law-enforcing authority would be entrusted relative to incorporating their traditional cultural laws. The proposed one and two year(s) term for elected Chief and Councilors in the self-government structure is very short, especially for the first cohort of people that will be occupying these offices. If realistic results are expected, it may be necessary for the Stellaquo to increase such officers' tenure.

There are currently four clan leaders that are expected to head the Stellaquo government. However, they have not indicated whether there will be one overall head from whom the other three will be taking directives. The possibility of all four having equal authority in managing the affairs of the government raises questions of how effective this arrangement would be, especially given that differences of opinion are expected to emerge.

A lack of trust currently exists between the Stellaquo and federal/provincial authorities, which will affect mutual intentions in pursuing reform processes. Trust is a key element in ensuring a healthy relationship between any governments, but central questions remain:

- Is there willingness on the part of both the federal and provincial governments to significantly devolve governing power over forestland and resources as proposed in the Stellaquo governance model?
- Will these governments be ready to relinquish such significant power and play a partnership role in a co-governance board?
- Is revenue-sharing possible, particularly in the areas suggested by the Stellaquo?

Another challenge is the integration of the differing value systems and orientations of the Stellaquo and non-First Nations peoples living in their territory. In addition, combining Aboriginal value systems with Canadian values will present similar challenges.

The Stellaquo made little mention of their commitment to remain a member of the Carrier Sekani Tribal Council (CSTC). Much of their vision for self-government was centered on internal structural reforms and their integration with provincial and federal governments. However, there is evidence of overlapping territorial boundaries with other First Nations, who are also members of the CSTC, which the federal and provincial governments recognize as the negotiating authority for land claims and treaty negotiations for the eight First Nations that comprise this body. Therefore, the nature of Stellaquo involvement in the CSTC will very much depend on the success of their proposed model of governance, especially when pre-contact manifestations of traditional governance probably involved their immediate neighbors in the feasting system.

Post-Research Development and Changes in the Stellaquo Vision

Since 2008, when research was concluded, the Stellaquo governance model has changed and evolved. The Stellaquo have developed programs with a series of projects currently being implemented to address the capacity development gap identified in Table 1. The nation has worked toward improving relations with the Nadleh Whut'en Band, their neighbour to the east, by addressing concerns regarding territorial overlap. A three-day workshop conducted in October 2009 reveals a strong commitment by the two nations to work together in the governance of forestlands and resources, especially in areas of overlapping jurisdiction. Participants in the workshop alluded to the fact that the two nations were co-resident and utilized natural resources harmoniously before colonial contact. Recommendations from the workshop suggest formation of a joint committee to further discussions leading to the formation of a possible joint protocol for the management of resources within areas of current overlap. One major recommendation that resonated with workshop participants was that the two nations were not going to allow the Crown to divide them any longer.

Programs to address capacity development gaps have also been initiated. In their drive to meet the need for qualified professionals, the Stellaquo are working with Vancouver Island University (VIU), University of Northern British Columbia and the College of New Caledonia to modify and develop VIU's current Fisheries Technician Diploma curriculum to meet the needs of interior BC fisheries ecosystems and watersheds. The Diploma training is proposed to commence in fall of 2010 and is designed to serve as a university bridging program. Courses offered would be credited to UNBC and VIU Natural Resources and Fisheries specialization bachelor studies. This would allow eight of Stellat'en's proposed graduates the opportunity to gain entry into the third year of a bachelor's degree program in any of the two Universities. Stellaquo has projected that by 2015, the community will/should have a pool of professional biologists, foresters or environmental officers. A current stock assessment project undertaken by the Department of Fisheries and Oceans (DFO) is employing eight Stellaquo seasonal Fisheries Technicians, but DFO has total control of all technical operations (Pinkerton 1989). Having qualified biologists is expected to change the operational control and management of this project in subsequent years, with operational authority revised to reflect higher SFN control and participation in the fisheries co-management relationship (Siar *et al.* 2006; Pinkerton 1993, 1999).

Another institutional development program that is being pursued is Geo-spatial and research methodology training. The first phase of this program commenced in April 2009 with the development of a *Draft Land Use Plan* and *Natural Resource Stewardship Plan*. The Stellaquo quickly recognized that having draft land use and natural resource stewardship plans was one thing, but having effective government capacity and structures to implement these plans was another. The next phase was the delivery of a joint Stellaquo and UNBC GIS training project scheduled for November 2, 2009–July 2010. Eight Natural Resource Department Staff (fisheries and forestry) were to be provided classroom and field assessment and inventory operational training. Trainees would be awarded a certificate at the end of the program. The GIS course/module is equivalent to six credit hours accredited at UNBC. The goal of the project is to build capacity within Stellaquo as GIS skills training, land use planning and Natural resource policy are vital natural resource governance tools identified in Table 3, which are needed by the Stellaquo for ecosystem stewardship.

Although not much has yet been done in implementing the proposed governance structure, the Stellaquo have proposed to address the need for the codification of their customary laws that would make way for the development of a constitution. A project has been proposed for 2010 to bring together community elders, Clan leaders and hereditary chiefs to commence discussion and planning of this process.

Conclusion

The aspirations of the Stellaquo to reinstate their traditional governance system and institute a comprehensive management plan for forestland and resources have led to several conclusions. To start, similar self-governance processes currently undertaken by First Nations governments in British Columbia and those proposed by the Stellaquo indicate that reconciliation between Aboriginal peoples and the Crown is possible under the province's *New Relationship Policy*. These new regimes are designing a holistic governmental apparatus similar to the Crown's, that reflects the emergence of legislatures, written constitutions, judicial systems and administrative institutions. These designs call for a tripartite inter-governmental relationship, with land use planning processes and debating power (rights, sovereignty, and jurisdiction) with the federal and provincial governments. This new power structure demonstrates Aboriginal peoples' desire for a new kind of relationship, constructed in mutual respect for differing cultural value systems and the recognition of overlapping geo-political jurisdictions and the jurisprudence on Aboriginal rights. The process of developing and reforming governmental relations must be realized according to First Nations-defined plans, priorities and processes, and be sustainable over a sufficient period of time so as to achieve a constitutionally defined mechanism of devolving authority. To conjure what forms these new kinds of governments will take and to undertake research as to how they will be constructed and implemented is as challenging as it is fascinating. Future studies will reveal their strengths and weaknesses based on practical experiences of Aboriginal peoples, the State and other stakeholders.

References Cited

Aasen, W., F.M. Craig, H. Hamilton, G. Hughes, C. Lewis, M. Lochhead, *et al.* (2006). *Carrier Sekani Tribal Council Aboriginal Interests and Land Use Study on the Enbridge Gateway Pipeline: An Assessment of the Impacts of the Proposed Enbridge Gateway Pipeline on the Carrier Sekani First Nations.* Document posted on the Carrier Sekani website, archived at http://www.cstc.bc.ca/downloads/Oil%20and%20Gas/AIUS%20Overview.pdf

Alfred, T. and J. Corntassel (2005). *Being Indigenous: Resurgences Against Contemporary Colonialism in Government and Opposition.* Victoria BC: Blackwell Publishing, Canada.

Azfar, O., S. Kahkoneh, A. Lannyi, P. Meagher, and D. Rutherford (2004). 'Decentralization, governance and public service: The impact of institutional arrangements,' pp 17-56 in M.S. Kimenyi and P. Meagher, eds., *Devolution and Development. Governance Prospects in Decentralizing States.* Aldershot, England: Ashgate Edited Publishing Limited.

Brown, D. (2002). Carrier Sekani self-government in context: Land and resources. *Western Geography* (Canadian Association of Geographers): 21-67.

Chartrand, P.L.A.H. (2002). *Who are Canada's Aboriginal Peoples?* Saskatoon, SK: Purich Publishing.

Dickason, O.P. (1992). *Canada's First Nations: A History of Founding Peoples from Earliest Times.* McClelland and Stewart Inc. The Canadian Publishers.

Edmunds, D. and E. Wollenberg (2004). *Local Forest Management: The Impact of Devolution Policies.* Earthscan Publications Ltd UK.

Fisher, R. (1976). 'Joseph Trutch and Indian land policy,' pp. 256-280 in J. Friesen and H.K. Ralston, eds. *Historical Essays on British Columbia.* Toronto: McClelland and Stewart.

Fisher, R. (1977). *Contact and Conflict.* Vancouver: UBC Press.

Flanagan, T. (2000). *First Nations Second Thoughts.* Toronto: McGill- Queens University Press.

Glück, P., j. Rayner, and B. Cashore (2007). *Changes in the Governance of Forest Resources.* Paper presented at the Global Forum conference (pp.79). Retrieved September 6, 2007 from http://www.yale.edu/forestcertification/ChangeintheGovernanceofForestResources-Gluck/pdf

Government of Canada (n.d.). *Parliament of Canada, Special Committee on Indian Self-Government, Indian Self-Government in Canada.* Ottawa: Ministry of Supply and Services) [Hereinafter, Penner Report].

Jensen, D. C. and Brooks (1991). *In Celebration of Our Survival. The First Nations of British Columbia.* Vancouver: University of British Columbia Press

Jorgensen, M. (2007). *Rebuilding Native Nations: Strategies for Governance and Development.* Tucson: University of Arizona Press.

Kaufman, J., F. Roberge, E.K. Fulton, J. Davies, and M. Nickerson (2003). Aboriginal governance in the Canadian Federal state 2015. Working paper 2003(3) IIGR, Queen's University. Retrieved January 21, 2008, from http://www.turtleisland.org/news/abgov2015.pdf

Lee-Johnson, E.B. (2006). *First Nations values that affect forest and land management in British Columbia.* Unpublished term paper for Forestry 528 course submitted April 2006, University of British Columbia, Vancouver.

Lee-*Johnson*, E.B. (2008). *The Governance of Forestland and Resources in British Columbia: Case Study of Stellat'en First Nation*. M.Sc. Forestry Thesis, University of British Columbia.

Mayers, J. (2006). 'Poverty reduction through commercial forestry: What evidence? What prospects?' *The Forests Dialogue*. Yale University School of Forestry and Environmental Studies. Research paper (July 2006, p.36). A TFD Publication. Retrieved March 11, 2007 from http://research.yale.edu/gisf/tfd/poverty_pub.pdf

Menzies, C. and M. Anthony (2005). Towards a Class-Struggle. *Anthropologica* 47:1-21.

McKee, C. (2000). *Treaty Talks in British Columbia: Negotiating a Mutually Beneficial Future*. Vancouver: UBC press.

McNeil, K. (2006). 'What is the inherent right of self-government?' Speaking notes Hul'qumi'num Treaty Group and National Centre for First Nations Governance, Governance Development Forum, Parkville, B.C. Retrieved April 7, 2007, from http://www.fngovernance.org/pdf/HulquminumSpeakingNotes.pdf

Molloy, D., C. White, and N. Hosfield (2002).*Understanding Youth Participation in Local Government*. A qualitative study prepared for DTLR. Retrieved July 23, 2007, from http://www.local.odpm.gov.uk/research/data/final.pdf

Moss, W. and E. Gardner-O'Toole (1991). *Aboriginal People: History of Discriminatory Laws* (BP-175E). Government of Canada document, retrieved March 21, 2007, from http://dsp-psd.pwgsc.gc.ca/Collection-R/LoPBdP/BP/bp175-e.htm

Natcher, D.C. and S. Davis (2007). Rethinking devolution: Challenges for Aboriginal resource management in the Yukon Territory. *Society and Natural Resources* 20: 271–279.

Newhouse, D.R. and Y.D. Belanger (2001). *Aboriginal Self Government in Canada. A Review of Literature Since 1960*. Native studies Trent University. Retrieved September 06, 2007, from http://www.iigr.ca/pdf/publications/205_Aboriginal_SelfGovernmen.pdf

Nietschmann, B. (1995). 'The fourth world: Nations versus state,' pp. 227-352 in G.J. Demko and W.B. Wood, eds., *Reordering the World: Geo-political Perspectives on the 21ˢᵗ Century*. Philadelphia: Western Press.

Orlowsk, P. and C.R. Menzies (2004). Educating about Aboriginal involvement with forestry: The Tsimshian experience—Yesterday, Today and Tomorrow. *Canadian Journal of Native Education* 28: 66-79.

Ottmann, J. (2005). *First Nations Leadership Development Within a Saskatchewan Context*. Unpublished master's thesis. University of Saskatchewan, Saskatoon, Canada, retrieved November 16 2007, from http://library2.usask.ca/theses/available/etd-04262005-094217/unrestricted/Ottmann.pdf

Palys, T.S. (n.d). Prospects for Aboriginal justice in Canada. Personal position statement, School of Criminology, Simon Fraser University. Retrieved November 28, 2007, from http://www.sfu.ca/~palys/prospect.htm

Peeling, A.C. (2004). Traditional governance and constitution making among the Gitanyow. Prepared for the First Nations Governance Centre. Retrieved September 11, 2006, from http://www.fngovernance.org/pdf/Gitanyow.pdf

Pinkerton, E., ed. (1989). *Co-operative Management of Local Fisheries*. Vancouver, BC: UBC Press.

Pinkerton, E. (1993). Co-management efforts as social movements. *Alternatives* 19(3): 33-38.

Pinkerton, E. (1999). Factors in overcoming barriers to implementing co-management in British Columbia salmon fisheries. *Ecology and Society* 3(2): 2.

Quinn, J.R. (2006). *Customary Mechanisms and the International Criminal Court: International Regulations and Local Processes in Post-conflict Society*. CPSA annual conference, Canadian Political Science Association. Toronto. Retrieved July 19, 2007, from http://www.cpsa-acsp.ca/papers-2006/Quinn-Customary.pdf

Rafoss, B. (2005). *First Nations' Government and the Charter of Rights: In Defense of the Charter: First Nations—First Thought.* A presentation to the Centre of Canadian Studies, University of Edinburgh, Scotland. Retrieved May 11, 2007, from http://www.cst.ed.ac.uk/2005conference/papers/Rafoss_paper.pdf

Reyntjens, D. and D.C. Wilson (2004). *Addressing the Environmental Challenges.* Conference briefing on Sustainable EU Fisheries, European Parliament, Brussels, Belgium. Working Paper No. 5. Institute for Fisheries Management and Coastal Community Development, Hirtshals, Denmark. Retrieved October 23, 2007, from http://www.ifm.dk/reports/110.pdf

RCAP—Royal Commission on Aboriginal Peoples (1996*). Report of the Royal Commission on Aboriginal Peoples.* Ottawa: Libraxus.

Riordan, C. (n.d). *Racial Discrimination Act 1975*: *A Review* by the Race Discrimination Commissioner: Book Review 85: 33. Retrieved May 5, 2007, from http://www.ilb.unsw.edu.au/docs/ilb_indexes_vol1-3.pdf

Saul, J.R. (2008). *A Fair Country: Telling Truths about Canada*, Penguin, Toronto.

Schedler, A. (1999). 'Conceptualizing accountability,' pp. 13-28 in A. Schedler, L. Diamond, and M. F. Plattner, eds., *The Self-Restraining State: Power and Accountability in New Democracies.* Boulder: Lynne Rienner.

Siar, S.V., M. Ahmed, U. Kanagaratnam, and J. Muir (2006). *Governance and Institutional Changes in Fisheries: Issues and Priorities for Research.* World Fish Center Discussion Series No. 3.

Stephen, C. and J. Taylor (2000). *Sovereignty, Devolution and the Future of Tribal-State Relations.* Tucson, AZ: Native Nations Institute, University of Arizona.

Tarnopolsky, W. and W. Pentney (1985). *Discrimination and the Law.* Don Mills, ON: DeBoo.

Tennant, P. (1985). Aboriginal rights and the Penner Report on Indian self-government. In M. Boldt and J.A. Long (Eds.). *The Quest for Justice: Aboriginal Peoples and Aboriginal Rights.* Toronto: University of Toronto Press.

Thom, B. (2001). Aboriginal Rights and Title in Canada after Delgamuukw: Part One, oral traditions and anthropological evidence in the Courtroom. *Native Studies Review* 14(1): 1-26.

UNDP—United Nations Development Program website (2006). *Definition of Capacity Development.* Retrieved February 27, 2007 from, http://www.capacity.undp.org

Weber, S., R. Trosper, and T. Maness (2009). 'Evaluation of Two Forest Governance Models based on Stellat'en First Nation Goals,' pp. 143-162 in by Marc G. Stevenson and David C. Natcher, eds., *Changing the Culture of Forestry in Canada: Building Effective Institutions for Aboriginal Engagement in Sustainable Forest Management.* Edmonton, AB: CCI Press and Sustainable Forest Management Network.

SECTION 3:
PLANNING TOOLS AND
CONSIDERATIONS FOR CO-EXISTENCE

Chapter Seven

Pikangikum Family Hunting Areas and Traplines: Customary Lands and Aboriginal Land Use Planning in Ontario's Far North

Nathan Deutsch and Iain Davidson-Hunt

Introduction

Aboriginal land use planning has taken on increased significance in Ontario through the first reading (2 June 2009) of Bill 191 entitled *An Act with Respect to Land Use Planning and Protection in the Far North*.[1] Of significance in this bill is that the Minister is required to work with First Nations who express an interest in preparing a community based land use plan. The bill proposes that the first step, following the expression of such an interest, is that "The parties must first approve terms of reference to guide the designation of a planning area and the preparation of the plan. Next, the Minister must make an order designating a planning area, to which the plan will apply" (Government of Ontario 2009:i). Given that a categorical imperative in state-initiated regional land use planning is the definition of a planning area (Hodge and Robinson 2001:95), it is not surprising that this would be a requirement for community-based land use plans. In this chapter we explore the potential of such planning areas to be rooted in First Nation conceptions of territory and space.

Our intent in this chapter is not to add to the debate concerning Aboriginal peoples and property rights. Rather our objective is to demonstrate that Anishinaabe land and resource use in northwestern Ontario is characterized by the widespread presence of spatial areas used by family groups for purposes of hunting, trapping, fishing and harvesting (Berkes *et al.* 2009; Tanner 2009). These areas of land and water have historically figured in the organization of the bulk of subsistence and commercial activities by extended family groups (Bishop 1974; Dunning 1959). These spatial areas received early scholarly attention as family hunting territories in a long-running debate over Aboriginal territoriality and property rights (Bishop 1974; Cooper 1939; Leacock 1954; Speck 1915; Speck and Eiseley 1939). While this debate has more or less waned in recent years, the organization of family hunting territories under changing conditions continues to intrigue scholars interested in land-use and land-use institutions through which indigenous hunting, value and knowledge systems are reproduced (Berkes 1987, 2008; Feit 1991). Traditional use areas have now taken on new importance as one way by which First Nation planning areas could be identified. It is in this context that this chapter describes the process of

---

[1] In Bill 191 'Far North' is described in detail but can be summarized as lands north of Woodland Caribou and Wabakimi Provincial Parks, and north of existing forest management units.

trapline registration by the Ontario Government in the traditional territory of the Pikangikum First Nation. We then explore ways in which trapline registration impacted customary rules related to resource use within the pre-existing system of family hunting areas.[2] We differentiate this latter system from the government trapline system by using the term 'customary,' to emphasize the continuity between the past, present and future despite alterations to Anishinaabe institutions such as kinship, worldview and language (Davidson-Hunt 2006). It is important to note that, although an imperfect fit with customary land use (Berkes *et al.* 2009), the registered traplines have played a significant role in delineating Pikangikum's planning area.

The Whitefeather Forest Land Use Plan & Regional Planning

There have been other regional land use planning processes in Ontario in which First Nations have been meaningfully involved, e.g., processes involving the Teme-Augama Anishnabai (Lane 2006) and Poplar River First Nation (Poplar River First Nation 2005). The Whitefeather Forest Initiative of Pikangikum First Nation was the first community-based land use planning process in the region now designated as the Far North of Ontario. The result was the Whitefeather Forest Land Use Strategy signed in 2006 by Pikangikum First Nation and the Ontario Ministry of Natural Resources (Pikangikum First Nation and OMNR 2006). As this was the first community-based land use plan created by a First Nation in partnership with Ontario it is interesting to note how these parties decided upon the boundaries of their planning area. In Anishinaabe, the planning area has been called *Ahkee Weeohnahcheekaywin Weembahbeepee Eekahnahn* and is distinct from the area they considered as their traditional territory (Pikangikum First Nation and OMNR 2006:26). *Weembahbeepeeeegwan* was used to convey the idea of an "area that stands out by itself within a larger area; a part of a landscape that forms a covering like a traditional tent." In the land use strategy, the following explanation is provided for the boundaries (Pikangikum First Nation and OMNR 2006:26):

> The boundaries of the Whitefeather Forest have been defined by the trapline areas of *Beekahncheekahmeeng paymahteeseewahch.* Long ago our people were not bounded by these traplines; however, in the present context the traplines provide a useful basis for Community-based Land Use Planning purposes. In addition, the Whitefeather Forest does not encompass the whole of Pikangikum First Nation's ancestral lands. Some of our trapline areas have been lost to non-Native trappers and to Woodland Caribou Park; one of our people's traplines is split by the 51st parallel which generally divides the Northern Boreal Initiative

---

[2] We use the term family hunting areas to refer to the area of land people use to hunt, trap, fish and gather resources. These areas are thus not only limited to commercial trapping activities, or for hunting, but are understood to be relevant for a broad range of land-based livelihood activities, both commercial and subsistence.

(NBI) from the Ontario Living Legacy planning area and hence Pikangikum First Nation is only planning for the portion in the NBI area.

Pikangikum First Nation has decided to use OMNR trapline areas as the basis for defining the Whitefeather Forest Planning Area out of respect for the customary custodial responsibilities of Pikangikum people and our neighbours. The head trappers on each trapline are some of our *keecheeahneesheenahbayg* (esteemed Elders) who provide guidance for our community-based decision-making processes in Pikangikum First Nation. Each of these head trappers exercises a customary custodial responsibility for their trapline areas and therefore these head trappers are central to both Pikangikum First Nation's relationship to the land and to our Community-based Land Use Planning process. The trapline areas have also been used to define the boundaries of the WFPA in order to respect neighbouring First Nations.

Regional planning begins from the idea that there are multiple, interacting planning units at different scales and each of those is a whole in and of itself, or what has been termed a 'holon.' As (Hodge and Robinson 2001:72) note the term holon finds its origins in the work of Koestler (1976) and Wilber (1997) and expresses the notion that within regional planning, a planning area is a holistic unit, or region, that encompasses smaller parts such as municipalities, while at the same time being part of a larger whole like a province. Conceptually, this term is at the root of similar ideas such as nested ecosystems or watersheds, while specifically including the dimension of decision-making. This idea has also been picked up in recent resilience literature that uses the notion of socio-ecological systems that both embed and are embedded within other systems and includes a consideration of decision-making (Folke *et al.* 2003). If we look back to the statement provided by Pikangikum, our interpretation is that they have put forward traplines as the basic planning unit and senior trappers as key participants in decision-making for such units.

Another key issue recognized in the regional planning theory is the need to obtain coordination amongst planning units at different spatial scales (Hodge and Robinson 2001:72). This idea again finds parallels in the resilience literature and the current focus on the need to coordinate across different scales for purposes of planning and management (Berkes 2002). In the approach taken by Pikangikum, this was achieved through an agreement amongst the senior trappers associated with Pikangikum who agreed to bring their trapline areas into the planning process. What is interesting in this approach is that there is horizontal coordination and cooperation amongst family groups and consensual decision-making for the pursuit of new land uses. This is distinct from regional planning theory in which a larger regional planning unit has been considered to 'transcend the capabilities (and authority)' of smaller units (Hodge and Robinson 2001:72). However, there is the notion of a territory doctrine in which a planning unit is a confluence of a people with a shared identity and a territory that can sustain the people. Whereas the dominant approach of regional planning is rooted in the function doctrine, whereby planning units are constructed for functional purposes using natural (e.g., ecosystem classification) or administrative (municipality, provincial, federal government) boundaries, it would seem that the territory doctrine is more analogous to the approach articulated by Pikangikum. We now turn to the notion of territory and the continuity of Pikangikum's customary institutions related to land and authority, and how they have been shaped through interactions with the Province of Ontario.

Figure 1: *The traplines of Pikangikum. The area in the lighter shade of gray is the planning area for the Whitefeather Forest Initiative. The dark gray polygon at bottom left is Woodland Caribou Provincial Park.*

Family Hunting Territories and Traplines
Beginning in the 1920s, provincial and territorial governments in Canada began registering parcels of land as traplines for a more coordinated approach to managing fur resources that would be in line with other government resource management institutions. Although trapline registration was purportedly initiated for protection of fur resources and Aboriginal interests (e.g., *see* Notzke 1994 and discussion on archival sources below), traplines also became instruments for the northern expansion of formalized state management institutions. In northwestern Ontario, 'state' forms of wildlife management made their debut through policies and legislation in fur management and specifically through the registration of trapping areas in the 1940s (*see* Fig.1).

In the course of investigating the evolution of Pikangikum's trapline system, we use materials from several archival collections, and interview and life-history material collected during field work with Pikangikum conducted between 2006 and 2009. Documentation from the Archives of Ontario and the

Library and Archives of Canada complemented oral histories of trapline registration from Pikangikum. In addition, reports from two anthropologists were used. A.I. Hallowell visited Pikangikum in the 1930s and recorded information on family hunting areas immediately prior to registration of the traplines. A second anthropologist, R.W. Dunning, writing in the 1950s, filled in gaps with his own archival and ethnographic investigation into changes in territorial organization of Pikangikum peoples. We used Hallowell's archives at the American Philosophical Society to explore the traditional land-use system, before the creation of the trapline system, and we found Dunning's archives at the University of Toronto to contain valuable material relating to pre-trapline organization to the workings of the government trapline system early in its implementation.

Land Use Prior to the Trapline Registration

The anthropologist R. W. Dunning noted that land-use and resource management off-reserve (i.e., on the traplines) in the decades prior to his fieldwork (1954-55) had been characterized by only weak relationships with both the federal and provincial governments. Also, though Pikangikum people were deeply intertwined in fur trade relationships, there was relatively little Euro-Canadian presence in fur-trade outposts, especially in Pikangikum or nearby Poplar Hill[3] (Lytwyn 1986). We thus found a corresponding scarcity of reports on land-use, hunting and trapping organization from both government and HBC sources. Our richest sources for archival documentation related to Pikangikum were the bodies of work produced by Hallowell and Dunning.

A.I. Hallowell traveled up the Berens River from Lake Winnipeg as far as Pikangikum over several summers in the 1930s (Brown and Grey 2009). Hallowell found large Anishinaabe groups gathered at sites on the shores of major lakes along the Berens River. Hallowell associated three summer fishing settlements with the Pikangikum band, at Pikangikum Lake, where the current reserve can now be found, and at two other sites further down the Berens River. One of these sites is now the location of the Poplar Hill reserve; the other Hallowell found at Barton Lake which is no longer occupied, yet remains a popular fishing area accessible from both reserves (Hallowell 1992:46) (*see* Fig. 2)

While summer settlements were located at the shores of major lakes, Hallowell described winter hunting and trapping areas formed of smaller family groupings, based on kinship ties. These groups took advantage of lake and river systems removed from the summer gathering places, and which formed travel routes to access furbearers and other resources needed for winter subsistence. Hallowell wrote of recognizable family hunting and trapping areas, which could be readily sketched on maps by certain members of various communities of the Berens River: "Their approximate boundaries could be easily traced by the Indians themselves" (Hallowell 1992:45) (*see* Fig. 3).

---

[3] In 1946, Pikangikum became a full post of the Hudson's Bay Company. From 1920 when it was created to 1946, Pikangikum was an outpost of the Little Grand Rapids trading post.

Figure 2: *Major summer settlements used by Pikangikum people in the 1930s, as described in Hallowell (1992).*

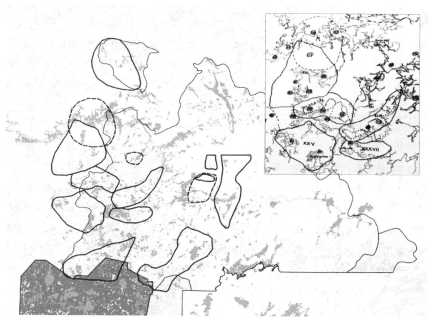

Figure 3. *Hunting territories redrawn from inset (upper right corner) by A.I. Hallowell (American Philosophical Society, Philadelphia). Dashed lines show 'old areas.' Solid lines show areas as of Hallowell's 1930s fieldwork.*

Later, in 1954-55, the anthropologist R.W. Dunning conducted his fieldwork, a major outcome of which was his book on social and economic changes linked to the introduction of government social assistance (Dunning 1959). Dunning's notes at the University of Toronto Archives contain little correspondence with government officials regarding traplines. However, they are rich with his own observations from fieldwork and archival investigations. Dunning (1959:55) noted that, from the time of the first government records for Pikangikum in 1876 to the period of his writing in the 1950s, the band had been made up of a number of hunting groups situated at various distances apart, based on the centers of their communally owned hunting territories (Dunning 1959:55). Table 1 provides a comparison of place names gathered by the Whitefeather Forest Management Corporation, Hallowell and Dunning for lakes used in summer and winter.

In describing their land use prior to trapline registration, Pikangikum people have emphasized ways in which the registration of traplines changed things. Paddy Peters noted some of these changes:

> After traplines were established... you couldn't trap with others like you could before. Elders talk about what it was like before... one big area. Traplines were planned in accordance with clan boundaries. These boundaries were made rigid. People were told to pick up traps that were across the line from their territories. (Paddy Peters, 15 November 2006)

Often, when asked about where people trapped before registration, we were informed that they were not restricted to any particular area. Elder Tom Quill Sr. (6 November 2007) recalls "before there were boundaries, we trapped anywhere.... My father would go all over the place." However, this statement must then be qualified as framing a set of other statements about land use. Flexibility of movement (i.e., within certain bounds) was conveyed to us in describing the area comprising all current Pikangikum traplines as 'one big area.' This statement was elaborated by Elder Charlie Peters, who recounted that, before trapline registration, there were indeed recognizable areas, but these were customarily, albeit flexibly, defined based on how far people would travel while pursuing their livelihood activities. "If we were at a lake, we would only trap so far" (Charlie Peters, 9 October 2008). Thus, boundaries appear to have been flexibly defined. Since boundaries were not fixed as they are at present, people were able to move between areas and join with different family groups following customary rules. Meetings were also an occasional occurrence between people and groups hunting and trapping in different areas during the winter. Although rules associated with the government trapline system focus on hindering the occurrence of trespass on others' trapping grounds, it is interesting to note that these meetings were not necessarily characterized by discordance or concern about trespass:

Table 1: Names of major lakes at the centre of Pikangikum family hunting areas.

English name	English translated name (WFI)	Anishinaabe name (WFI)	English name (Hallowell)	Anishinaabe name (Hallowell)	English name (Dunning)	Anishinaabe name (Dunning)
Barton Lake	White Spruce Narrows/Little Duck Lake	*Obeemeenahiigohkang/ Shi-shi-pe-si-wi-sa-ka-i-kang* (from syllabics)	Duck Lake	Not given	Duck Lake/ Spruce Point	*Opimena-eko-kangg/Sesep-saka-egun*
Berens Lake	Not translated	*Pi-kwa-ta-me wi-sa-ka-i-kang* (from syllabics)	Sturgeon Lake	*Name*	Sturgeon Lake	*Pekwutamaywe-saka-egun*
Cairns Lake	Not translated	*Ki-chi-wa-she-ka-mi-shing* (from syllabics)	Big Clear Water Lake	*K'tchiwac'egomic sagahigan*	Not given	Not given
Hawk Cliff/ Eagle Rock Lake	Hawk Cliff	*Keeneew Wahbik*	Not given	Not given	Eagle Rock Lake	*Kiniw-wapikwunk*
Kirkness Lake	Clear Lake	*Wahshaygahmeeshing*	Clear Water (?) Lake (Kirkness)	*Kawasigamisi sagahigan*	Clear Water Lake	*Washaykumisheng*
Mud Lake	Shallow Mud Lake	Not available	Shallow Mud Lake	*Kapakwakiwagakosagahi-gan*	Mud Lake	*Kapakoshkeewakake-saka-egun*
Pikangi-kum Lake			Pikangikum	Not given	Pekangi-kum Lake	*Pegunchegum*
Roderick Lake	Not translated	Not given	Little clear water Lake (Roderick Lake)	*Wac'egamic'is sagahigan*	Clear Lake	*Washay-gamesesink*
Silcox Lake	Not translated	*Pahkayahgahmak*	Not given	Not given	Bay Lake	*Pukay-yakamak*
Throat River	Not translated	*Ohkohtahgaizeebee*	Throat River	Not given	Throat River	*Okotangunesepe*

> We used to see McDowell Lake People when trapping in this area for muskrat... These were not planned meetings. Sometimes we would camp overnight with them (George M. Suggashie, 7 November 2007).

People also observed certain rules regarding how beaver lodges—the most important and scarce resource in the fur trade at the time of trapline registration—were appropriated:

> It wasn't like the way it is today. Like within your own trapline area, if there's a beaver lodge, the person within the trapline area would assume ownership. Anyone who would find the lodge would assume ownership over that lodge (George B. Strang, 25 October 2007).

In the 1930s, Hallowell (1992) found no rigid principle of intergenerational succession of hunting/trapping areas in the traditional system. Dunning observed in the 1950s that areas were typically passed from father to son (Dunning 1959), yet records in his archives on the occupancy of areas shows significant variation from this pattern. As maps available in archives were compared to trapline history data collected during fieldwork with Pikangikum, we found that the new administrative rules that came with trapline registration had important implications and consequences. Even so, that traditional land use institutions continued to a large extent to guide Pikangikum peoples' relationship to their family hunting and trapping areas. Maps drawn from notes and sketches in Dunning's archives show movement of extended family groups within the traditional system (*see* Figs. 4 and 5). These maps illustrate residential shifts in the major family groups residing at major lakes. *Nindoodemag*[4] are animal names used by Anishinaabe people to indicate affiliation with patrilineal clans. Clan affiliation is inherited patrilineally, so it is possible to observe processes of kin-based succession of areas during a period when Christian surnames were rare and inconsistent across generations.

There appears to have been slow change in membership at several of the major lakes, indicating that at times, areas were not simply passed down to male heirs. It also seems that new areas were occupied over time, although this may simply reflect the quality of the record more than actual expansion. It is quite possible that relatively large areas were sub-divided through time as the population increased, as is indicated by Dunning for the later years covered in his archives (1959). In addition, Hallowell recorded the clan groups residing at each of the major summer settlements (Table 2). Barton Lake is no longer settled in summer, and people from this settlement have since found their way to other locations.

---

[4] For a discussion of the meaning of the Anishinaabe word *indoodem* and the organization of the clan system, see Bohaker (2006).

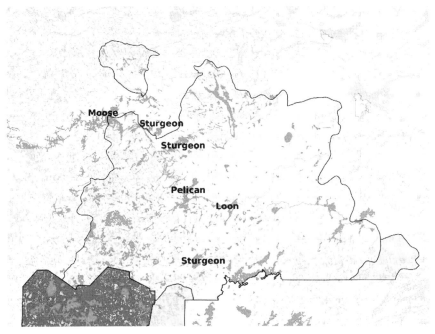

Figure 4. *Locations of family groups 1876*
(adapted from Dunning, University of Toronto Archives).

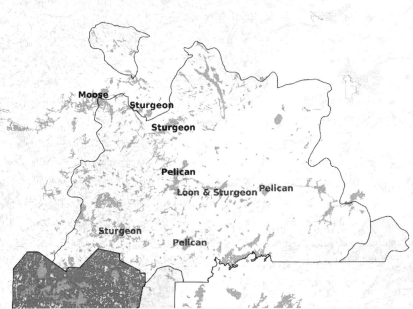

Figure 5. *Locations of family groups 1900. Changes from figure 4 are printed dark gray type* (adapted from Dunning, University of Toronto Archives.

Table 2: Clan names at major summer settlements[5]

Lake name	Clan names
Little Duck Lake (Barton Lake)	Sturgeon (chief), Pelican, Moose, Loon
Poplar Hill	Moose, Sturgeon
Pikangikum	Moose, Sturgeon, Pelican, Loon

Establishment of the Registered Trapline System

In the 1930s, federal and provincial authorities began to negotiate the setting aside of exclusive hunting and trapping areas for Aboriginal bands in Ontario. Their actions were driven by a concern that over-trapping had caused serious furbearer depletions across Ontario, and much of the northern forests of Canada. The exclusive band areas proposed in the 1930s were never fully realized in northern Ontario. However, the province adopted a policy of prohibiting new white trappers north of the Canadian National Railway (CNR) line by the 1940s. The registration of traplines seems to have been the means by which this prohibition policy was put into action. According to Pikangikum people, officials justified the creation of the trapline system as a means to stop imminent white incursions into their traditional areas. However, Pikangikum people explained that a real threat from white trapping had never materialized in areas known to Pikangikum trappers. Matthew Strang recalled stories of only two white trappers in Pikangikum territory (Matthew Strang, 7 September 2007). Likewise, a 1941 report from the Sioux Lookout treaty party found minimal activity of white trappers:

> Except for possibly four or five white trappers along the Manitoba boundary between Berens River and Cobham River, Indians have the whole country to themselves for trapping and hunting. Very few halfbreeds [Métis] or non-treaty Indian trappers are resident in the district.[6]

---

[5] Hallowell papers (American Philosophical Society, Philadelphia).
[6] Sioux Lookout Treaty Party Report, 23 July 1941. Ontario Archives, RG1-427.

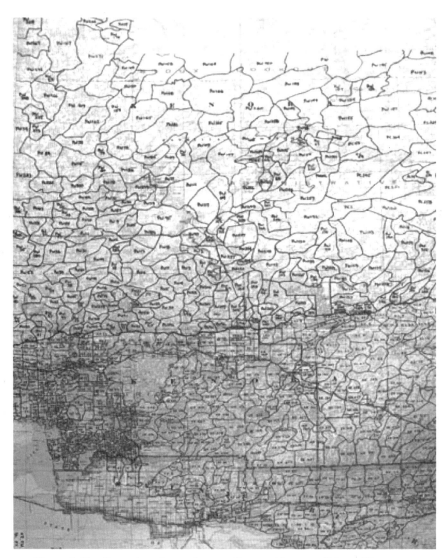

Figure 6. *Traplines of northwest Ontario, Ontario Ministry of Natural Resources Library, Peterborough, Ontario.*

An internal circular from the Deputy Minister of Fish and Wildlife, Ontario, stated in 1946 that:

> ...the region north of the northernmost transcontinental railway line has been allotted to Treaty Indians by agreement with the Indian Affairs Branch and is therefore not open to licensed trapping except by persons with established rights who have been

accustomed to trap in this region and whose right has previously been recognized by the Department.[7]

The issue of how trapline areas were actually designated, and the role of Pikangikum people in their design, is critical for understanding continuities and changes in the customary system. To the south of Pikangikum, on Crown land that had been previously surveyed and divided into townships, there was already a system of assigning one township per trapper as a trapline area (Fig. 6).[8] This was never was the case in Pikangikum as the advent of registered traplines in the late 1940s was the first time that Pikangikum had presented a spatial representation of its customary land use system to the province.

Evidently, the traplines of Pikangikum were based on pre-existing family hunting areas. The advantages to this technique seem to have been justified because of the enormous area that had to be covered by few officers in the short period they were given to complete the registration process. On February 4, 1948, Harkness, chief of the Fish and Wildlife Division at the Department of Lands and Forests, Ontario wrote to the HBC:

> North of the CNR, our officers met all the Indian trappers at treaty time [summer] and the boundaries of their traditional trapping grounds were plotted on maps of the district. Each band was then broken down into family groups, each group with its own trapping area, and the names of the members of each group listed. This work was accomplished with the help of the Indian Agents, the traders and the missionaries at the various settlements throughout the district including Severn and Weenusk. No doubt some adjustments in the area boundaries will have to be made next summer and if your post managers can secure information pertaining to any changes of that kind it will be appreciated.[9]

Dunning (1959:27) confirms that "government registration of the trappers and territories was based on the 1947 grouping of trappers and their own definition of existing trapping areas as drawn on large-scale maps." However, the origin of the boundaries seems to be somewhat more ambiguous and complicated due to the presence of other factors at play.

Twenty areas were registered for Pikangikum (Dunning 1959:27). Although prior preparations seemed to have been made by district officers of the federal Indian Affairs Board, the entire mapping process north of the CNR line

---

[7] F. A. MacDougall, Circular #14, 1946. Ontario Archives, RG1-427.

[8] It is possible to see in Figure 6 in some parts of the south, the residual rectangular township shapes. These township-based traplines in the south were also re-drawn to take into account to some extent natural features about the time of registration of the trapline system in northern Ontario.

[9] Letter, C.H. Harkness, 4 February 1948. Ontario Archives, RG1-427, File no. 2207.

was carried out within a single summer.[10] Because of limited staff, it seems to have taken some time before the registered trapline system was actually operational—as opposed to a system on paper:

> In the Patricia District we have not as yet enough field personnel to adequately cover this large territory and bring all the various groups of trappers, who are almost all Indian trappers, into the fur management program immediately.[11]

Although mapping was done in the summer of 1947, the system itself was not completed until later. In a description of the process, Hugh Conn, the fur supervisor at the Indian Affairs Branch in Ottawa regarded the mapping exercise as "more or less a fact finding expedition, since little or no information is available in Toronto concerning trapping in the isolated districts of the Province, and we discussed tentatively the possibility of holding a conference in Toronto at the conclusion of the summer season, which would set the stage for the actual organization next summer."[12]

It is also relevant that there was already some institutional experience in terms of working with trappers and with the traditional family hunting areas system: "The custom in the area is to trap in family groups. In order to organize fur management among them an understanding of their trapping methods is necessary."[13] The Province borrowed its rationale from a collaborative beaver preserves program involving the HBC and the Indian Affairs Branch: "In fact the new regulations provide for much the same management plan of Band area and family group trapping grounds as have been in operation on the Beaver Preserves."[14] In addition, it would appear that the province made an effort to manipulate trapline boundaries on maps to force them to conform roughly to watersheds for management purposes. Crichton, writing the province's perspective on the matter stated that, "In more northern parts of the province, watersheds may include quite a large area upon which a number of Indian families may trap together, which is very desirable."[15]

Although it seems the federal and provincial governments engaged in extensive discussion and negotiations from the 1930s to the establishment of the trapline system in the 1940s, the period of communication with Pikangikum people regarding traplines was relatively brief. One reason for this seems to be linked to rising awareness of threats posed by the traplines to treaty rights,

---

[10] Correspondences between Dunning and Harkness, Dunning Papers, University of Toronto.
[11] Correspondence between Harkness and H. L. Keenleyside, 20 April 1948, Ontario Archives, RG1-427.
[12] Correspondence between Conn and Orford, Ottawa, 7 June 1947, National Archives of Canada, RG10 c-8105
[13] Fur Advisory Committee report, 6-7 February 1950, Ontario Archives, RG1-427, File no. 2207.
[14] Letter, F.A. MacDougall, 19 December 1947, Ont. Arch. RG1-427, File no. 2207
[15] Crichton, V. 1948 "Registered traplines." Reprinted from *Sylva* Vol. 4, No. 2. pp 3-15, in National Archives of Canada, RG10, c-8104.

which was most readily felt by Indian Affairs Department. Trapline registration in other parts of Canada were also understood by some First Nations to be a breach of their treaty rights (Notzke 1994). Ontario developed an interest in assuming responsibility for fur management even though it was previously not a concern.[16] It therefore moved quickly to register a vast area in a very short period of time. On the other hand, the province's choice to establish traplines mirroring pre-existing family hunting areas also received much attention and debate, as evidenced in various communications among government departments. The selection of traditional family hunting and trapping areas, albeit modified to suit bureaucratic requirements of governmental legibility, seems to have been motivated by a concern for conflict-avoidance and the desire for a relatively frictionless transfer of power to the provincial natural resource management authorities.

Records held by the provincial and national archives allow for a cursory sketch of what the trapline registration process might have entailed. Visits were made to communities in northern Ontario and traplines were sketched out in 1947, but work was not complete until somewhat later. Provincial authorities and the Indian Affairs Branch were assisted by whomever was available, including Indian Affairs district officers, missionaries and HBC personnel. It was noted that the first maps only showed "the approximate boundaries of such traplines."[17] Registration of the traplines cannot be solely understood on the basis of statements from bureaucrats and government authorities. Pikangikum perspectives on trapline registration reveal the importance of this transition in terms of decisions made by elders when registration was done. Registration of traplines was not easily compatible with the system of customary family hunting areas, yet the traplines are today understood by Pikangikum elders to be an outcome of the recognition of traditional family areas by the government. In addition traplines are recognized as having played a role in sustaining beaver populations. Pikangikum elders recalled the sense in which their parents set up the trapline system, in conjunction with the province:

> The trapline boundaries were not written by a white person. These people were asked where their trapline areas were, and that's what their response was: Those boundaries. The boundaries that are in my trapline area were written by my father and [his brother]. That's why [trapline registration is] so important; because the elders at the time drew these boundaries. And these boundaries will not change. They will stay as they are. And the MNR will not change these boundaries... Elders long ago, they're the ones who set those boundaries... Long ago when these boundaries were written, there were times when someone else would sneak into someone else's trapline and trap. There would be actions taken. That's how they

---

[16] The beaver preserve system, a precursor to trapline organization, was administered jointly by the Hudson Bay Company and the Federal Government.
[17] *Ibid.*

would honour these boundaries. (George B. Strang, 25 October 2007)

George Strang emphasizes in this statement that traplines were registered in accordance with Elders' understandings of their customary areas. He and others saw the registration of traplines as desirable because it offered a solution to some problems in the use of commercial fur resources: "Long ago the people would trap unlimitedly. They would harvest fur without any limits." (George B. Strang, 25 October 2007). However, trapline registration also brought about changes which the government used to limit the mobility of Pikangikum people who chose to follow their own customary system of land use.

Land Use Following Trapline Registration

Several important features of the new registered trapline system made an impact on the ways in which Pikangikum people moved about on the land. The government was a non-factor in fur management before traplines were introduced. After 1947, trapline leaders were assigned the role of senior trapper in relation to other trappers using traplines, and to the province with regard to wildlife management matters. "There was no such thing as a trapline boss [senior trapper] back when I was born. Everyone just went anywhere. [My father] told the MNR exactly where he trapped, where he went up to. My father was the first [senior trapper]" (George B. Strang, 25 October 2007). There was leadership of areas before traplines were registered, but this kind of leadership in relation to Provincial wildlife management was new.

Although traplines became fixed spatial entities, the province allowed some flexibility; traplines and boundaries could be changed according to the needs of Pikangikum people. Since 1947, several major changes have been made to trapline boundaries. In one case, an area was split into two smaller areas. "The people in this area made an agreement and divided the area in two." (Senior trapper, Jimmy Keeper). Interestingly, divisions do not appear to have been made clearly along family lines. In fact, in numerous cases, adjacent trapline holders have continued to co-operate, camp together and hunt together, even though they occupy different registered traplines. However, once boundaries were in place, the wildlife management authorities brought in new rules which further restricted peoples' movements:

> MNR was very strict. They took people to court as a result of going to someone else's trapline. That's the way it happened. In my trapline area, when I would see someone else's trap there, I would personally take ownership of that trap (George B. Strang, 25 October 2007).

While Pikangikum people felt forced to stay within their allotted trapping areas, in partial compensation, they were also told that settlers couldn't trap in their areas either.

People only trapped in their areas. They didn't go into other people's traplines. I was told that the white man couldn't come in to trap because it's someone else's trapline (Tom Quill Sr., 6 November 2007).

Figure 7. *Trapline boundaries redrawn from maps in Ontario Ministry of Natural Resources Library, Peterborough, Ontario. Family groups in 1954 redrawn from Dunning, University of Toronto Archives.*

Despite changes enforced by the province, important features of the family hunting areas endured. In earlier times, groups of relatives tended to trap together:

We were in groups of relatives. We had our own trapline with a group of relatives... Non-relations went to different places. We never knew each other in groups. Not really well. So we had to go with relatives. It was hard for us to get to know different groups. That was a long time ago (George M. Suggashie, 30 April 2007).

It is quite possible that the registered trapline system accentuated the pattern of groups of relatives staying together. Writing soon after the creation of the government trapline system, Dunning (1959) concluded that, with few exceptions, succession of trapline areas occurred patrilineally, and that residence on trapline areas in the time of increasing government influence has tended to favour increased rigidity of this system of succession. The set of maps in Figures 7 and 8 compare trapline membership in 2008 by major family groups with those present in 1954-55 during the winter trapping season of Dunning's field work. Although families have Christian surnames, the clan names again have been used in the maps for continuity with the previous set of maps, and to understand broader movements at the level of extended family units as Pikangikum's population continues to grow. The maps also show some subdivision of areas according to trappers' wishes that are not reflected in the government traplines.

Figure 8. *Trapline boundaries and family groups in 2008.*
Changes from Figure 7 are in gray type

Dunning notes that changes in personnel on the traplines were made in accordance with trappers' wishes at annual trapline meetings on First Nations reserves (Dunning 1959, 27). This format continued into the early 1980's when the last few trapline meetings were held on reserves between trappers and Ontario Ministry of Natural Resources (OMNR) conservation officers. After this, changes in trapline membership were communicated less formally, and on a case-by-case basis (Claire Quewezence, OMNR Assistant Park Superintendent and Deputy Conservation Officer, personal communication, October 2008). Throughout the latter half of the 20th century, senior trappers continued to invite others onto their traplines, in accordance with Pikangikum's customary system:

> My father was invited to go to another trapline, other than the one he was born at. From then on, that's where he went. You have to be invited to go to another trapline. It's possible to ask people to trap on their trapline, or if you want to use a section of the trapline. (Paddy Peters, 15 November 2006)

The policy of the province was to encourage steady, long-term occupation of a single area to foster good wildlife stewardship as well as to guard against poaching of valuable fur resources:

> A trapper is required to spend as much time on his trapline as possible in order to thoroughly look after and maintain it. In this way, the District office can very well see just who are looking after their traplines and who are not (Crichton 1948).

This provision seems to have been consistent with the way trapline succession has worked over time. Senior trappers tended to be those most knowledgeable about their area.

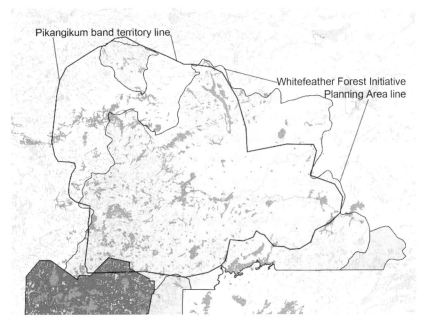

Figure 9. *Pikangikum 'band territory' in Dunning (1959) compared with contemporary Whitefeather Forest Initiative planning area.*

Long-term occupation of an area by extended family groups was customary before trapline registration. However, with the advent of registered traplines, areas were at increased risk of being alienated from individuals and families, much like private property. This latter change seems to have had some impact in Pikangikum's case as senior trappers wanted to reaffirm their presence on the land for fear that the land and resources would be re-allocated by government to others, rather than following Pikangikum's customary process of succession. Senior trapper John-Pierre Keesic explained that senior trappers may invite others to trap in their registered trapline area "Just to know that people are there, just to make signs that you were there" (J. P. Keesic, 23 October 2007). This policy has also become relevant in the context of current land-use planning activities. Because Pikangikum manages traplines at the band level, they are not open to private dealings between individuals as, for example, when traplines held by outsiders adjacent to Pikangikum's traplines become vacant. Traplines, formerly part of family areas belonging to Pikangikum people had in the past been allocated to white trappers. Now, as white trappers retire and their trapline tenures are given over to First Nations in Ontario, it is the Band that makes decisions about who takes charge of these areas, a process which has been agreed on between First Nations and the province. Traplines or parts of traplines have also been passed from one First Nations community to another following negotiation, customary arrangements, or movement of trappers between communities. To illustrate some of these shifts, Dunning's (1959) map of the Pikangikum band territory (the sum of Pikangikum hunting group areas as observed by Dunning), is compared to the present Whitefeather Forest Initiative planning area in Figure 9.

Traplines and Aboriginal Land Use Planning

As First Nations communities become increasingly involved in Aboriginal/community-based land use planning, there will be a need to consider and understand the processes by which planning areas will be determined. This chapter has provided a mix of archival and oral accounts of the process of creation of the registered trapline system for Pikangikum First Nation. It has examined broadly the emergence of a new system of land use based on continuities with the customary hunting areas.

In our overview, it appears that there has been a sense of who speaks for certain areas of land. The Whitefeather Forest Initiative of Pikangikum envisages ways that customary land use, tied to the traplines/system of family hunting areas can be re-invigorated through new land-based planning activities. Through the role of the Whitefeather Forest Initiative, a partnership with the Ontario Ministry of Natural Resources has been built upon dialogue, by which customary authority linked to knowledge of family hunting areas is being negotiated in a new planning context. The Land Use Strategy (http://www.whitefeatherforest.com) is a product of Pikangikum's continuing interaction with the province and provides evidence of the new importance of traplines in planning.

In terms of the land use planning area for Pikangikum, it was created through agreement among the family groups to work together and create the planning area bringing together only those traplines held by the families associated with Pikangikum First Nation. Decisions taken for the planning area come from a council of stewards (*Ohnahshohwayweeneeng*), which in the case of Pikangikum has evolved into the Whitefeather Steering Group. Whereas natural resource management decisions from the OMNR were formerly brought to the attention of head trappers, the WFMC now brings together head trappers and elders (*keecheeahneesheenahbayg*) with OMNR officials to discuss new land uses over a planning area larger than the single trapline. Head trappers and elders from particular family groups have authority to bring knowledge forward in the planning process for areas they know through experience on the land. In terms of regional planning practice, this is most consistent with the territory doctrine and would suggest that the basic planning unit for Aboriginal land use planning is the family hunting territory or the trapline. Stewards would be involved in decisions for those areas.

The question in terms of spatial planning is whether Pikangikum people can pursue new land use planning initiatives without falling prey to imposed government planning processes, and the technologies and ideologies that support them. That is, can Pikangikum people maintain their relationship to the land through the use of holistic planning units within the larger umbrella of regional planning, while being able to propose their own forms of spatial authority, negotiate boundaries, and create new relations with the state? That is a story that Pikangikum people have yet to live. However, it is important to note that both Hallowell and Dunning did note that people could draw out their areas, and this has also been noted by officials responsible for the mapping of the trapline system. Of relevance to Bill 191, we should then expect that Aboriginal peoples, when asked, could come up with a process to define their planning regions. When given the chance to participate in defining regional planning areas, the Pikangikum First Nation has shown they are quite capable of doing so. Thus, planners must be cautious not to rely solely on historical records or contemporary circumstances, and aim to bring the people living today into the process in order to develop their own approach to land use planning.

References Cited

Berkes, F. (1987). 'Common-property resource management and Cree Indian fisheries in subarctic Canada,' pp. 66-90 in B.J. McCay and J.M. Acheson, eds. *The Question of the Commons: The Culture and Ecology of Communal Resources.* Tucson: The University of Arizona Press.

Berkes, F. (2002). 'Cross-scale institutional linkages: Perspectives from the bottom up,' pp. 293-321 in E. Ostrom, T. Dietz, N. Dolšak, P. Stern, S. Stonich, and E.U. Weber, eds., *The Drama of the Commons.* Washington: National Academy Press.

Berkes, F. (2008). *Sacred Ecology,* 2nd ed. New York: Routledge.

Berkes, F., I. Davidson-Hunt, N. Deutsch, C. Burlando, A. Miller, C. Peters, P. Peters, R. Preston, J. Robson, M. Strang, A. Tanner, L. Trapper, R. Trosper, and J. Turner (2009). 'Institutions for Algonquian land use: change, continuity and implications for sustainable forest management,' pp. 35-52 in M.G. Stevenson and D.C. Natcher, eds., *Changing the Culture of Forestry in Canada: Building Effective Institutions for Aboriginal Engagement in Sustainable Forest Management.* Edmonton, AB: CCI Press and Sustainable Forest Management Network.

Bishop, C.A. (1974). *The Northern Ojibwa and the Fur Trade: An Historical and Ecological Study.* Toronto: Holt, Rinehart and Winston of Canada, Limited.

Bohaker, H. (2006). *Nindoodemag*: The significance of Algonquian kinship networks in the eastern Great Lakes region, 1600–1701. 3rd Series. *William and Mary Quarterly* 63(1): 51-80.

Brown, J.S.H., and S.E. Grey, eds. (2009). *Memories, Myths and Dreams of an Ojibwe Leader William Berens, as told to A. Irving Hallowell.* Montreal: McGill-Queens University Press.

Cooper, J.M. (1939). Is the Algonquian family hunting ground system pre-Columbian? *American Anthropologist* 41(1): 66-90.

Crichton, V. (1948). 'Registered traplines.' Reprinted from *Sylva* 4(2) pp 3-15, in National Archives of Canada, RG10, c-8104.

Davidson-Hunt, I. 2006. Adaptive learning networks: developing resource management knowledge through social learning forums. *Human Ecology* 34(4): 593-614.

Dunning, R. (1959). *Social and Economic Change Among the Northern Ojibwa.* Toronto: University of Toronto Press.

Feit, H.A. (1991). 'Gifts of the land: Hunting territories, guaranteed incomes and the construction of social relations in James Bay Cree society,' pp. 223-268 in N. Peterson and T. Matsuyama, eds., *Cash, Commoditisation and Changing Foragers.* No. 30 in *Senri Ethnological Studies.* Osaka: National Museum of Ethnology.

Folke, C., J. Colding, and F. Berkes (2003). 'Synthesis: building resilience and adaptive capacity in social-ecological systems,' pp. 352-387 in F. Berkes, J. Colding, and C. Folke, eds., *Navigating Social-ecological Systems: Building Resilience for Complexity and Change.* Cambridge, UK: Cambridge University Press.

Government of Ontario (2009). Bill 191: *An Act with Respect to Land Use Planning and Protection in the Far North.* Toronto: The Legislative Assembly of Ontario.

Hallowell, I.A. [J.S.H. Brown, ed.]. (1992). *The Ojibwa of Berens River, Manitoba: Ethnography into history.* Fort Worth: Harcourt Brace Jovanovich College Publishers.

Hodge, G. and I.M. Robson (2001). *Planning Canadian Regions.* Vancouver: UBC Press.

Koestler, A. (1976). *The Ghost in theMachine.* Danube edition. London: Hutchinson. Reprint, New York: Random House.

Lane, M.B. (2006). The role of planning in achieving Indigenous land justice and community goals. *Land Use Policy* 23: 385-394.

Leacock, E. (1954). The Montagnais hunting territory and the fur trade. *American Anthropological Association Memoir* 78:1-59.

Lytwyn, V. (1986). *The Fur Trade of the Little North: Indians, Pedlars, and Englishmen East of Lake Winnipeg, 1760-1821.* Winnipeg: Rupert's Land Research Centre, University of Winnipeg.

Notzke, C. (1994). *Aboriginal Peoples and Natural Resources in Canada.* North York: Captus University Publications.

Pikangikum First Nation in cooperation with Ontario Ministry of Natural Resources (2006). *Keeping the Land: A Land Use Strategy for the Whitefeather Forest and Adjacent Areas.* Pikangikum, Ontario.

Poplar River First Nation (2005). *Asatiwisipe Aki Management Plan.* Winnipeg.

Speck, F.G. (1915). The family hunting band as the basis of Algonkian social organization. New series, *American Anthropologist* 17(2): 289-305.

Speck, F.G. and L.C. Eisley (1939). The significance of hunting territory systems of the Algonkian in social theory. *American Anthropologist* 41(2): 269-280.

Tanner, A. (2009). 'From fur to fir: In consideration of a Cree family territory system of environmental stewardship,' pp 53-62 in M.G. Stevenson and D.C. Natcher, eds., *Changing the Culture of Forestry in Canada: Building Effective Institutions for Aboriginal Engagement in Sustainable Forest Management.* Edmonton, AB: CCI Press and Sustainable Forest Management Network.

Whitefeather Forest Initiative (2009). http://www.whitefeatherforest.com. Accessed January 12, 2009.

Wilber, K. (1997). *The eye of spirit: An Integral Vision for a World Gone Slightly Mad.* Boston: Shambhala.

Chapter Eight

Road Rash: Ecological and Social Impacts of Road Networks on First Nations

Daniel D. Kneeshaw, Mario Larouche, Hugo Asselin, Marie-Christine Adam, Marie Saint-Arnaud, and Gerardo Reyes

Introduction

First Nations communities are confronted with many environmental and social issues associated with natural resource extraction and other land use practices on natural ecosystems. Timber harvesting, oil and gas exploration and development, hydroelectric development and mining have in some areas led to the creation of extensive road networks on forested lands used by Aboriginal peoples. Surprisingly, the ecological and social impacts of road network development and maintenance on forested lands used and occupied by First Nations communities have seldom been considered and, as such, are poorly understood. Yet such knowledge is critical for land use planning in these forested environments.

Road networks have direct impacts on wildlife and ecosystem processes that, in turn, impact First Nations communities by altering hunting, gathering, trapping, fishing and other cultural activities. Roads can modify the diversity and abundance of various fish and wildlife species as well as cause changes to reproduction, migration and foraging behaviours. Roads have modified Aboriginal relationships with the land by increasing public access to previously isolated areas and by changing modes of travel. An often ignored effect is the changing relationships that road networks cause with other nearby communities (Aboriginal or not). The impacts of road networks linked to resources extraction continue to affect First Nations communities' relationships with the land. Improved planning and management that minimizes negative impacts is needed, especially since not all roads are equivalent in terms of their effects.

Developing and maintaining efficient road networks is a strategic component for forest land management, economic growth and land use planning. Most forest roads are constructed to facilitate access to remote regions for the exploitation of natural resources (primarily wood, minerals and fossil fuels). However, subsequently and often concurrently, they are used for outdoor recreation, scientific research and other endeavors (Kneeshaw 2008). Human population growth and the associated intensification of natural resource extraction have led to the rapid expansion of road networks (Asselin 2007). As a result, regions without road access have become increasingly rare (Bourgeois *et al*. 2005; Forman 2000).

Table 1. Impacts of roads on natural habitats and consequences for First Nations.

Road impact	Source	Effects	Consequences for First Nations
Chemical input to atmosphere, soils and water	Vehicle emissions; de-icing and other road maintenance chemicals and products	Smog; degradation of air quality; potential loss of soil fertility; bioaccumulation of toxins in biota; degradation of water quality; eutrophication; acidification	Health of exposed communities and workers; loss of food sources or negative impact on health
Modified watercourses	Road construction, maintenance, and use	Changes in water flow patterns and levels, sedimentation rates, floodplain ecology, and nutrient cycling; increased turbidity; degradation of fish spawning habitat conditions; shifts or changes in species composition and abundance	Changes or losses in fishing grounds and of semi-aquatic food sources such as trapped animals
Modified terrestrial habitat	Road construction, maintenance, and use	Habitat loss and fragmentation; soil compaction; loss of biodiversity; shifts or changes in species composition and abundance, plant-pollinator interactions, dispersal rates and patterns, reproductive success, and animal behaviour; spread of exotic species	Modifications in animal migration; changes in hunting grounds and human travel corridors
Access	Maintained roads	Poaching, multiple uses by many communities	Conflicts or tensions between communities
Increased connectivity	Connected road network; unplanned development	Species invasion; spread of predators, disease, etc.	Changes in the landscape and species composition; new understandings; loss of traditional ecological knowledge (TEK)
Speed of travel	Changed modes of travel	Less time traveling and learning of the land	Less time away from home; less time invested in understanding the land and relationships between events on the landscape

While the expansion of road networks has been identified as having multiple benefits for remote communities in terms of possibilities for economic growth and increased accessibility to markets and social services (schools, hospitals, etc.), there are growing concerns about threats to biodiversity and the environment. Many studies have reviewed the ecological impacts of road networks on various ecosystems (Forman and Alexander 1998; Lindenmayer and Franklin 2002; Trombulak and Frissell 2000). The principal concerns include altered local climate conditions, habitat loss and fragmentation, degradation of air and water quality, altered nutrient cycling, loss of flood and drought mitigation capabilities, soil fertility loss, modification of animal behaviour and reproductive success, increased animal mortality, increased intra-specific isolation, and greater anthropogenic pressure on remaining natural landscapes and wildlife populations (Table 1).

The effects of roads on ecosystems are not limited to road construction and maintenance. In the United States, for example, public roads physically occupy only about 1% of the total land surface area, yet they affect approximately 15 to 20% of the total land base (Forman and Alexander 1998). Given the significant impacts of road networks on ecosystems and ecological processes, there is need for improvement in the planning, design, technology and materials used, construction methods, as well as a thorough understanding of the negative impacts on biodiversity and the environment so that measures to mitigate and reduce these impacts are developed and employed.

In addition to ecological impacts, road networks have significant socio-economic effects on local communities. Roads can have positive effects on livelihoods through improved access to markets, increased social interactions, migration and settlement, as well as greater economic opportunities (Nelson *et al.* 2006). Lack of mobility and high transportation costs are key impediments that lead to the formation of 'spatial poverty traps' (Deichmann 1999; Pender *et al.* 1999; Bigman and Fofack 2000). In fact, road network density has been shown to be positively correlated with national per capita income (Sarkar and Ghosh 2000).

However, roads have also caused detrimental changes to socio-environmental institutions and dynamics. Most of these changes have been documented in developing countries where poaching, deforestation and indifferent or apathetic social attitudes toward conservation and environmental protection have contributed to fragmentation and biodiversity loss (Allen *et al.* 1985; Keller and Berry 2007). More evidence is showing that the negative impacts of roads are not limited to developing countries. Road networks are, in fact, becoming a significant issue for many First Nations communities in North America.

In Canada, there are more than 700,000 First Nations peoples belonging to more than 600 different bands, most of who live within one of the 2,283 reserves in the country (Dickason 1996), and more than 1 million Aboriginal peoples when including the Métis and Inuit. The majority of these communities remain closely tied to forest ecosystems for subsistence, employment as well as for cultural and spiritual rituals, customs and beliefs (Berkes 2008; Berkes *et al.* 1994; Dickason 1996). The development of extensive road networks for forestry purposes and oil/gas exploration and development has affected intra- and intercommunity relationships, which have subsequently contributed to the deterioration of socio-environmental relationships and thus Aboriginal ties to forest ecosystems. Such effects are obviously not only linked to the roads themselves, but are compounded by large-scale modifications to the forest ecosystems on traditional lands.

Assuring that natural resource extraction does not exceed an ecosystem's productive capacity is of paramount importance for sustainable development (Berkes *et al.* 2002; Chapin *et al.* 2004). Yet, what is often overlooked or ignored is that these same resources, and the lands from which they are harvested, are critical to First Nations ways of life. This has led to conflicting demands on natural resources by western society and First Nations peoples (e.g., Blakney 2003; Wyatt 2004; Saint-Arnaud 2009), adding further pressure to resource extraction industries already under close scrutiny due to declining natural resources and growing public concerns regarding corporate environmental and social impacts (Anderson 1997; Lertzman and Vredenburg 2005). Moreover, as a means of achieving greater control of activities on their traditional lands for self-determination, and a desire to be independent of government assistance, First Nations peoples are becoming more involved in economic development enterprises (Anderson 1997; Trosper *et al.* 2008, Wyatt 2008). Subsequently, the influences of western society have permeated into First Nations ways of life. Modern, faster means of access to resource extraction and use require less intimacy with the natural landscape often undermining traditional relationships with and attitudes about the land. Although First Nations peoples seek social, economic and infrastructural improvement, developments need to be acceptable to the communities and be consistent with their worldviews.

The substantial impact of industrial practices on First Nations peoples and their ancestral lands has led to a sense of urgency for the development and implementation of policies that ensure the survival of indigenous cultures: the loss of large expanses of contiguous terrain and associated biodiversity has been linked to reductions in the diversity of human culture (Anderson 1997; Lertzman and Vredenburg 2005). Knowing that development of road networks is a major cause of forest fragmentation and biodiversity loss (Fig. 1; Forman and Alexander 1998), and that it has significant social and cultural impacts on First Nations communities, has forced land managers, land use planners and policy makers to address these impacts prior to regulatory approvals.

This chapter examines the effects of forest road networks on 'natural' ecosystems and discusses the collateral effects of road development on First Nations communities. Our first objective is to evaluate the impact of roads on ecosystem processes and wildlife populations, focusing on game species important to traditional activities such as hunting and trapping. Subsequently, we evaluate the effects of roads on First Nations communities and their socio-environmental dynamics. Although roads provide access to the land and affect travel patterns, they have far more complex social and cultural repercussions for First Nations communities. These effects contrast with the expected benefits of roads, e.g., increased mobility and access to potential economic opportunities outside of isolated communities, while bearing in mind that the impacts of different types of roads are not equal.

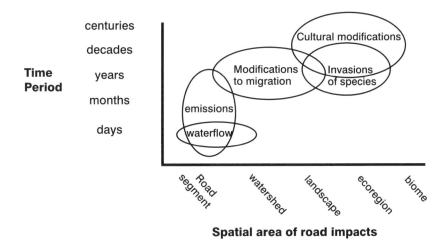

Figure 1. *Spatial and temporal dimensions of ecological effects of roads, modified from National Research Council (2005).*

Impacts of Forest Road Networks on Natural Ecosystems and their Use by Aboriginal Peoples

Extensive road network development in Canada's forests has led to habitat loss, fragmentation and drastic changes to landscape structure and composition. Increases in the quantity of edge habitat, decreases in total surface area and homogenization of remaining forest patches (Trombulak and Frissell 2000; Gucinski *et al.* 2001; Lindenmayer and Franklin 2002) have resulted in an ecosystem matrix markedly different from pre-industrial conditions (Fahrig 2003). Indeed, road network density and the extent of linear disturbance now serve as indicators of habitat quality and integrity at the landscape scale (Forman and Alexander 1998; Heilman *et al.* 2002).

Habitat fragmentation substantially modifies the demographic and genetic structure of forest vegetation by isolating populations, altering plant-pollinator interactions, and disrupting dispersal and reproduction mechanisms (Reh and Seitz 1990; Trombulak and Frissel 2000; Honnay *et al.* 2005). In combination, these effects can lead to major changes in plant species diversity and relative species abundance. Changes to vegetation communities can have negative impacts on the abundance of medicinal plants, as well as culturally important wildlife species.

Road networks can alter or obstruct behaviours, home ranges, migration patterns and reproductive success of animal species important to First Nations peoples either directly because of other economic, nutritional, social, cultural, and spiritual values, or indirectly as some animals serve as food sources for target species or play other important ecological roles. Examples include: invertebrates (Mader 1984; Baur and Baur 1990; Haskell 2000), herpetiles

(Ashley and Robinson 1996), small mammals (Oxley *et al.* 1974; Burnett 1992), large mammals (Garland and Bradley 1984; McLellan and Shackleton 1988; Dyer *et al.* 2002), birds (Anthony and Isaacs 1989; Fernandez 1993; Reijnen and Foppen 1994) and fish (Meehan 1991).

Changes in animal behaviour and migration routes brought about by road construction can be disastrous to Aboriginal communities that structure activities around the presence and abundance of fish and wildlife at given places and times of year (e.g., Payette *et al.* 2004). While such situations may not lead to food shortages due to the availability of alternative food sources (i.e., from grocery stores), road construction is yet another potential factor that is challenging First Nation access to traditional food sources. Furthermore, reduced hunting and trapping opportunities and scarcity of game caused by disruptions to migratory patterns and game trails prevent younger generations of Aboriginal peoples from forming and maintaining strong connections to the land.

The increased connectivity created by roads has been linked to the spread of pathogens, disease and insect pests (Boulet and Darveau 2000; Trombulak and Frissel 2000; Gucinski *et al.* 2001). Roads have also facilitated colonization of invasive species such as trembling aspen (*Populus tremuloides*) into boreal and sub-boreal ecosystems (Laquerre 2007; Zelazny 2007). White-tailed deer (*Odocoileus virginianus*), a species historically associated with temperate forests, is also taking advantage of road networks in its expansion northward into boreal regions. Anishinabe hunters have noted the recent immigration of this species into their hunting grounds and consider their increase to be inevitable (Saint Arnaud, unpublished data). Native species such as the beaver (*Castor canadensis*) have also proliferated with road expansion across the landscape. This species responds not only to the increased presence of trembling aspen (a preferred food source), but also to culverts which are easily dammed to flood adjacent areas. However, since roads themselves are often flooded, forest companies hire trappers (often Aboriginal in northern areas) to kill beavers during summer months to prevent damage to roads (Asselin, personal observation). This is a problem as First Nations peoples usually do not trap beaver during the summer because it is the young-rearing season, the meat is not palatable and the fur is of poorer quality.

Road networks have been linked to range expansions of predators such as the coyote (*Canis latrans*). Previously restricted to western ecosystems, coyotes are now ubiquitous throughout many ecosystems that have lost indigenous predators such as wolves (*Canis lupus*) (Berger and Geese 2007). The effect of increased predation associated with road networks has also been shown to be an important factor causing declines in populations of woodland caribou (*Rangifer tarandus*) (James and Stuart-Smith 2000), a species of importance to a number of northern boreal peoples. Population declines due to natural predators are also compounded by increased hunting pressure.

Access to formerly inaccessible lands facilitates legal hunting and trapping, but also poaching. Excessive hunting and poaching has caused decreases in populations of culturally important species such as bear (*Ursus* spp.), wolf (*Canis lupus*), ungulates (*Alces alces, Rangifer tarandus, Odocoileus* spp.), pine marten (*Martes americana*), fisher (*Martes pennanti*), lynx (*Lynx canadensis*) and wolverine (*Gulo gulo*) (Ferreras *et al.* 1992; Ruggiero *et al.* 1994; McLellan *et al.* 1999; Simon *et al.* 1999; Trombulak and Frissel 2000; Gucinski *et al.* 2001). The problem is exacerbated when territories are used for subsistence, recreational and industrial purposes.

Road networks increase wildlife mortality rates through collisions with automobiles (Mech *et al.* 1988; Trombulak and Frissell 2000; Strittholt and Dellasala 2001). Most collisions occur with white-tailed deer (*Odocoileus virginianus*), moose (*Alces alces*) and black bear (*Ursus americanus*), all culturally important species for First Nations. In the province of Quebec alone, between 2,500 and 5,500 deer, moose and black bear are killed annually due to collisions with automobiles (MTQ 2000; de Bellefeuille and Poulin 2004). Although these collisions probably do little to reduce the overall populations of these species, they may reinforce the stereotype of western culture as a wasteful and wanton destroyer of life. Although not usually reported since there is often no vehicular damage, many collisions also occur with turtles. However, such collisions disrespect First Nations' culture and spirituality; for many Aboriginal peoples, turtles play a key role in world creation. Many other unreported collisions with 'non-valuable' species may also reinforce this stereotype and differences between holistic and anthropocentric worldviews.

Road networks negatively affect water quality through inputs of dust, de-icing salts and other chemicals, and by modifying natural water flow patterns and levels, and by increasing turbidity and sedimentation (Beschta 1978; Reid and Dunne 1984; Jones *et al.* 2000, Myers-Smith *et al.* 2006). Germain and Asselin (unpublished data) found that the vast majority of sites of interest to the Anishinaabe are located close to water bodies or water courses (<60 m). Furthermore, changes in water quality can adversely affect fishing grounds. Even if some of these effects are limited temporally, they can still have large impacts on First Nations communities since a loss of traditional fishing grounds even for a few years may last generations.

Social Impacts of Forest Road Networks on First Nations Communities

Road networks have radically changed the landscape in which First Nations communities live (Sherry and Johnson 1999); previously inaccessible areas can now be accessed by motorized vehicles. In the recent past, this was viewed primarily as positive since roads passing through or in proximity to communities improved access to outside resources and provided economic opportunities and greater access to goods and social services. Economic possibilities are critical since unemployment and low incomes are endemic to Aboriginal communities across the country (Anderson 1997). Roads can have a direct impact by bringing tourists, and thus money, to communities, and can foster economic and social interactions with other communities, agencies and organizations. Roads can also provide community members access to employment outside their communities without forcing workers to relocate or leave their communities for extended periods of time. Community members partaking in traditional hunting, gathering, fishing and trapping activities can now rapidly travel to sites within traditional lands that would have previously required days or weeks of absence from one's family and the community (Saint-Arnaud 2009). Road networks also permit elders to maintain a link with the land when travel by traditional means and trails becomes problematic. In addition, roads permit greater community access to social services such as health care, particularly for the very young, old or vulnerable. Even so, they can also cause direct negative effects, e.g., health problems related to road dust (Gordian *et al.* 1996; Segerstedt and Forsberg 2006).

Roads in many First Nations lands were originally constructed to harvest forest resources at minimum cost (Chung *et al.* 2008). While their purpose was for timber extraction, it cannot be assumed that they will subsequently provide access to social services and foster economic development needs of the local communities. For example, a forest road network density of 0.8 km for each square kilometre (km^2) was found in an Anishinaabe community in Quebec (Saint-Arnaud *et al.* 2005), while in the lower foothills of the Rocky Mountains in Alberta, Frair et al. (2008) observed a road density of 0.9 km^2. Only a minimal proportion of the roads in either study are used regularly by local communities. Road network development for forest resource extraction, therefore, should serve as a reminder of how Aboriginal territories are often substantially modified by outside interests with little return or real benefits to local First Nations communities. Indeed, more and more local communities do not view road construction as a positive development opportunity and therefore question the value of infrequently used forest road networks constructed near their communities.

The environmental impacts of forestry activities have created tensions between forestry companies and First Nations communities. For many northern First Nations, roads are associated with timber extraction, and the extensive nature of forestry road networks may actually be viewed as an indicator of a forest company's occupation and control over the territory. Resource extraction may ensure the viability of many boreal towns, however communities are increasingly required to organize themselves according to the location of existing road infrastructure rather than planning a road network to meet their needs. The absence of a co-management approach to forestry, road planning and land-use planning generally means that road network development may be viewed by First Nations as yet another colonial endeavor to extract natural resources while forcing First Nations to make the best of the imposed infrastructure. Other users are also demanding better planning of forestry roads, although this is primarily to control access (Kneeshaw 2008), whereas First Nations needs are multiple, varied and interrelated.

An oversimplified view of roads as all having the same spatial-temporal impacts has emerged. However, the benefits and disadvantages of a paved road or a primary gravel road *versus* a tertiary road built for forestry purposes and subsequently abandoned are not equal. While primary roads provide continuous access along main travel corridors, tertiary roads provide little benefit to people beyond opening up the territory for hunting. Such openings also result in fragmentation of the forested land base (Forman and Alexander 1998) and provide increased access to non-local users as well as natural predators.

Road networks can create and exacerbate inter- and intra-community conflicts. This is particularly problematic when unrestricted access leads to overexploitation of resources on traditional lands. Management of territorial resource access was historically provided by local community members, with resource access mediated through family control of certain areas and through formal and informal institutions monitoring activities in the territory. It is becoming evident that the extensive nature and rapid development of roads in the boreal region has significantly affected the capacity of informal Aboriginal institutions to monitor activities on their lands, a criticism shared by some non-Aboriginal groups (Kneeshaw 2008). For First Nations, these impacts also have cultural repercussions. First, control of access by local institutions is made difficult by road saturation of the territory. It is, for example, unrealistic to expect a population of 400 to effectively monitor and control more than 4,800 km of roads (Saint-Arnaud *et al.* 2005). Different communities and different user groups now have easy access to the territory. Traditionally, hunting and trapping required an expertise in the activity itself, but also knowledge of the environment and territory (Davidson-Hunt and Berkes 2003). Community experts would accompany and teach individuals; and thus, the activities were valued and not readily available to all (Adam, personal observation). Road construction in forests used by First Nations communities continues to occur at an alarming rate. In some cases, substantial changes have taken place within only one generation (Asselin 2007).

The development and expansion of road networks has, for many First Nations peoples, changed the manner in which they access the land. In the past, knowledge of the land was required to safely navigate through parts of the territory (Davidson-Hunt and Berkes 2003; Saint-Arnaud 2009). In abandoning traditional means and routes of travel, younger Aboriginal peoples, in particular, have lost the intimate contact previous generations maintained with their lands and resources. In discussions with members of the Pikogan First Nation community, it was noted that hunting, fishing and trapping are now concentrated along main road arteries where access and travel is by pick-up truck. Occasions where hunters would spend weeks alone or in small groups moving across the landscape in rhythm with natural forest cycles are disappearing. The current use of automobiles for transportation is not conducive to developing knowledge of the land equivalent to that of previous generations, for whom travelling across the land was an important part of being (Davidson-Hunt and Berkes 2003). The greater speed of travel, focused attention on the road itself and inner distractions (e.g., music, cell phones, other passengers) reduce attention levels and the amount of time spent observing the land and acquiring intimate ecological knowledge. No longer is the individual part of the land through which he or she is traveling. Rather, s/he is a visitor traveling in an enclosed environment, largely disassociated from the natural landscape.

Traditionally, networks of interconnected lakes and rivers and footpaths were important for voyaging across the land. The slower speed of travel and the greater time spent on the land meant that all places were named and known to members of the community (Saint-Arnaud, unpublished data). These places were often associated with knowledge that was important for physical and cultural survival. Locations for wildlife hunting also varied with the season and the occurrence of natural disturbances, e.g., fire (Davidson-Hunt and Berkes 2003). According to a study on hunting preferences by Haener *et al.* (2001), First Nations hunting preferences were often associated with older-aged forests and water access, suggesting that the interest was in the activity as much as in the target of the hunt. Summer and winter hunting grounds, as well as village locations were a function of resource access associated with seasonal changes, instead of being planned as part of road infrastructure and access to services found in towns and cities.

Conclusions

Transportation systems are crucial for communication, economic growth, and human settlement. Extensive road networks do, however, have many negative environmental and social repercussions in northern regions. Habitat loss and threats to biodiversity are clearly major concerns, as are the marked changes to First Nations cultures. Traditional land use and subsistence practices were not only important for First Nations peoples' dietary needs and well-being, but were centrally linked to cultural identity and social structures. Road networks have contributed to the gradual shift in the way traditional lands are being accessed and used, and arguably have weakened inter-generational links to the land. Road networks are only one of the many changes to the natural landscape that has led to profound disruptions to the social, economic, political and spiritual life of First Nations peoples (Fohlen 1985; Dickason 1996; Niezen 2000). Drastic changes to traditional ways of life and societal behaviours have contributed to undermine First Nations peoples' self-confidence, self-reliance and cultural identity.

Today, the integrity and sustainability of First Nations communities continue to be challenged and disrupted by many factors, including forest road networks and the numerous activities and practices associated with their use, maintenance and development. Decreased traditional use of lands, strong feelings of loss of connectedness with the past and to the natural landscape, loss of cultural knowledge, identity and language, as well as increased social tensions within and among communities are occurring with increasing frequency, all of which are sensitive issues requiring careful attention in future land use planning. Forest road networks have many negative as well as positive influences on First Nations communities. Moreover, they are increasingly abundant in hitherto inaccessible landscapes. These factors alone suggest that planning, construction, use and decommissioning of road networks—whether for forestry, fossil fuel extraction, or general land-use planning—are critical, and are exercises with which affected Aboriginal peoples and communities must be involved from the start.

References Cited

Allen, J.C. and D.F. Barnes (1985). The causes of deforestation in developing countries. *Annals of the Association of American Geographers* 75: 163-184.

Anderson, R.B. (1997). Corporate/indigenous partnerships in economic development: the First Nations in Canada. *World Development* 25: 1483-1503.

Anthony, R.G. and F.B. Isaacs (1989). Characteristics of bald eagle nest sites in Oregon. *Journal of Wildlife Management* 53: 148-159.

Ashley, E.P. and J.T. Robinson (1996). Road mortality of amphibians, reptiles and other wildlife on the Long Point Causeway, Lake Erie, Ontario. *Canadian Field-Naturalist* 110: 403-412.

Asselin, H. (2007). Emplois en dents de scie. Exploration des facteurs invoqués pour expliquer les crises dans l'industrie forestière québécoise. Greenpeace Canada, Toronto. <http://web2.uqat.ca/asselin/Asselin_Emplois_en_dents_de_scie.pdf>

Baur, A. and B. Baur (1990). Are roads barriers to dispersal in the land snail *Arianta arbustorum*? *Canadian Journal of Zoology* 68: 613-617.

Berger, K.M. and E.M. Gese (2007). Does interference competition with wolves limit the distribution and abundance of coyotes? *Journal of Animal Ecology* 76: 1075-1085

Berkes, F. (2008). *Sacred ecology*. Second edition. Routledge, New York.

Berkes, F., J. Colding and C. Folke, eds. (2002). *Navigating Social-ecological Systems. Building Resilience for Complexity and Change.* Cambridge University Press, Cambridge.

Berkes, F., P.J. George, R.J. Preston, A. Hughes, J. Turner, and B.D. Cummins (1994). Wildlife harvesting and sustainable regional Native economy in the Hudson and James Bay Lowland, Ontario. *Arctic* 47: 350-360.

Beschta, R.L. (1978). Long-term patterns of sediment production following road construction and logging in the Oregon coast range. *Water Resources Research* 14: 1011-1016.

Bigman, D. and H. Fofack (2000). *Geographical Targeting for Poverty Alleviation, Methodology and Applications.* Washington: The World Bank.

Blakney, S. (2003). Aboriginal forestry in New Brunswick: conflicting paradigms. *Environments* 31: 61-78.

Boulet, M. and M. Darveau (2000) Depredation of artificial bird nests along roads, rivers, and lakes in a boreal balsam fir forest. *Canadian Field-Naturalist* 114: 83-88.

Bougeois, L., D. Kneeshaw, and G. Boisseau (2005) Les routes forestières au Québec: les impacts environnementaux, sociaux et économiques. *Vertigo* 6 <http://www.vertigo.uqam.ca/vol6no2/art16vol6no2/vertigovol6no2_laurence_bourgeois.pdf>

Burnett, S.E. (1992). Effects of a rainforest road on movements of small mammals: mechanisms and implications. *Wildlife Research* 19: 95-104.

Chapin, F.S. III, G. Peterson, F. Berkes, T.V. Callaghan, P. Angelstam, M. Apps, C. Beier, Y. Bergeron, A-S. Crépin, K. Danell, T. Elmqvist, C. Folke, B. Forbes, N. Fresco, G. Juday, J. Niemelä, A. Shvidenko, and G. Whiteman (2004). Resilience and vulnerability of northern regions to social environmental change. *Ambio* 33: 344-349.

Chung, W., J. Stuckelberger, K. Aruga, and T.W. Cundy (2008). Forest road network design using a trade-off analysis between skidding and road construction costs. *Canadian Journal for Forest Research* 38: 439-448.

Davidson-Hunt, I. and F. Berkes (2003). Learning as you journey: Anishinaabe perception of social-ecological environments and adaptive learning. *Conservation Ecology* 8:5 <http://www.consecol.org/vol8/iss1/art5>

de Bellefeuille, S. and M. Poulin (2004). *Mesures de mitigation visant à réduire le nombre de collisions routières avec les cervidés.* Ministère des Transports du

Québec. Rapport No. RTQ-04-05.
<http://www.mtq.gouv.qc.ca/portal/page/portal/Librairie/Publications/fr/ministere/recherche/etudes/rtq04-05.pdf>

Deichmann, U. (1999). *Geographic Aspects of Inequality and Poverty* The World Bank <http://povlibrary.worldbank.org/files/5319_povmap.pdf>

Dickason, O.P. (1996). *Canada's First Nations: A History of Founding Peoples from Earliest Times*. Toronto: Oxford University Press.

Dyer, J.S., J.P. O'Neill, S.M. Wasel, and S. Boutin (2002). Quantifying barrier effects of roads and seismic lines on movements of female woodland caribou in northern Alberta. *Canadian Journal of Zoology* 80: 839-845.

Fahrig, L. (2003). Effects of habitat fragmentation on biodiversity. *Annual Review of Ecology Evolution and Systematics* 34: 487-515.

Fernandez, C. (1993). The choice of nesting cliffs by golden eagles *Aquila chrysaetos*: the influence of accessibility and disturbance by humans. *Alauda* 61: 105-110.

Ferreras, P., J.J. Aldama, J.F. Beltran, and M. Delibes (1992). Rates and causes of mortality in a fragmented population of Iberian lynx *Felis pardina* Temminck, 1824. *Biological Conservation* 61: 105-110.

Fohlen, C. (1985). *Les indiens d'Amérique du Nord*. Coll. Que Sais-je? 1ère édition. Paris: Presses Universitaires de France.

Forman, R.T.T. (2000). Estimate of the area affected ecologically by the road system in the United States. *Conservation Biology* 14: 31-35.

Forman, R.T.T. and L.E. Alexander (1998). Roads and their major ecological effects. *Annual Review of Ecology and Systematics* 29: 207-231.

Frair, J.L., E.H. Merrill, H.L. Beyer, and J.M. Morales (2008). Thresholds in landscape connectivity and mortality risks in response to growing road networks. *Journal of Applied Ecology* 45: 1504-1513.

Garland, T. and W.G. Bradley (1984). Effects of a highway on Mojave Desert rodent populations. *American Midland Naturalist* 111: 47-56.

Gordian, M.E., H. Ozkaynak, J. Xue, S.S. Morris, and J.D. Spengler (1996). Particulate air pollution and respiratory disease in Anchorage, Alaska. *Environmental Health Perspectives* 104: 290-297.

Gucinski, H., M.J. Furniss, R.R. Ziemer, and M.H. Brookes (2001). *Forest Roads: A Synthesis of Scientific Information*. U.S. Department of Agriculture, Forest Service, Pacific Northwest Research Station. Portland, OR. Gen. Tech. Rep. PNW-GTR-509.

Haener, H.K., D. Dosman, W.L. Adamowicz, and P.C. Boxall (2001). Can stated preference methods be used to value attributes of subsistence hunting by Aboriginal peoples? A case study in Northern Saskatchewan. *American Journal of Agricultural Economics* 83: 1334-1340.

Haskell, D.G. (2000). Effects of forest roads on macroinvertebrate soil fauna of the Southern Appalachian Mountains. *Conservation Biology* 14: 57-63.

Heilman, G.E., J.R. Strittholt, N.C. Slosser, and D.A. Dellasala (2002). Forest fragmentation of the conterminous United States: assessing forest intactness through road density and spatial characteristics. *BioScience* 52: 411-422.

Honnay, O., H. Jacquemyn, B. Bossuyt, and M. Hermy (2005). Forest fragmentation effects on patch occupancy and population viability of herbaceous plant species. *New Physiologist* 166: 723-736

James, A.R.C. and A.K. Stuart-Smith (2000). Distribution of caribou and wolves in relation to linear corridors. *Journal of Wildlife Management* 64: 154–159.

Jones, J. A., F.J. Swanson, B.C. Wemple, and K.U. Snyder (2000). Effects of roads on hydrology, geomorphology, and disturbance patches in stream networks. *Conservation Biology* 14: 76-85.

Keller, G.R. and J. Berry (2007). Reduced impact logging road issues in tropical forests. *Transportation Research Record* 1989: 89-106.

Kneeshaw, D.D. (2008). 'L'écotourisme dans les territoires forestiers : une source de conflit ou une opportunité de meilleure gestion,' in B. Sarrasin, and M. Lequin, eds., *Le tourisme sur les territoires forestiers*. Quebec: Les presses de l'Université du Québec.

Laquerre, S. (2007). *Analyses multi-échelles du phénomène d'enfeuillement du couvert forestier de la région de l'Abitibi*. M.Sc., Université du Québec en Abitibi-Témiscamingue, Rouyn-Noranda.

Lertzman, D.A. and H. Vredenburg (2005). Indigenous peoples, resource extraction and sustainable development: an ethical approach. *Journal of Business Ethics* 56: 239-254.

Lindenmayer, D.B. and J.F. Franklin (2002). *Conserving biodiversity: a comprehensive multi-scaled approach*. Washington: Island Press.

Mader, H.J. (1984). Animal habitat isolation by roads and agricultural fields. *Biological Conservation* 29: 81-96.

McLellan, B.N. and D.M. Shackleton (1988). Grizzly bears and resource-extraction industries: Effects of roads on behaviour, habitat use and demography. *Journal of Applied Ecology* 25: 451-460.

McLellan, B.N., F.W. Hovey, R.D. Mace, J.C. Woods, D.W. Carney, M.L. Gibeau, W.L. Wakkinen, and W.F. Kasworm (1999). Rates and causes of grizzly bear mortality in the interior mountains of British Columbia, Alberta, Montana, Washington, and Idaho. *Journal of Wildlife Management* 63: 911-920.

Mech, L.D., S.H. Fritts, G.L. Radde, and W.J. Paul (1988). Wolf distribution and road density in Minnesota. *Wildlife Society Bulletin* 16: 85-87.

Meehan, W.R. (1991) *Influences of Forest and Rangeland Management on Salmonid Fishes and their Habitats*. Special Publication no. 19. Bethesda: American Fisheries Society.

MTQ—Ministère des Transports du Québec (2000). *Les transports au Québec*. Recueil de données statistiques.

Myers-Smith, I.H., B.K. Arnesen, R.M. Thompson, and F.S. Chapin III (2006). Cumulative impacts on Alaskan arctic tundra of a quarter century of road dust. *Ecoscience* 13: 503-510.

National Research Council (2005). *Assessing and Managing the Ecological Impacts of Paved Roads*. Committee on Ecological Impacts of Road Density, National Academic Press, Washington DC, 324 p.

Nelson, A., F. Pozzi, and A. de Sherbinin (2006). Towards development of a high quality public domain global roads database. *Data Science Journal* 5: 223-265.

Niezen, R. (2000). Recognizing indigenism: Canadian unity and the international movement of indigenous peoples. *Comparative Studies in Society and History* 42: 119-148

Oxley, D.J., M.B. Fenton, and G.R. Carmody (1974). The effects of roads on populations of small mammals. *Journal of Applied Ecology* 11: 51-59.

Payette, S., S. Boudreau, C. Morneau, and N. Pitre (2004). Long-term interactions between migratory caribou, wildfires and Nunavik hunters inferred from tree rings. *Ambio* 33: 482-486.

Pender, J.L., S.J. Scherr, and G. Durón (1999). *Pathways of development in the Hillsides of Honduras: Causes and Implications for Agricultural Production, Poverty, and Sustainable Resource Use*. Washington: International Food Policy Research Institute (IFPRI).

Reh, W. and A. Seitz (1990). The influence of land use on the genetic structure of populations of the common frog *Rana temporaria*. *Biological Conservation* 54: 239-249.

Reid, L.M. and T. Dunne (1984). Sediment production from forest road surfaces. *Water Resources Research* 20: 1753-1761.

Reijnen, R. and R. Foppen (1994). The effects of car traffic on breeding bird populations in woodland. 1. Evidence of reduced habitat quality for willow warblers (*Phylloscopus trochilus*) breeding close to a highway. *Journal of Applied Ecology* 31: 85-94.

Ruggiero, L.F., K.B. Aubry, S.W. Buskirk, L.J. Lyon, and W.J. Zielinski (1994). *The Scientific Basis for Conserving Forest Carnivores: American Marten, Fisher, Lynx and Wolverine in the Western United States*. Department of Agriculture, United States Forest Service, Rocky Mountain Forest and Range Experiment Station. Ft. Collins, CO. Gen. Tech. Rep. RM-254.

Saint-Arnaud, M., L. Sauvé, and D. Kneeshaw (2005). Forêt identitaire, forêt partagée: trajectoire d'une recherche participative chez les Anicinapek de Kitcisakik. *VertigO* 16 <http://www.vertigo.uqam.ca/vol6no2/art20vol6no2/vertigovol6no2_st_arnaud_et_coll.pdf>

Saint-Arnaud, M. (2009). *Contribution à la définition d'une foresterie autochtone: le cas des Anicinapek de Kitcisakik*. Ph.D. thesis, Université du Québec à Montréal, Montreal.

Sarkar, A.K. and D. Ghosh (2000). *Meeting the Accessibility Needs of the Rural Poor*. IASSI Quarterly, Indian Association of Social Science Institutions.

Segerstedt, B. and B. Forsberg (2006). Health effects from road dust. *Epidemiology* 17: S233.

Sherry, E.E. and C.J. Johnson (1999). The forgotten forest: revisiting the forestland allocation strategy. *Forestry Chronicle* 75: 919-927.

Simon, N., F. Schwab, M. LeCoure, F. Philips, and P. Trimper (1999). Effects of trapper access on a marten population in central Labrador. *Northeast Wildlife* 54: 73-76.

Strittholt, J.R. and D.A. DellaSala (2001). Importance of roadless areas in biodiversity conservation in forested ecosystems: Case study of the Klamath-Siskiyou Ecoregion of the United States. *Conservation Biology* 15(6): 1742-1754.

Trosper, R.L., H. Nelson, G. Hoberg, P. Smith, and W. Nikolakis (2008). Institutional determinants of profitable commercial forestry enterprises among First Nations in Canada. *Canadian Journal of Forest Research* 38: 226-238.

Trombulak, S.C. and C.A. Frissell (2000). Review of ecological effects of roads on terrestrial and aquatic communities. *Conservation Biology* 14: 18-30.

Wyatt, S. (2004). *Co-existence of Atikamekw and Industrial Forestry Paradigms. Occupation and Management of Forestlands in the St-Maurice River Basin, Québec*. Ph.D. thesis, Université Laval, Quebec.

Wyatt, S. (2008). First Nations, forest lands, and 'Aboriginal forestry' in Canada: from exclusion to co-management and beyond. *Canadian Journal of Forest Research* 38: 171-180.

Zelazny, V.F. (2007). *Our Landscape Heritage: The Story of Ecological Land Classification in New Brunswick*. Second edition. New Brunswick Department of Natural Resources.

Chapter Nine

Aboriginal Land Use Mapping: What Have We Learned From 30 Years of Experience?

Stephen Wyatt, David C. Natcher, Peggy Smith and Jean-François Fortier

Introduction

For the past 30 years, Aboriginal peoples in Canada have documented the extent to which they have used traditional lands and resources, both prior and subsequent to European settlement. These studies have taken a variety of forms that include traditional land use and occupancy studies that document the traditional territories of Aboriginal communities, map biographies that record an individual's use of the land over the course of his or her lifetime, and resource use studies that quantify the amount of wildlife resources harvested from the land over a specific period of time. For Aboriginal peoples, it is often hoped that these studies will assist in their assertion of *a priori* claims to lands that they have long occupied, while helping to create 'spaces' in which Aboriginal communities can negotiate more equitable roles in land use planning and management. For government agencies and resource developers (including the forest industry), land use studies are often supported in efforts to identify and avoid sites of Aboriginal interest and concern, while also meeting fiduciary obligations for consultation. However, for all parties, and despite 30 years of experience, it remains unclear whether any of these objectives are being met.

This chapter, derived from the experience and contributions of Aboriginal community representatives, industry, governments and academics from across Canada,[1] considers some of the challenges and issues associated with land use studies and mapping, and the ways in which this information is being used in forest management and land use planning in Canada.

---

[1] This chapter is the result of an SFM Network funded 'State of Knowledge Project' that reviewed and synthesized Canadian experiences in harmonizing the interests of Aboriginal peoples and the forest industry through the use of traditional knowledge and land use occupancy studies (Wyatt *et al.* 2010). Many of the ideas presented here were discussed at a workshop in Saskatoon on January 15 -16, 2009.

Three Decades of Land Use Studies in Canada

Aboriginal land use and occupancy studies in Canada are generally considered to have arisen in the 1970s, especially with work among the Cree of Fort George (Weinstein 1976) and the Inuit (Freeman 1976). However, their beginnings can be traced to the traditions of Boas (1888) who recognized that the recording of locally-used place names could articulate the link between Aboriginal cultures and the landscapes that they used and occupied. Shortly afterward, Mauss (1905) described seasonal variation in the life of the Eskimos (Inuit), linking lifestyles to group and individual practices, seasons and resource use. Speck (1915), a student of Boas, recorded the ethnographic details of northern Algonquian hunting territories, leading to an interest in recording hunting territories and debate whether or not these were a form of 'private property.' Although this work was initially of purely academic interest, it was subsequently realized that such studies could support Aboriginal territorial claims.

The Inuit Land Use and Occupancy Project (Freeman 1976), undertaken in preparation of Inuit land claims, established the basic model of land use mapping that is still used today (Robinson and Ross 1997). This is based on the 'map biography' in which respondents are asked to locate and map harvesting or related land use activities during their adult lives (i.e., hunting, fishing, gathering), as well as other important elements such as burial sites, travel routes, and spiritual locations. Community land use patterns are then aggregated by map categories with outer areas representing boundaries and high-density areas representing the spatial intensity of community land use. This approach was subsequently used in Labrador (Brice-Bennett 1977), and has since become a standard method in Canada, in part due to straightforward documentation, visually effective maps and a perception of scientific validity (Usher *et al.* 1992). However, map biographies also have their limitations, particularly as it is impossible to recall, record and map all information about land use, and maps are inappropriate for documenting explanations of why or how land is used (Thom and Washbrook 1997).

Concurrent with the Inuit studies of the 1970s, the Fort George Cree Resource Use and Subsistence Economy Study (Weinstein 1976) used similar techniques to link land use and the contribution of these to subsistence economies. Researchers determined the spatial distribution of harvesting activities of roughly 1,500 community residents covering a geographical area of approximately 60,000 sq. km. Harvesting data were used to determine equivalent values for commercial foodstuffs, thereby estimating the economic effects of a proposed hydroelectric project. Difficulties associated with this approach include a failure to fully account for variability and seasonality of wildlife resources or for historical changes in land use and residency by Aboriginal groups. In a later review of his own work, Weinstein (1993:13) observed that the geographic extent of Cree land use was at an all time low when the study was undertaken in the mid-1970s.

A third variant has also been used to combine local land use patterns with proposed or existing industrial activity in order to assess the spatial aspects of conflicts on traditional lands (e.g., Brody 1981). A map biography, consistent with the approach developed by Freeman, is linked to the mapping of competing resource development and recreational harvesting activities (Weinstein 1993). This method was originally developed by the Union of British Columbia Indian Chiefs and is particularly useful for a spatial demonstration of the effects of industrial development on subsistence activities. From their origins in anthropological research, land use studies became increasingly common during the 1980s as Aboriginal peoples sought to obtain recognition of their land use in the face of development (as in the Kayahna Land Use conducted during the Ontario Royal Commission on the Northern Environment) and their rights through litigation.

Until recently, Canada's courts have obliged Aboriginal claimants to demonstrate or prove their occupancy of land, and land use and occupancy studies have become an accepted way of doing so (Elias 2004). However, with the Supreme Court of Canada's decision in *Haida* (2004),[2] Aboriginal peoples no longer need to prove the existence of their resource use and land title rights in the eyes of the courts for Crown governments to consult and accommodate their interests; "the duty (to consult) arises when the Crown has knowledge, real or constructive, of the potential existence of the Aboriginal right or title and contemplates conduct that might adversely affect it."[3] For governments to wait until Aboriginal rights are proven in court while allowing resource development to proceed is 'not honourable' or consistent with the Crown's fiduciary obligation to Aboriginal peoples.

More recently (1990s), land use research has found its way into resource management fields such as forestry, conservation and mining, oil and gas development. Government policies and sustainable forest management guidelines now encourage mapping or studies as a means of documenting Aboriginal use of lands as a contribution to effective management. However, the response of researchers has been more cautious. Natcher (2001 identified a series of potential problems in resource management associated with the application of land use research. In particular, he emphasized that Aboriginal Land Use and Occupancy Studies (ALUOS) should be only one of a number of tools used to include Aboriginal knowledge in the management process and should never be used to the exclusion of knowledge holders themselves. Karjala and Dewhurst (2003) went beyond simply mapping land use and sought to involve Aboriginal communities in developing alternative management scenarios. A different direction has been established by Berkes and colleagues (e.g., Armitage *et al.* 2007; Berkes and Folke 1998) who emphasize the social context of Aboriginal knowledge and land use, arguing that understanding the link between social and ecological systems is a prerequisite to maintaining and promoting sustainability and resilience.

ALUOS are now common throughout Canada. Elias (2004:62) estimated that more than 100 studies have been completed across Canada at a cost of over $100 million. However, this is probably conservative; many studies are undertaken under terms of strict confidentiality as Aboriginal communities and their advisors prepare for negotiations or litigation over land rights. Cheveau *et*

---

[2] *Haida Nation v. British Columbia (Minister of Forests)*, [2004] 3 S.C.R. 511, 2004 SCC 73
[3] *Ibid.* at para. 35.

al. (2008) noted a relative paucity of published studies of traditional ecological knowledge in forestry in Canada, and called for greater diffusion in accessible papers. Elias (2004:62) summarized the dilemma for land managers stating that "land use and occupancy studies could provide information managers need" but that "so long as … studies are primarily a legal tool, they and the wealth of cultural information they contain will languish."

Robinson and Ross (1997) attribute the willingness of Aboriginal peoples to document land use knowledge to their ongoing struggle to gain respect, equity and empowerment in the land management process. Documenting this knowledge can lead to increased self-respect and self-reliance. However, while the social and political influence of cartographic representation as a means of community empowerment has been well documented, land use studies in themselves may not be enough to ensure the protection of Aboriginal rights to the land (Natcher 2001). While such research represents a positive step toward articulating the rights and land use needs of Aboriginal communities, empowerment requires that Aboriginal communities express their concerns and aspirations within institutional frameworks that recognize their rights as users.

Legal Bases for Aboriginal Land Use and Occupancy Studies

Since Aboriginal and treaty rights were enshrined in the Canadian Constitution in 1982, the Supreme Court of Canada (SCC) has ruled on a number of cases that outline what it takes to establish Aboriginal rights. These rights may exist in the form of 'aboriginal title,' a unique form of land ownership that differs from simple fee ownerships or rights based on use, and may be defined in negotiated agreements with the Crown, either historic treaties or modern day land claims. Whatever the form of right, the onus has been on Aboriginal peoples to prove their land use and occupancy (McNeill 1999). To do this requires compiling evidence that can be held up to expert scrutiny and whatever legal or resource management challenges the Crown advanced. Many of these claims involve accommodation or compensation, or the development of arguments to transform land management regulations or operational practices—there is much at stake.

With SCC cases defining Aboriginal rights in the 1990s and 2000s, standards for proof of Aboriginal rights were set out. In the landmark decision of the Supreme Court in *Delgamuukw* (1997),[4] the court stated that Aboriginal peoples must demonstrate: 1) occupancy prior to European settlement, 2) continuity between present and pre-European occupancy, and 3) exclusive occupancy through physical evidence on the ground, such as dwellings and regular use of resources, including the delineation of boundaries. As well, Aboriginal people must demonstrate that the use and occupancy of their defined territory was governed by forms of customary law (Thom 2001). Interviewing community members and mapping land use became one, if not the main, form of accepted proof.

---

[4] *Delgamuukw v. British Columbia* [1997] 3 S.C.R. 1010,

In *Delgamuukw* (1997), the SCC described the determination of occupancy:

> ...by reference to the activities that have taken place on the land and the uses to which the land has been put by the particular group. If lands are so occupied, there will exist a special bond between the group and the land in question such that the land will be part of the definition of the group's distinctive culture.

In the *Marshall* and *Bernard* cases (2005),[5] the Supreme Court emphasized the importance of the cultural connection:

> Therefore, anyone considering the degree of occupation sufficient to establish title must be mindful that aboriginal title is ultimately premised upon the notion that the specific land or territory at issue was of central significance to the aboriginal group's culture. Occupation should be proved by evidence not of regular and intensive use of the land but of the tradition and culture of the group that connect it with the land.

With Aboriginal peoples bearing the burden of proof to support their claims for Aboriginal rights, whether in or out of court, many have turned to ALUOS as a means of documenting their land use areas and activities. Integrally tied to the demonstration of traditional land use and occupancy has been the exploration of Aboriginal customary law, which supports and governs traditions of land use. Although land use has been referenced by the courts as having an Aboriginal customary law and governance component, ALUOS are not always used in ways consistent with customary law or the approval of Aboriginal communities. For example, under the Crown's duty to consult, the Crown may take information out of context to speed up resource development decisions, leaving Aboriginal communities with impressive land use maps, but little to show in the way of negotiated agreements for accommodation of land use rights as customary governance arrangements.

ALUOS also trigger issues around ownership of and access to Indigenous knowledge. As a result, information-sharing agreements or protocols that address intellectual property rights have become a necessary part of sharing the information gathered through ALUOS. The use of ALUOS as evidence in land claims has also necessitated standards that will hold up to legal and scientific scrutiny. Although *Delgamuukw* (1997) affirmed the use of oral testimony for Aboriginal societies, the courts still scrutinize this oral evidence closely and require confirmation in the form of written records and archaeological evidence (Thom 2001). In *Tsilhqot'in Nation v. British Columbia* (2007),[6] Justice Vickers laid out his criteria for assessing oral evidence, including how oral history is preserved, how that history is transmitted from one generation to the next, how the truth of oral history is protected, who is entitled to learn and pass on the

---

[5] *R. v. Marshall, R. v. Bernard*, [2005] 2 S.C.R. 220, 2005 SCC 43.
[6] *Tsilhqot'in Nation v. British Columbia* 2007 BCSC 1700.

history, and if there are people who are more trusted than others to remember and transmit this history. The judge also set out guidelines for judging the reliability of witnesses called to convey oral history (PIP 2009).

Thom and Washbrook (1997) note the impact of requirements for legal evidence when planning an ALUOS. While a study may be intended for use in negotiations or in resource co-management, Aboriginal peoples may revert to litigation if these processes break down. Therefore, "research methods should follow the standards of the law of evidence in the pursuit of data integrity, and research strategies should aim to meet the tests that courts have applied to aboriginal claims to rights and title cases" (Thom and Washbrook 1997:5). Many Crown processes for documenting Aboriginal land use in resource management, such as those for forest management in Ontario (described below), have not prescribed standards that will hold up in a court of law. The need for such standards has led to the popularity of several guidebooks on conducting ALUOS (e.g., Tobias 2000, 2010; Garvin *et al.* 2001; WCMF 2007).

Whatever their challenges, ALUOS have the potential to provide not only legal evidence for claims of Aboriginal and treaty rights, but also evidence to inform resource development and management practices to protect traditional land uses. Beginning in the early 1990s a series of decisions by the SCC established the 'duty to consult and accommodate,' obliging government agencies, forestry companies and other resource developers to consult with Aboriginal peoples concerning the effect of proposed activities on their traditional lands and rights. Importantly, consultation is understood to include providing Aboriginal communities with a meaningful role in the decision-making process. Within this context, land use mapping is seen as a means for government agencies and forestry companies to demonstrate that consultation has occurred. An ALUOS can certainly provide important and useful information to managers, but the interests, expectations and rights of Aboriginal peoples are usually more extensive than the type of information that can be captured on a land use map (Natcher 2001). Hence, consultation processes must also, if the extent of the potential infringement of rights is significant, involve accommodation, including taking steps to avoid harm and minimize the infringement of rights. This requires other mechanisms to effectively integrate Aboriginal concerns into forest management and land use planning.

Policy Drivers for ALUOS

The growing number of traditional land use and occupancy studies conducted in Canada in recent decades is partly a response to the legal issues identified above, but also to several other policy factors, both governmental and non-governmental. These can vary across provinces, between forestry companies and from situation to situation, and so each ALUOS should be considered on its own merit, within its own unique context. Some provinces have developed specific requirements for mapping Aboriginal land use as a part of their policies and processes for either forest management planning or for relations with Aboriginal communities. For example, the Ontario Forest Management Planning Manual obliges forest managers to prepare an Aboriginal Background Information Report and an Aboriginal Values Map as a part of consultation processes

(OMNR 2004; McGregor, Chapter 10 this volume). Management plans must also include information on how identified values have been protected. However, it is important to note that many First Nations in Ontario do not accept this process as responding adequately to their rights (*see* McGregor this volume). Provincial governments in British Columbia and Alberta have established programs to encourage and support Aboriginal communities in conducting land use mapping with the view to using this information to facilitate resource management and planning (Elias 2004). In other provinces, such as New Brunswick and Nova Scotia, governments do not require either government forestry agencies or private forestry companies to engage in ALUOS.

Although provinces have constitutional authority over natural resources, federal government roles include Aboriginal issues and coordination of provincial efforts, leading to several initiatives concerning sustainable forestry across Canada. The Canadian Council of Forest Minsters includes the "Extent to which forest management planning takes into account the protection of unique or significant Aboriginal social, cultural or spiritual sites" as an indicator of the sustainability of forest management (CCFM 2003). The 2003 National Forest Strategy includes the rights and participation of Aboriginal peoples as one of eight strategic themes (NFSC 2003). Although land use studies and mapping are not specifically mentioned, the strategy includes commitments to "incorporate traditional knowledge in managing lands and resources" and developing institutional arrangements that give effect to "land claim settlements, treaties and formal agreements on forest resource use and management." In contrast, the most recent version of the strategy avoids discussion of Aboriginal and treaty rights and identifies only two priorities (transforming the forest sector and climate change), making only minor mention of historical Aboriginal relationships with forests and their potential role in the forest economy (CCFM 2008).

Over the last decade, sustainable forest management certification has become increasingly important in Canadian forestry. Managers have a choice of three performance-based standards, but requirements concerning Aboriginal land use are quite different among them (Collier *et al.* 2002). The Forest Stewardship Council (FSC) standard uses the strongest language in relation to Aboriginal peoples, requiring that forest management respect Aboriginal rights, providing several clear indicators for this. The Canadian Standards Association (CSA) standard requires consultation with Aboriginal communities, while the Sustainable Forestry Initiative (SFI) standard simply obliges managers to confer with Indigenous peoples.

Five Dynamics that Affect the Application of Land Use Studies

There is great variation in the practices used to conduct land use studies and different models exist for applying this information in land use planning. Given this diversity, it is inappropriate to propose a single set of characteristics or processes for successful studies. Instead, we identify a series of inter-related dynamics—questions that Aboriginal communities and their partners need to consider before commencing a study. How a specific situation is positioned in relation to these dynamics will affect how the study should be planned (Fig. 1). These five dynamics affect the way in which a land use study needs to be planned and conducted, and the contribution that it could make to forest management and land use planning. Each of the dynamics also has an impact on others. Hence, we represent them as a star (Fig. 1).

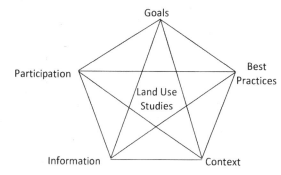

These five dynamics affect the way in which a land use study needs to be planned and conducted, and the contribution that it could make to forest management. Each of the dynamics also has an impact on others. Hence, we represent them as a star.

Figure 1. *Inter-related dynamics affecting land use studies.*

Dynamic 1—Goals: Why Study Land Use?

Each party in a land use study has its own goals and objectives. Understanding the goal of a proposed study helps to establish information needs and appropriate methods. A study aimed at improving forest management or building relations may imply a collaborative project with government and industry, while a study by a company aimed at demonstrating compliance with regulations could result in a token effort with little interest by managers in using this information. The goal also influences the quality of information gathered—a legal proceeding may require a research methodology that can be defended against expert witnesses called by opposing parties, while information for members of a First Nation may be subjected to scrutiny and evaluation by elders.

Land use studies and mapping are especially important in establishing Aboriginal title and ancestral rights in the courts or in political negotiations. This can create an expectation that any study or map could eventually be used in court or legal processes, leading participants to adopt strategic behaviour such as favouring areas of importance for upcoming legal battles or withholding funding or support. Research methods and standards used for court proceedings may also result in information that is inappropriate for forest management and land use planning or that is more expensive than necessary.

Dynamic 2—Information: What to Collect and What to Share?

Aboriginal knowledge about land and occupation is both extensive and complex, and it is unlikely that any single project would be able to collect all available and relevant information. Hence choices need to be made about what to collect and how to do so; taking account of goals as well as the interests of people who hold knowledge. However, Aboriginal information is often subject to limitations on use and transfer. Site-specific information such as fish spawning sites or burial grounds may be considered too sensitive to be released outside the community. Other information or knowledge relating to the use or history of the land and the people may be passed only to those who have demonstrated that the information will be used with respect. Land use study processes that are based on making information available may not be acceptable to communities that wish to keep information confidential. Conversely, not sharing information may result in Aboriginal interests being ignored while management decisions and actions go ahead.

Dynamic 3—Participation: Is it Worthwhile?

Although Aboriginal peoples often participate in land use and occupation studies, some question whether these studies really serve the best interests of the community and its members. Providing information for forest and land use management may actually serve as a substitute for direct Aboriginal participation in planning and management. A study funded by government or industry may be understood as fulfilling the 'duty to consult,' replacing the need for further discussions with the community. It is also a process that usually requires significant investment by Aboriginal communities in money and time, especially for elders and knowledge holders, as well as community leaders and negotiators. Producing a study and maps for use in legal proceedings or land use planning processes obliges Aboriginal peoples to present their culture and knowledge in terms that conform to the views and concepts of Western jurists. Although Indigenous maps have contributed to advancing Aboriginal interests, mapping is rapidly being assimilated into institutionalized processes steeped in colonial power relations (Bryan 2009). For Nietschmann (*cited in* Bryan 2009:24), it is a matter of 'map or be mapped!'

Aboriginal peoples may therefore prefer to refuse to participate in a land use study, accepting the risk of having their interests overlooked by others. For industry and government, such a refusal could create difficulties in planning and in compliance with consultation requirements. Aboriginal peoples need to consider the place of land use studies as part of a strategy to achieve their long-term goals. For their part, governments and forest industries should recognize that their expectations of consultation processes may not fit with an Aboriginal community's own strategy.

Dynamic 4—Best Practices: Best for What and for Whom?

A number of guides to best practices in land use and occupation studies exist, having been prepared by a variety of individuals and agencies. However, Peter Elias, a recognized expert in land use and occupancy studies, said "today there is no way of knowing which practices might be deemed 'best practices'" (Elias 2004:v). Despite the multitude of studies conducted, leading practitioners and researchers are rarely able to freely share their experiences and methodologies—partly because of concerns about confidentiality and legal implications. Hence, most guides represent principally the authors' personal views of what constitutes 'best practice,' rather than a sector-wide evaluation of the strengths and weaknesses of different techniques. While some methodological diversity enables studies to be adapted to the goals and characteristics of each project, it remains important to ensure the quality and consistency of information.

A second aspect of 'best practice' concerns how the information is to be used, especially in relation to forest management. Few existing guides consider this, while the research literature identifies problems such as different forms of knowledge, preferences for quantitative data, a bias toward site-specific information, and a range of legal and institutional considerations. However, communities, managers and researchers across the country have developed a variety of models. 'Best practice' in using land use information appears to be encouraging parties to develop individual approaches for their own situation.

Dynamic 5—Context: Considering a Range of Options for Aboriginal Engagement

No matter how well land use and occupation studies and maps are prepared, it is still necessary to recognise their limitations. Studies are a source of data and information that can help contribute to attaining the various goals of Aboriginal Nations and/or their partners. However, studies themselves will not achieve recognition of Aboriginal title, create employment, change the way management and planning are conducted on forested and Aboriginal lands or ensure the survival of culture and values. Instead, land use and occupation should be considered as part of a wider collection of activities, processes and tools that can help develop Aboriginal engagement in forest and land use planning and management. This will depend upon the context—the opportunities that exist for First Nations and their partners in a specific situation. Aboriginal nations, and other actors, need to consider their overall goals as well as the characteristics of their own situation in order to decide how land use and occupation studies can correlate to a wider strategy of social, economic and environmental sustainability.

Conclusion: Limitations and Challenges Associated with Land Use Studies

Based on these five dynamics, it is clear that land use studies have a number of shortcomings, particularly when they are promoted as a solution to all land use problems for Aboriginal communities. These challenges can be grouped according to the methodologies employed, the practical needs for conducting a study, and the institutional issues associated with using this information to make land use and forest management decisions in Indian Country (Table 1).

Resource developers, forestry companies and government agencies often provide funds for land use mapping and studies as a part of development planning and approval processes. This may be seen as way to meet obligations to consult with Aboriginal peoples, as a means of facilitating access to resources or as a step in protecting certain values associated with forest landscapes. For some organisations, financing a study leads to an expectation that land use information will be freely available to them, or may even become their property to be used or distributed as necessary. In contrast, Aboriginal peoples consider that their information remains their intellectual property and that they should retain control of it in order to ensure that it is used in appropriate ways. Mapping and land use studies are often perceived as a means of consultation. However, Aboriginal peoples typically have a broader range of issues that they wish to raise with land managers, such as title and rights, management objectives, employment or benefit sharing. As a result, consultation processes risk becoming a debate about what is and is not open for discussion. Forest industries are generally unable to address issues of policy or rights, and land use mapping is not a substitute for consultation around these issues.

Table 1. Challenges associated with Aboriginal land use studies.

Research methodologies	Practical needs	Institutional issues
• **Dependence on memory ethnography**, meaning that studies present a snapshot of what happened, based on peoples' memories. • **Studies are always incomplete** because there will always be missing information and, as land use changes over time, land use data will need to be regularly updated. • **Bias toward sites** because these are easy to mark on maps. Other cultural information is often ignored. • **'Blank spaces are unused'** is a common reaction when people from outside the community look at maps. • **Dynamism in land use is overlooked** because maps present static information.	• **Funding** is essential and Aboriginal groups need support from industry and governments. • **Trust of elders and the community** is critical if they are to share information. • **Trained and experienced people** are needed within communities rather than consultants. • **Technology** is useful, but should not overpower the social aspects. • **Coordination** with other communities helps build skills and techniques for studies and for using the information.	• **Recognizing the validity of Aboriginal knowledge** remains a challenge for many resource professionals. • **Short-term proposals** often oblige communities to react immediately, rather than prepare long-term plans. • **Consultation** is understood in different ways by Aboriginal peoples and by industry and governments. • **'Lurking legislation'** is an expectation that studies will be used in court and so should be kept secret.

Clearer government policies and planning guidelines should support the application of land use studies and mapping for sustainable land use and forestry. Government action to resolve major issues of rights and access to resources could enable industry and Aboriginal communities to discuss more specific issues related to management and practices. Guidelines and modifications to provincial forest planning frameworks should address the characteristics and utility of Aboriginal knowledge, and clarify ways in which information is collected, controlled and used. Finally, a certain degree of flexibility is necessary to enable forest and land use planners, whether they work for industry, governments or other agencies, to modify management planning methods and prescriptions to better incorporate information collected during land use studies. In terms of management implications, the following recommendations were identified:

1. Aboriginal land use is dynamic and complex, and includes cultural values, practices, history and social relationships for living on the land. It is not simply a set of maps showing specific sites.
2. Mapping sites or conducting a land use study is NOT a substitute for consulting Aboriginal communities about forest management and land use planning.
3. Clear goals are needed for land use studies and mapping projects, especially where several parties with different interests are involved.
4. Consultation frameworks and legal issues such as Aboriginal title are best addressed by governments, while planning and operational matters are more approporiate for management authorities (whether industry, government or others).

Although ALUOS had its origins in ethnographic research, the expansion of the field across Canada over the last 30 years has been largely due to litigation and legal processes concerning claims of Aboriginal rights and title. ALUOS have been an effective method for demonstrating Aboriginal use and occupancy of land as a means of establishing rights and title. This creates a situation where Aboriginal communities may view an ALUOS first as a confidential resource that could be used as part of a current or eventual legal strategy, and secondly as information to be shared with land use planners and forest managers as part of a process to protect their activities and uses of the land.

For Aboriginal peoples, land use, occupancy and knowledge are all linked to their culture and to a holistic view of the environment. A typical land use map is unable to document this complexity and is only a partial representation of Aboriginal interests in their traditional lands. Forest industry managers who view an ALUOS as simply another layer of information to be incorporated into a forest management plan overlook the cultural importance attached to the information by the Aboriginal community. Using an ALUOS in such a way may contribute to increasing tensions between Aboriginal peoples and industry, rather than an opportunity to build bridges. ALUOS can provide a wealth of information for both members of the Aboriginal community and for outsiders, contributing to better management of forests, lands and natural resources. We believe, however, that this can only be effective when an ALUOS is part of a broader collaborative arrangement that respects the rights and interests of both Aboriginal peoples and other parties.

References Cited

Armitage, D., F. Berkes, F., and N. Doubleday, eds. (2007). *Adaptive Co-Management: Collaboration, Learning, and Multi-Level Governance*. Vancouver, B.C.: UBC Press.

Berkes, F. and C. Folke (1998). *Linking Social and Ecological Systems: Management Practices and Social Mechanisms for Building Resilience*. Cambridge, UK: Cambridge University Press.

Boas, F. (1888). 'The Central Eskimo,' pp. 399-669 in *Sixth Annual Report of the Bureau of American Ethnology for the Years 1884-1885*. Washington, DC.: The Smithsonian Institute.

Brice-Bennett, C. (1977). *Our Footprints Are Everywhere. A Report of the Labrador Inuit Land Use and Occupancy Study*. Labrador Inuit Association.

Brody, H. (1981). *Maps and Dreams*. Vancouver: Douglas and McIntyre.

Bryan, J. (2009). Where would we be without them? Knowledge, spaces and power in indigenous politics. *Futures* 41: 24-32, www.elsevier.com/locate/futures

CCFM—Canadian Council of Forest Ministers (2003). *Defining Sustainable Forest Management in Canada; Criteria and Indicators*. Natural Resources Canada: Canadian Council of Forest Ministers.

CCFM—Canadian Council of Forest Ministers (2008). *A Vision for Canada's Forests: 2008 and Beyond*. Natural Resources Canada: Canadian Council of Forest Ministers.

Cheveau, M., L. Imbeau, P. Drapeau, and L. Belanger (2008). Current status and future directions of traditional ecological knowledge in forest management: a review. *The Forestry Chronicle* 84(2): 231- 243.

Collier, R., B. Parfitt, and D. Woollard (2002). *A Voice on the Land: An Indigenous Peoples' Guide to Forest Certification in Canada*. Ottawa and Vancouver: National Aboriginal Forestry Association and Ecotrust Canada.

Elias, P. (2004). *Standards for Aboriginal Cultural Research in Forest Management Planning in Canada*. Canada's Model Forest Network, Aboriginal Strategic Initiative. Project ASI-03/04-003.

Freeman, ed. (1976). *Inuit Land Use and Occupancy Project*. Ottawa: Minister of Supply and Services Canada.

Garvin, T., S. Nelson, E. Ellehoj, and B. Redmond (2001). *A Guide to Conducting a Traditional Knowledge and Land Use Study*. Edmonton, Alberta: Natural Resources Canada, Canadian Forest Service.

Mauss, M. (1905). *Essai sur les variations saisonnières des sociétés eskimo. Étude de morphologie sociales*. L'Année Sociologique 9.

McNeill, K. (1999). The onus of proof of Aboriginal title. *Osgoode Hall Law Journal* 37: 775-803.

Natcher, D. (2001). Land use research and the duty to consult: a misrepresentation of the Aboriginal landscape. *Land Use Policy* 18(2): 113-122.

NFSC—National Forest Strategy Coalition (2003). *National Forest Strategy 2003-2008*. National Forest Strategy Coalition.

OMNR— Ontario Ministry of Natural Resources (2004). *Forest Management Planning Manual for Ontario's Crown Forests*. Edited by Ontario Ministry of Natural Resources. Toronto: Queen's Printer for Ontario. 440p.

PIP—Powley Implementation Project (2009). *Canadian Legal Standards of Evidence*. Available from http://www.abo-peoples.org/Metis/Legal.html . [accessed 4 June 2009.

Robinson, M.P. and M.M. Ross (1997). Traditional land use and occupancy studies and their impact on forest planning and management in Alberta. *The Forestry Chronicle* 73(5): 596-605.

Speck, F. (1915). The Family Hunting Band as the basis of Algonkian Social
     Organization. *American Anthropologist* 17: 289-305.
Thom, B. (2001). Aboriginal rights and title in Canada after Delgamuukw: Parts I and II.
     Native Studies Review 14(1&2): 1-26 & 21-42.
Thom, B. and K. Washbrook (1997). 'Co-Management, Negotiation, Litigation.
     Questions of Power in Traditional Use Studies,' in *Annual Meetings of the
     Society for Applied Anthropology*. Seattle, Washington.
Tobias, T.N. (2000). *Chief Kerry's Moose: A Guide to Land-use and Occupancy
     Mapping, Research Design and Data Collection*. Vancouver, BC.: Union of
     B.C. Indian Chiefs and Ecotrust Canada.
Tobias, T.N. (2010). *Living Proof: The Essential Data-Collection Guide for Indigenous
     Use-and-Occupancy Map Surveys*. Union of B.C. Indian Chiefs and Ecotrust
     Canada, Vancouver, BC.
Usher, P., F. Tough, and R. Galois (1992). Reclaiming the land-Aboriginal title, treaty
     rights and land claims in Canada. *Applied Geography* 12(2): 109-132.
WCMF—Waswanipi Cree Model Forest (2007). *Ndoho Istchee: An Innovative Approach
     to Aboriginal Participation in Forest Management Planning*. Waswanipi Cree
     Model Forest.
Weinstein, M.S. (1976). *What the Land Provides: An Examination of the Fort George
     Subsistence Economy and the Possible Consequences on it by the James Bay
     Hydroelectric Project*. Grand Council of the Crees (of Quebec).
Weinstein, M.S. (1993). 'Aboriginal land use and occupancy studies in Canada,' in
     Workshop on Spatial Aspects of Social of Social Forestry Systems. Chiang Mai,
     Thailand: Chiang Mai University.
Wyatt, S., J.-F. Fortier, G. Greskiw, M. Hébert, D. Natcher, S. Nadeau, P. Smith, D.
     Théberge, and R. Trosper (2010*). Can Aboriginal Land Use and Occupancy
     Studies be Applied Effectively in Forest Management?* Edmonton, Alberta:
     Sustainable Forest Management Network.

Chapter Ten

Aboriginal Values Mapping in Ontario's Forest Management Planning Process

Deborah McGregor

Introduction

In recent decades there has been, both among the general public and those involved in forestry and environmental issues, a growing sense that change is needed in dominant society's values regarding how we relate to forests. In part, this realization stems from international pressure from such sources as the World Commission on Environment and Development (WCED 1987), and the United Nations (e.g., through *Agenda 21* produced at the 1992 United Nations Conference on Environment and Development) (UNDSD 2004). Prominent organizations the world over are calling for improved forest practices and increased protection of forest resources as part of global efforts to avert world-wide environmental catastrophe. This has led many North American governments, including Canada's federal and Ontario's provincial governments, to closely examine existing forestry policies and practices, and enact modifications and new guidelines intended to put Canada and Ontario on a path toward greater sustainability in forest use and management.

Governments are also experiencing demand for change in forest practices from more locally-based sources. In Canada and the US, registered foresters, forest scientists, forest technicians, and environmental and social activists have been undertaking increasingly demonstrative and coherent efforts to produce a wide-scale shift from conventional methods of industrial forestry to more 'ecologically responsible' or sustainable systems of forest use (Drengson and Taylor 1997; Hammond 2008). The volume and logic of the arguments being presented, as well as the appearance in reality of some of the large-scale and long-term negative impacts of forestry predicted by critics, have placed significant pressure on governments to make real changes to the conventional system, for political, ecological and economic reasons (Merkel 2007; Rayner and Howlett 2007).

An important component of the suggestions for change has been that Aboriginal peoples need to be both consulted for their knowledge of forest ecosystems and included meaningfully in forest related decision-making, particularly where such decision-making affects Aboriginal communities, as it most often does, especially in Canada (Hickey and Nelson 2005; Smith 2007; Wilson and Graham 2005). Furthermore, the nature and scope of Aboriginal and treaty rights are still being defined in the courts, and it is increasingly expected that Aboriginal rights should be recognized and accounted for in the resource management arena (Government of Canada 2008). Not only is there the outstanding issue of land claims to be resolved from a moral and legal perspective, but there remains a significant body of Aboriginal Traditional Knowledge (TK) which could guide new forestry systems toward increased sustainability (Davidson-Hunt and O'Flaherty 2007; Houde 2007; O'Flaherty *et al.* 2008; Stevenson 2005).

In recent decades, the reconstruction of Aboriginal nations, including the call for the settlement of land claims, the assertion of Aboriginal and treaty rights and the meaningful inclusion of Aboriginal peoples in decision-making processes has received increasing support in the international political arena. This is expressed well in the United Nations *Declaration on Indigenous Rights*, which recognizes "...that respect for indigenous knowledge, cultures and traditional practices contributes to sustainable and equitable development and proper management of the environment" (UNGA 2007:2).

Canadian federal and provincial governments have thus been under intensifying pressure, not only from international sources, but from Indigenous peoples themselves. Subsequently, the Ontario provincial government developed policy frameworks aimed at moving toward sustainable environmental and resource management decision-making which includes Aboriginal participation. Such policies in Ontario now extend beyond forest management planning into broader land use planning, as evidenced by undertakings such as the Northern Boreal Initiative (NBI) and the 'Far North' Planning Initiative (*see* Deutsche and Davidson-Hunt, this volume). Both these initiatives involve First Nations communities who are determined to protect and maintain ancient relationships with the land, and who are voicing their input into land use planning and decision-making, of which forest management planning is a major component.

This chapter discusses the central focus of Ontario's response to the call for more meaningful Aboriginal engagement in forest land use planning, 'Aboriginal Values Mapping.' The province views Aboriginal Values Mapping (AVM) as a key mechanism for the incorporation of TK into its forest management plans (Brubacher and McGregor 1998; Cheveau *et al.* 2008; McGregor 2000). As such, the degree of success of this process will largely determine the extent to which Aboriginal views on forestry are heard in Ontario.

Background: The Need for Aboriginal Involvement

There is more at stake for First Nations and forest management in Canada and Ontario than simply forestry. There are numerous related and underlying factors which cannot be ignored if one is to achieve an effective understanding of the topic. These include pre-and post-contact Aboriginal history as well as issues around self-government, land claims, treaties and Aboriginal rights and title. It is these factors which give rise to Aboriginal goals for forest management, but also to many of the current conflicts among stakeholders, governments and First Nations. It is important to develop an historical understanding of the relationship between Aboriginal and non-Aboriginal peoples in Canada in order to appreciate how it continues to influence the current state of Aboriginal participation in forest management and planning. As Georges Erasmus (1989:92), former Grand Chief of the Assembly of First Nations, commented, "To understand the native or indigenous point of view on conservation or environmental matters, one must understand our history, our culture, and the way we see our relationship with nature."

In the history of Canada we can see a parallel relationship between forest management and the colonization of Aboriginal peoples. Absolute sovereignty prior to European contact involved the duties and activities that any sovereign nation would undertake, including ways of relating to and engaging with other nations as well as managing internal affairs and relations among its citizens (RCAP 1996a). Inherent in this sovereignty were sophisticated methods of relating to and utilizing the natural environment, including forests. First Nations peoples in North America had highly refined ways of understanding and explaining the world and its resources. Complex systems of resource use and stewardship were developed long before Europeans realized this continent existed. Discussion of the details of such systems is beyond the scope of this paper, but the reader can turn to works such as Berkes (2008) and Menzies (2006) for more detail. Since the arrival of Europeans, however, Aboriginal peoples have become excluded from managing the forests on their territories (McGregor 2006; NAFA 1993; RCAP 1996a). As in the rest of Canada, colonization and settlement in Ontario has wrested from Aboriginal peoples most of their access to and use of their traditional territories. Aboriginal peoples have been denied authority and jurisdiction over their territories in a variety of ways, including the enactment of legislation (e.g., *British North America Act, 1867, Indian Act, 1876*), policy and treaties (Borrows 2005; RCAP 1996a; Teillet 2005). This exclusion has occurred as a result of deliberate efforts to make way for European settlement and resource development interests (NAFA 1993; Rude and Deiter 2004; Teillet 2005) and remains a highly contentious issue.

Despite numerous prolonged attempts by Canadian governments to assimilate Aboriginal peoples into mainstream society, thereby eradicating the 'Indian problem,' Aboriginal peoples have continued to retain much of their cultural heritage, and have even embarked on a path of cultural resurgence in traditional activities in many cases (Houde 2007; Smith 2007). Wyatt's (2008) review of Aboriginal forestry over the past 30 years chronicles such evolution of Aboriginal involvement in the forest sector. Wyatt (2008:176) observes that:

> It is clear that Canadian forestry is evolving to provide a greater role for aboriginal peoples. Aboriginal rights are being defined and recognized, First Nations are taking their place in forest industries, and forest management increasingly takes their interests into account.

Thus, it is not only governments that are making changes in regards to the inclusion of Aboriginal peoples in forestry, but also industry and other partners (Hickey and Nelson 2005) .

Ontario's Timber Management Hearings: The Basis for Current Aboriginal Involvement

In Ontario, a significant turning point in Aboriginal participation in forestry occurred with the 'Timber Management Hearings,' which took place between

1988 and 1992. These hearings were required at the time by the Minister of the Environment as part of the class environmental assessment for timber management being conducted by the Ministry of Natural Resources (MNR). During the hearings, Aboriginal peoples provided testimony as to their historical and current exclusion from forestry in Ontario. The Environmental Assessment Board (EAB), which was overseeing the process and whose *Decision on the Class Environmental Assessment for Timber Management in Ontario* was published in 1994, states that:

> The evidence we received on employment, poverty and access to off-reserve timber convinced us of the historical and present day exclusion of native communities from sharing in the social and economic benefits enjoyed by non-native communities from timber operations on Crown Land (EAB 1994:13).

While MNR recognized this situation to some extent, the EAB (1994:357) stated clearly that resolution to such problems would require much more than simple recognition:

> MNR told us that it recognizes the very high rates of unemployment and limited opportunities for developing a stable economic base in many Aboriginal communities. Although their witnesses gave us examples of MNR's attempts to encourage Aboriginal involvement in timber management operations, the results are not likely to be impressive without access to timber and creative thinking.

Not only did the EAB's decision require a much broader approach to forestry than the conventional timber management framework would allow, but it also specifically dictated that Aboriginal concerns be addressed in all new forest management plans. The Board's decision stated that (1994:13):

> In the first we approve the Timber Management Native Consultation Program, which parallels the overall planning process but affords opportunities for First Nations and Aboriginal communities to get recognition for their unique concerns into Timber Management Planning. In our opinion, this program can offer the same protection for the values of native communities against the adverse impacts of timber operations that we are approving for other northern Ontario communities and interests.

The 1994 EAB decision was reviewed in 2002 and input was received from the public, stakeholders and Aboriginal peoples. In 2003, the MOE approved *Declaration Order MNR-71*, including a number of conditions relating to Aboriginal peoples (Ministry of the Environment 2003). The Declaration Order exempts the MNR's forest management activities from certain sections of the *Environmental Assessment Act*, provided the MNR meets a series of over 50 conditions. Among the conditions in the Declaration Order are requirements for

carrying out an Aboriginal consultation process and creating two main types of
Aboriginal-related reports: the *Aboriginal Background Information Report* and
the *Report on Protection of Aboriginal Values.*

Significantly, a key required component of the Aboriginal Background
Information Report is the Aboriginal Values Map (Larose 2009, Sapic *et al.*
2009). The 1994 EAB decision and *Declaration Order MNR-71* have helped
shift the focus of provincial forest policy and improve opportunities for
Aboriginal peoples in forestry (Bombay 1995; LaRose 2009; McKibbon 1999).
The question then becomes "Are Aboriginal peoples satisfied with these
improvements?" The goal of this chapter is to begin answering one aspect of this
question by evaluating a key Aboriginal component of the new forest
management planning process: Aboriginal Values Mapping.

Aboriginal Values Mapping

Aboriginal Consultation Under MNR's New Forest Management Planning Manual

A major change in Ontario's forest management policy in relation to Aboriginal
peoples is that the province has now formalized Aboriginal involvement in
forest management planning. As part of this process, the new *Ontario Forest
Management Planning Manual* was developed "to implement the Board's
decision and the newly appointed *Crown Forest Sustainability Act* (CFSA)
(McKibbon 1999:9). The new manual, most recently updated for June 2004,
"...is the pivotal document that provides direction for all aspects of forest
management planning for crown lands in Ontario..." (MNR 2004:i). Whereas the
CFSA reflects the decisions of the EAB, the planning manual reflects the legal
requirements of the CFSA, the EAB decision and more recently the
requirements derived from conditions laid out in the *Declaration Order MNR-
71*. The planning manual contains the regulations for the CFSA and implements
the terms and conditions of the EAB decision and *Declaration Order MNR*-71,
thus laying out steps for how Ontario is to achieve sustainable forestry.

Ontario's new forest management planning process, as embodied in the
FMP manual, is regarded as one of the most environmentally rigorous in the
world. It addresses sustainability and develops indicators of such, and is also
designed to respond to a multitude of non-timber values. However, it remains
highly debatable as to how well the planning process addresses Aboriginal
concerns.

According to Naysmith (1996:6), the new planning manual represents, "a
substantial opportunity for First Nations to incorporate their concerns, values
and aspirations into the earliest stages of developing forest management plans."
As quoted above, the EAB decision presents opportunities for Aboriginal
peoples to 'get recognition of their unique concerns' addressed in the planning
process (EAB 1994:13). In practice, however, this 'recognition' of Aboriginal
concerns occurs at the MNR district *administration* level in the planning
process. Due to their unique historical situation, First Nations insist on a nation-
to-nation relationship where they are partners in meaningful government level
decision-making, and reject token participation in previously established

planning processes. The new planning process does not accommodate this position (Bombay 1995).

Whereas the MNR had previously described Aboriginal peoples as stakeholders in their planning process, the EAB contested that, "We are persuaded by the evidence we heard that it is incorrect to characterize the interests of First Nations people and Aboriginal people as the same as other stakeholders" (EAB 1994:346). The EAB decision, the Declaration Order and MNR's Forest Management Planning Manual each recognize the need for a separate parallel process for addressing Aboriginal needs and values.

In some respects, the Aboriginal public consultation process recognizes the unique needs and values of Aboriginal peoples through the provision of a voluntary process in which identified Aboriginal communities can choose to participate. The 'Aboriginal Consultation Approach' is a parallel, yet separate, planning process representing MNR's interpretation of the EAB condition for separate Aboriginal involvement. Although the current FMP manual *does* reflect the EAB decision and Declaration Order conditions required for Aboriginal participation in the process, it has *not* necessarily represented what Aboriginal peoples *want* from participation in forestry (Larose 2009).

Research has shown that the approach to consultation identified in the FMP manual (MNR 1996, 2004) is lacking from an Aboriginal perspective, particularly with respect to the Native Values Mapping component (Larose 2009; McGregor 2000; Sapic *et al.* 2009). Recognizing that the process requires improvement, MNR undertook a review process to produce a new *Forest Management Guide for Cultural Heritage Values* (MNR 2007). This Guide, which replaces the 1991 *Timber Management Guide for the Protection of Cultural Resources* (MNR 1991), is intended to provide specific direction for forest managers as they implement the new 'Aboriginal Consultation Approach.' The new Guide provides mandatory standards and guidelines as well as best management practices to be followed by forest managers in the protection of all cultural heritage values. Unlike its predecessor, the new Guide, created with the input of at least one Aboriginal member of the 'revision team' along with input from two Aboriginal workshops held in 2004 and 2005, now specifically includes Aboriginal values among its list of cultural heritage values, and details the steps required of forest managers to ensure that Aboriginal values are protected. The new guide seeks to describe Aboriginal values in forest management planning. It includes sections that provide more detail (and images) of Aboriginal cultural values, including a section devoted to best management practices for Aboriginal values identification (MNR 2007).

Aboriginal Values Maps

Despite its flaws, the current Aboriginal consultation approach aims to acquire information on Native communities via the Aboriginal Background Information Report (which includes an Aboriginal Values map), as well as the Preliminary and Final Reports on the Protection of Identified Aboriginal Values. The Aboriginal Background Information Report includes (MNR 2004:A-135):

(a) a summary of the use of natural resources on the management unit by Aboriginal communities, in particular hunting, fishing, trapping and gathering;

(b) forest management-related problems and issues for those Aboriginal communities;

(c) an Aboriginal values map which identifies the locations of natural resource features, land uses and values which are used by, or of importance to, those Aboriginal communities.[1] In particular, the following features, land uses and values will be mapped:

  (i)   areas of significance to local Aboriginal communities, such as areas used for traditional or recreational activities;

  (ii) boundaries of trapline management areas of those Aboriginal communities (i.e., all registered trapline areas associated with individual Aboriginal communities);

  (iii) Reserves and Aboriginal communities;

  (iv) areas that have been identified as being required as Reserve lands or for economic or capital development projects of those Aboriginal communities;

  (v)   areas used by those Aboriginal communities for fuel wood or building materials;

  (vi) sites of local archaeological, historical, religious and cultural heritage significance to those Aboriginal communities, including Aboriginal cemeteries, spirit sites and burial sites; and

  (vii) areas of archaeological potential as a result of Aboriginal involvement in the archaeological predictive modeling; and

(d) a summary of the negotiations between MNR and Aboriginal communities.

A key component in the successful creation of an Aboriginal Values Map is the participation of Aboriginal communities in the forest management planning process. The Manual requires MNR to contact Aboriginal communities in or adjacent to the management unit under consideration in an attempt to secure their participation in the development of an Aboriginal consultation approach (MNR 2004:A-80). However, while an Aboriginal Background Information Report (containing an Aboriginal Values Map) is *required* for approval of a forest management plan, Aboriginal participation *per se* is not. This can put MNR in an awkward position if Aboriginal communities do not wish to participate in the forest management process. While there is a clear need

---

[1] Publicizing the location of certain values may be detrimental to conservation, in which case information would not normally be shown on the Aboriginal values map(s).

for Aboriginal participation in the forest management planning process (via the consultation programs or otherwise) in order to achieve successful production of the Aboriginal Values Map, MNR will nevertheless proceed with the Aboriginal Background Information Report in the absence of Aboriginal participation (MNR 2004:A-132).

Research Methods

Research was undertaken to provide an initial analysis in response to the specific question of how successfully the MNR's Aboriginal Values Mapping process incorporates Aboriginal Traditional Ecological Knowledge (TEK) into forest management planning. Incorporating TEK includes meeting the needs and protecting the values of Aboriginal peoples as defined by Aboriginal respondents. In order to address this topic, 52 interviews were conducted with representatives from MNR, the forest industry and Aboriginal communities (the three 'response groups') who had participated in the new Aboriginal consultation process as originally required by the EAB decision. A 'Grounded Theory' approach (Glaser 1994; Strauss 1987) was used to classify and analyse the interview response data. Grounded Theory involves the identification of patterns from qualitative data (in this case, interview responses) which are then classified into 'Response Categories,' followed by a smaller number of broader 'Themes,' and then the identification of two or three 'Core Variables' which explain the primary characteristics of the data (see McGregor 2000 for detailed methodology and research results). The interviews conducted for this study yielded eight Response Categories, five Themes and two Core Variables which explain the major issues around Aboriginal Values Mapping as experienced by participants in the process. These results are presented below.

Research Results

Response Categories
The interview data collected were initially sorted into eight response categories. These categories, along with highlighted characteristics or key points, are summarized in Table 1 (further details are presented in McGregor 2000:96-128).

Themes
Once similarities and differences among the responses provided by the three response groups were identified, it was necessary to begin to explain these results. This was achieved by first collapsing the response categories into central 'Themes,' of which five emerged from the data. The major thrust of each theme is highlighted below (detailed explanations are found in McGregor 2000:129-174).

1. The Relationship between Aboriginal Values and TEK.
While all three response groups acknowledged that TEK is hard to define, there was also general agreement that Aboriginal values with respect to forests represent merely a small subset or fragment of the holistic life knowledge which

constitutes TEK. From an Aboriginal perspective, trying to isolate and incorporate TEK 'fragments' out of context and/or without the meaningful involvement of the original knowledge holders results in loss and/or distortion of intended meanings.

Table 1. Response Categories derived from interviews with participants in MNR's Aboriginal involvement process (22 representatives from MNR, 10 from Industry, 20 from First Nations).

Response Category	*Key Characteristics/Points from the Perspective of Each of the 3 Response Groups* (**FN**=First Nations; **Ind**=Industry; **MNR**=Ministry of Natural Resources)
1. Definitions of Aboriginal Forest Values	**FN**: emphasis on spiritual as well as physical values; broad, holistic, flexible; not confined to physical location. **Ind**: emphasis on physical values; readily identifiable on a map. **MNR**: preference for physical, map-friendly values; recognition of existence of hard-to-define values; uncertainty as to how to protect hard-to-define values.
2. Understanding Differences Between Personal Views of Aboriginal Values vs Those of the Other Response Groups	All 3 response groups recognized the existence of world view differences between Aboriginal and non-Aboriginal respondents, particularly the much greater Aboriginal focus on holistic, spiritual, and non-physical values.
3. Data Collection for the Aboriginal Values Mapping Process	**FN**: community-driven process; sources known to data collector, so information could be verified or queried; interviews, public events, widespread community involvement, but process seen as inappropriate for representing Aboriginal realities. **Ind**: not involved in data collection. **MNR**: maps, computer modelling; sources not necessarily known to data collector; information from individual sources may be used to represent entire community; First Nations viewed as lacking capacity to participate effectively
4. Degree of Assurance that all Values are Protected	**FN**: generally felt their values were not protected. **Ind**: all identified (i.e., generally point-specific) values were protected, but recognized not all values were identified. **MNR**: all identified values are protected; recognized difficulty protecting non-point-specific values; recognized First Nations dissatisfaction with level of values protection.

5. Potential of Aboriginal Participation to Contribute to FMP	**FN**: believe they can offer unique perspectives, especially involving a philosophy of forest stewardship; can provide highly specific information on local forest ecosystem, its ecological functioning and its history. **Ind**: similar to what First Nations felt they had to offer, but concerns about the validity of Aboriginal contributions which are not quantifiable, difficult to verify scientifically. **MNR**: same as for Industry.
6. Definitions of Traditional Ecological Knowledge	**FN**: broad, holistic definition; includes spirituality, world view, ethics, necessity of "living the knowledge" daily, as well as ecological information. **Ind**: focused on ecological information. **MNR**: broader definitions than industry, but not as broad as First Nations (e.g., didn't include spirituality); knowledge has predictive capacity; experiential; includes management aspects.
7. The Potential Contribution of TEK to Forest Management Planning	Despite having different definitions of TEK, all 3 response groups were remarkably similar in their views of what TEK might offer forest management; including helping to identify and protect a variety of Aboriginal and forest values, improve forest management and planning, and provide current and historical ecological information.
8. Degree of Assurance that Forest Management Planning Currently Incorporates TEK	**FN**: TEK is being used to some extent; major barrier to TEK use is its devaluing in relation to western science; relationship-building the best way to further TEK use. **Ind**: TEK, while having a potential contribution to make, is rarely used; TEK not valued as highly as western science; relationship-building the best way to further TEK use. **MNR**: TEK, while having a potential contribution to make, is rarely used; only potential existing formal process for TEK incorporation is Aboriginal Values Mapping, but this is designed for identifying Aboriginal Values, not incorporating TEK; relationship-building the best way to further TEK use.

2. Aboriginal Values in Forest Management Planning.
Responses obtained confirmed that Aboriginal and non-Aboriginal views of what constitute 'Aboriginal values' differ considerably. From an Aboriginal perspective, spiritual values, for example, form an integral part of *every* forest value, while non-Aboriginal peoples tend to view spiritual values as separate from other forest values. Another major difference concerns the fundamental appropriateness of mapping Aboriginal values at all. Many Aboriginal people involved in the process feel that some values cannot and should not be mapped, while the procedural guidelines for Aboriginal Values Mapping (which do concede that some values cannot be mapped) state nevertheless that in order to be protected, Aboriginal values *must* be geographically located (Sapic *et al*. 2009). Cultural discrepancies around what can/cannot or should/should not be mapped have been observed by other researchers. Bryan (2009:29), for example,

observes that mapping processes potentially obscure what is really important to Aboriginal peoples, or as he puts it, mapping does not "...disclose a fully formed indigenous geography." What is excluded from mapping can be just as revealing of the differences in viewpoints as what information *is* mapped. With such world view differences, it is not surprising to find that resistance or refusal to participate in Aboriginal Values Mapping is not uncommon. Resistance to mapping that is required by state management regimes has also been noted elsewhere, as in the 'counter mapping' efforts enacted to "increase the power of people living in a mapped area to control the representation of themselves and to increase their control over resources" (Johnson *et al.* 2006:87).

3. The Representation of Knowledge in the Mapping Process.
Used within a community-derived framework and incorporating local goals in areas such as education and recording oral history, maps can be a powerful and positive force for Aboriginal peoples (Tobias 2000). They can be used to protect geospatial manifestations of 'traditional knowledge' (Poole 1995a:1). They can also be useful in responding to external pressures and exerting an influence in such matters as land claims, self government and resource development proposals (Robinson and Ross 1997; Tobias 2000). Whatever the purpose, if an Aboriginal community is to benefit, it is important that, "...the content remains traditional ecological and cultural knowledge and practice" (Poole 1995b:74). It is necessary to ensure that the *context* for sharing knowledge remains as well as the intent (Johnson *et al.* 2006). Aboriginal Values Mapping as it is currently practiced, however, remains preoccupied with meeting the pre-defined requirements of an externally initiated and perpetuated process. This project seeks to capture, in map form, the geospatial aspects of Aboriginal peoples' knowledge. However, forcing traditional knowledge into the fixed categories and boundaries of western maps "...further perpetuates the loss of Indigenous geographic knowledge" (Johnson *et al.* 2006:87). Standardizing and universalizing Indigenous geographic knowledge is really a colonizing process, as it makes "...the world known through the standardized knowledge of western cartography" (Johnson *et al.* 2006:89). Bryan (2009:30) notes that, "The map remains unchallenged as a means of expressing the way things are, passing itself off as a picture rather than a self-consciously fashioned representation that must conform to certain conventions in order to be taken seriously." Maps have thus been used to justify "...indigenous peoples exclusion from state societies or as an argument for their inclusion through recognition of culturally specific forms of use and occupancy" (Bryan 2009:25).

The fact that Aboriginal Values Mapping is an externally driven process with highly specific purposes remains true even if a First Nation chooses to conduct the exercise within their own community utilizing their own knowledge and resources. The values data eventually have to be provided in suitable form (maps) to be utilized in the externally imposed framework (e.g., a FMP). The community is coerced into fragmenting and redefining its knowledge, understandings and values for outsiders who hold preconceived notions of what knowledge is appropriate for forest and land use planning (Stevenson 206).

First Nations find themselves struggling to identify values in a process which may or may not recognize those values as valid or credible, in addition to mapping certain values which may not be appropriately mapped. This is perceived as a loss of control of traditional knowledge, in the form of 'values,' to a dominating regime. The entire process of Aboriginal Values Mapping is therefore troublesome from an Aboriginal perspective. Aboriginal Values Mapping under the Ontario forest management planning framework can easily be viewed as just another way for outsiders to gain more control of traditional lands, and may actually serve to exclude Aboriginal peoples from any meaningful role in decision-making.

4. Aboriginal Values Protection

In this research, Aboriginal representatives did not feel their values were adequately represented or protected, even in cases where a high level of Aboriginal participation had been achieved in the forest management planning process. This is due primarily to the externally-derived and driven nature of the Aboriginal Values Mapping process. As noted above, Aboriginal participants in the mapping process were under pressure to somehow make their values and understandings 'fit' with predetermined non-Aboriginal definitions of what those values should look like. For example, Aboriginal respondents stated that they felt their values were considered 'acceptable' as long as they didn't 'tie up a lot of land' slated for logging operations. As well, only point-specific values (i.e., ones that could be readily mapped) were even considered for protection. This 'dots on a map' approach to traditional land use mapping, which is discussed elsewhere in this volume (e.g., see Passelac-Ross, Webb and Stevenson), has recently been explicitly rejected by Judge Vickers in *Tsilhquot'in Nation vs. British Columbia* (2007 BSSC 1700). Nonetheless, faced with such criteria, Aboriginal peoples involved in values mapping have often found themselves in the 'Catch 22' situation of having to either 'map' their values even if they could not be appropriately mapped—how do you pinpoint spiritual values on a forest map, when in your view everything in the forest has spirit?—or not have their values mapped at all. Such situations have led to an inherent tension in the process, described by the late geographer Bernard Nietschmann, as a case of having to 'map or be mapped' (cited in Bryan 2009).

This rather forced fragmentation and extraction of knowledge from its original holders in some cases could be seen as having effects directly in opposition to the stated intention of the Aboriginal Values Mapping process. Instead of affording a level of protection to Aboriginal values, Aboriginal participation in the mapping process merely lends credibility to forestry operations which were seen by those Aboriginal participants as destructive of their values. This is a prime example of how Aboriginal knowledge, in the control of others, loses its intended meaning. Indeed, such misappropriation can even become dangerous from an Aboriginal perspective when culturally constructed knowledge is taken from the original knowledge holders, fragmented, and reapplied in an unfamiliar context (Stevenson 2006). In order to protect Aboriginal values adequately, the traditional knowledge which informs such values—although it certainly can be shared in appropriate contexts—needs to remain under the control of its original holders. This means that adequate protection of Aboriginal values will require the protection not only of knowledge, but of Aboriginal peoples' rights as well. The moral and legal issues around what are now commonly referred to as 'intellectual property rights' are beyond the scope of this paper, but are the subject of increasing international attention (*see* e.g., WIPO 2000). Working effectively with and protecting Aboriginal peoples and their rights will require substantial improvements in the understanding of Aboriginal peoples and their issues, as discussed in the final theme.

5. Lack of Understanding of Aboriginal Peoples.
When considered from an Aboriginal perspective, the flaws in Aboriginal Values Mapping projects become readily apparent. The seriousness of these flaws and the ease with which they can be identified begs the question of how or why the process was established in its current form in the first place. A deep-seated chronic misunderstanding of Aboriginal peoples, and their needs, rights and aspirations by non-Aboriginal peoples, goes a long way to explaining this situation. This certainly is not the first time in this country that non-Aboriginal endeavours have backfired as a result of a lack of understanding of Aboriginal people; witness the Oka crisis (Dickason 1997; RCAP 1996a) and the Ipperwash tragedy (Linden 2007), for example. Susan Berneshawi (1997:121) observes that, "...there is a general lack of understanding of how dispossession of land, either through dismissed land claims or land transfers for development and/or resource harvest, has a substantial impact on First Nation communities." A comprehensive historical description of Aboriginal/non-Aboriginal relations in Canada, including colonization and its devastating impacts on Aboriginal peoples, cannot be presented here, but there are already numerous texts on this subject (e.g., Berger 1991; Furniss 1999; Getty and Lussier 1983; Miller 1991; Richardson 1993; Wright 1992). The bottom line is that there has been a long history of misunderstandings and sustained attempts, both conscious and unconscious, to colonize and assimilate Aboriginal peoples by the dominant society in this country. Despite this, there is a movement to try and learn from Aboriginal peoples about the unique knowledge they have developed and managed to maintain throughout. It becomes, then, crucial and more productive to support Aboriginal peoples in their recovery and their attempts to realize their rightful place in Canada. In light of this, the Aboriginal Values Mapping project takes on new importance; it can either be a help or a hindrance to Aboriginal peoples. The next section highlights possibilities for the emerging relationship that Aboriginal peoples are attempting to forge with other Canadians in the context of confederation.

Core Variables
With few exceptions, the research data indicate a lack of success of the Aboriginal Values Mapping project in carrying out those tasks for which it was supposedly intended. This situation has not changed in Ontario since this process formally began in 1996, almost a generation ago. Although the new FMP process and improved technical guidelines for incorporating Aboriginal values promise numerous potential benefits to the process, there are just as many reasons as to why they are currently not providing such benefits in the eyes of those who have participated thus far (Larose 2009; Sapic *et al.* 2009). In particular, Aboriginal respondents have indicated that they do not feel that the

Aboriginal Values Mapping process has been successful in protecting or advancing their interests. Interestingly, industry and MNR representatives have also indicated concerns around certain aspects of the program. The identification of Core Variables in this research serves to provide key insights as to why Aboriginal Values Mapping has not yet met the expectations of the participants, and offers suggestions as to how to improve this situation. Research indicates that there are two Core Variables that influence the meaningful participation of Aboriginal peoples in Ontario's Aboriginal Values Mapping process:

1. World View, Spirituality and Aboriginal Values.
An underlying theme flooding the literature and confirmed by this research is the fact that Aboriginal and non-Aboriginal peoples simply have different world views (Houde 2007; Larose 2009). The existence of different world views need not translate into conflicts in resource management if both world views are accepted and accommodated. Unfortunately, most, if not all, state environmental and resource management regimes in Canada do not recognize Aboriginal world views and the knowledge it produces as valid. This knowledge is ignored, de-valued and sometimes deliberately excluded (Stevenson 2005, 2006). Past and present national and provincial forest management policy regimes reflect this attitude. As such, it can be expected that, in Ontario's FMP process and the Aboriginal Values Mapping exercise it requires, world view differences will continue to run as a conflicting undercurrent.

2. Relationships and Power.
A theme which has appeared throughout the research and which is also found in the relevant literature is the power imbalance that underlies the relationship between Aboriginal and non-Aboriginal peoples in Canada. In fact, domination and assertion of power over Aboriginal peoples have been defining characteristics of historical and contemporary Aboriginal/non-Aboriginal relations in Canada (RCAP 1996a). This relationship has a long history and is deeply rooted in the later manifestation of colonialism in this country.[2] It is a revisionist history which has oppressed Aboriginal peoples and has caused tremendous disruption in all spheres of Aboriginal life. Such inequitable power distribution manifests itself in Ontario's FMP process and is thus inherent in the Aboriginal Values Mapping exercise. Four key components contribute to this Core Variable:

i. Lack of Knowledge About Aboriginal Peoples.
There is a lack of understanding among resource management professionals and policy makers as to the historical dispossession of traditional lands from Aboriginal peoples, even though the agencies these professionals work for have frequently been key players in this process. The impacts of this dispossession

---

[2] See J.R. Saul's (2008), *A Fair Country: Telling Truths About Canada*, for a popularized account of how Canada, during the late colonial period, came to deny its Aboriginal origins and the role that Aboriginal peoples played in the formation of Canadian culture, identify and economy.

are also poorly understood, as demonstrated in prevalent attitudes toward Aboriginal land use, management and governance.

ii. Exclusion of Aboriginal Peoples from the Dominant Resource Management Paradigm.
Exclusion of Aboriginal views from the dominant discourse on environmental and resource management represents a violation of Aboriginal and treaty rights (Smith 2007). Although there are exceptions, the policy and legislative frameworks which govern Canada's forest industry continue to alienate and exclude Aboriginal peoples from forest management (Wilson and Graham 2005). This involves restricting access to forest resources (e.g., harvesting timber) and limiting participation in decision-making such that Aboriginal cultural and traditional uses and values continue to be unaccounted for (NAFA 1993). There has been, of course, considerable conflict over forest resources between Aboriginal and non-Aboriginal societies as a result (Smith 2007). Aboriginal assertions of rights and court decisions in their favour have recently led to a somewhat more favourable climate for Aboriginal involvement in decisions impacting their lands (Wyatt 2008). Despite these small inroads into the current system, the state of Aboriginal forestry in Canada is unfortunately still characterized by exclusion and the centralization of power.

iii. Forced Conformity to Dominant Modes of Western Resource Management.
State systems of resource management are not neutral or objective; they are products of the world view, society and culture that produced them (Ellis 2005, Houde 2007; McGregor 2008; Stevenson 2006). Chapeskie (1995:12) states that:

> ...state institutions...assume that the Euro-Canadian technical paradigm of resource management possesses a superior intrinsic rationality and predictive capacity. Such power is assumed to endow this paradigm with a universal applicability that should transcend cultural boundaries.

Aboriginal peoples are therein forced to conform to the dominant resource management paradigm in order to be 'involved' or to have a say in what happens on their lands. Dominant systems are imposed and in some cases Aboriginal peoples have been forced to accept them to gain access to lands and resources (Stevenson 1999). Situations like this are symptomatic of inequitable power relations. Chapeskie (1995:18) adds that, "Anishinaabe people find themselves in a position of having to accept that this discourse inevitably governs discussions concerning land use issues." This results in further suppression of Aboriginal/land relationships and any attempts to maintain them. Further, the forced adoption of environmental resource management concepts and procedures excludes Aboriginal concepts of land stewardship and tenure, which are rooted in their languages, from any meaningful role in the discourse (Stevenson 2006).

iv. TEK in Resource Management: Finding Adequate Expression.

Given issues of 'power, dominance and resistance' (Feit 1998:2) within which the use of Aboriginal knowledge in resource management tends to occur, Stevenson (1999:163) asks:

> ...why Aboriginal and First Nations people would want to....expose their knowledge, much of which is considered proprietary if not sacred, to decontextualization and misappropriation by outsiders....(and/or) participate in processes that systematically "hamstring" their contributions, abilities and rights to manage human activities and impacts on their lands?

Stevenson (2006) adds that where Aboriginal world views are dominated by Western paradigms, it follows that adequately expressing Aboriginal concepts such as those involved in TEK become extremely difficult. While Stevenson's (2006) evaluations of TEK use in resource management showed limited progress despite the interest and burgeoning research in this area, these disappointing results are in no way indicative of the importance of TEK and the potential contribution of Aboriginal peoples (Cheveau *et al.* 2008; Ellis 2005). Rather, fully appreciating and utilizing Aboriginal knowledge must occur in the context of positive, equal and healthy relationships. "Traditional Ecological knowledge is absolutely essential. Crafting a *relationship* between us is absolutely essential" (LaDuke 1997:36). From an Aboriginal perspective, positive relationships hold part of the key for moving toward sustainability and the fair and equitable use of TEK in environmental and resource management.

A new relationship between Aboriginal and non-Aboriginal parties involved in resource management is urgently needed. While attempts at various forms of power-sharing have been tried in the past ('co-management' being a primary example), "...almost all arrangements envisage provincial, territorial or federal governments having the final say on matters of central concern" (RCAP 1996b:666). A new approach, therefore—one in which final decisions are negotiated rather than enforced—continues to be sought (*see* NAFA 2007 for a discussion of this topic). Aboriginal perspectives on this issue arise from the fact that since the time of contact Aboriginal peoples have regarded their relationship with newcomers as one of nations interacting. The call for a nation-to-nation relationship between Aboriginal peoples and Canadians is thus not new or unknown among federal and provincial governments. Indeed, it was the foundation of the Royal Proclamation (1763) and the modern doctrine of the recognition and reconciliation of Aboriginal rights. In the years following first contact, fragile relations of peace, friendship and rough equality were given the force of law in treaties (*see* RCAP 1996a:5-18).

Forging a path forward into a more sustainable and mutually beneficial future will require the development of a new type of relationship between Aboriginal and non-Aboriginal peoples. In 2007, the Government of Ontario created the stand alone Ministry of Aboriginal Affairs in an effort to build stronger relationships with Aboriginal peoples (MAA 2009). This approach advocates for "charting a new course for constructive, cooperative relationships with Aboriginal people in Ontario" (MAA 2009:1). It remains to be seen whether the work of this new Ministry will yield the type of relationships desired by Aboriginal peoples in Ontario.

The creation of the Ministry of Aboriginal Affairs was a recommendation that emerged from the Ipperwash Inquiry (Linden 2007) which, like the Royal Commission on Aboriginal Peoples, stemmed from a violent conflict with tragic consequences. Like the Royal Commission on Aboriginal Peoples, the Ipperwash Inquiry called for new relationships with Aboriginal peoples. The next section describes in more detail what this new 'relationship' might look like.

Nation to Nation Relationships: The Co-existence Approach

In 1996, the Royal Commission hinted at a new type of relationship, referred to as 'co-existence.' RCAP (1996b:428) stated unequivocally that the current relationship:

> ...has not worked and cannot work. The Aboriginal principles of sharing and co-existence offer us the chance for a fresh start. Canadians have an opportunity to address the land question in the spirit of these principles.

Co-existence is a concept that has been passed on through oral tradition and is symbolically represented in the Two Row Wampum (treaty belt) of the Haudenosaunee and other Aboriginal nations who subscribe to similar ideas (Borrows 1997). It can be viewed as a way to have two world views and knowledge systems interact in an equitable fashion. This notion is summarized by the Royal Commission on Aboriginal Peoples (RCAP 1993:45):

> The widespread concerns for authentic Aboriginal voice, for authentic representation of Aboriginal experience and history, are continuing legacies of the colonial past. They underline the power relationship between Aboriginal and non-Aboriginal people in Canada. A related concept recurring throughout is the necessity of parallel development, perhaps best captured symbolically in the Two Row Wampum belt. Hamelin advocates a process of intercultural convergence and cohabitation:
>
> There is symbolism in the train that enhances its value-added by using two rails that are independent yet associated for the task. Writers will think of independent canoes moving along the same body of water without colliding. Still others will envision a dog sled team on the tundra, each animal using its own track to jointly pull the sled. These metaphors imply that the mutual regime would include both independent and communal traits.

In the co-existence approach, neither Aboriginal nor western systems of knowledge are considered subordinate or dominant, but run parallel, each with their own forms of validity and reliability, their own standards for legitimizing knowledge. The concept suggests that, "Together, side by side, we go down the river of life in peace and friendship and mutual co-existence" (Lyons 1988:20). Hill (1990:27) describes it as an "...image of the two water vessels, each containing laws and beliefs of the two distinct peoples, [which] creates a visual symbol of separate nations equal in rights, traveling in the same direction, but not crossing each other's paths." John Mohawk, Haudenosaunee scholar, observes that, "The wampum represents the peoples' best thinking put into belts....it was a symbol of a people's successful accomplishment of coming to one mind about how they were going to go on...in a permanent relationship of peace and tranquility between the two sides" (Mohawk 1988:xiv). The Two Row Wampum is thus a metaphor for the relationship historically desired by Aboriginal peoples with non-Aboriginal peoples. Hope for the establishment of this relationship continues to the present day (Borrows 1997).

The concept of co-existence explains how First Nations and non-Aboriginal peoples can peacefully interact, how two distinct peoples can share lands, resolve differences and exchange knowledge. This idea has relevance and application in many areas. One of these areas is resource management (Brubacher and McGregor 1998; Chapeskie 1995; Ransom and Ettenger 2001; Stevenson 2006). As Stevenson (2005:7) suggests, "The effective contributions of Aboriginal peoples, and their knowledge and management systems...will not be realized until environmental policy-makers and resource managers consider them equally with those of scientific researchers...in forest decision-making."

Co-existence is a potentially promising bridge between sustainable forest management as conceptualized by mainstream Canada and as held by Aboriginal peoples (TK). Brubacher and McGregor (1998:18-19) anticipate that the co-existence approach can serve as a starting point for renegotiating an old relationship in a contemporary context:

> ...a co-existence approach would promote a focus on formally acknowledging Aboriginal people as legitimate partners in resource management. It would ensure their rightful place in the development and implementation of management policies and decision making....By drawing upon principles which express the values and perspectives of both Aboriginal and non-Aboriginal cultures, there is potential for developing an effective co-existence model, one that bridges distinctions by building upon shared values.

The co-existence approach does not devalue western or Aboriginal resource management practices or the knowledge that informs them. It does not allow for the domination of one over the other. Both systems are valued, and most importantly for Aboriginal peoples, their cultural survival is assured. Aboriginal world views and all they have to offer will no longer be threatened, dominated or distorted. Such a relationship requires the important work of 'decolonizing' policies and institutions in an effort to move from mere Aboriginal participation in forestry to 'true' partnerships with Aboriginal peoples (McGregor 2004). The re-ordering of environmental and resource management regimes is required as an important decolonization strategy

(Natcher and Davis 2007; Stevenson 2006; White 2006; Wyatt 2008). Due to the long history of colonization, it becomes critically important to support the emergence and establishment of Aboriginal environmental and resource management governance (including institutions) based on the worldviews, traditions, values and knowledge of Aboriginal peoples themselves.

Summary and Conclusions

The aim of this research has been to determine, at least in qualitative terms, the degree to which MNR's Aboriginal Values Mapping project has been successful in incorporating Aboriginal knowledge and values into forest management planning, as experienced by those who have participated in the process. The interview responses obtained revealed that participants, by and large, were not satisfied with the process. Two Core Variables were distilled from the interview data which explain the primary reasons for this lack of success to date. In essence, these two variables express the conclusion that the lack of success in the Aboriginal Values Mapping program is due to two primary factors: differences in the fundamental world views and value systems of Aboriginal and non-Aboriginal peoples, and the fact that one party (western society) is in a position to dominate the other. It seems apparent that it will not be possible to resolve the former variable without first addressing the latter: as long as the non-Aboriginal parties in any negotiation continue to hold most or all of the real power, little progress toward meaningfully addressing Aboriginal values will be made. The main recommendation being made here, then, is that Aboriginal and non-Aboriginal parties involved in forest management planning in Ontario begin—and some have already begun, but much remains to be done—to develop and maintain long-term, mutually beneficial relationships.

This finding is similar to those arrived at in other land use and resource management planning initiatives in Canada and internationally (Bryan 2009; Johnson *et al.* 2006; Peace and Louis 2008). In 2008, the Canadian Institute of Planners released its *Celebrating Best Practices of Indigenous Planning* issue, which spoke to the benefits of grounding planning in the traditional culture and values of Aboriginal peoples. Certainly the potential for mapping as part of larger planning initiatives for addressing Aboriginal goals and aspirations exists, and Aboriginal peoples continue to explore such options (Berris *et al.* 2008; Ray and Harper 2008). The fundamental challenges presented by inequitable power relationships and differences in world view, values and spirituality will continue to play themselves out in broader land use planning exercises and are not unique to forest management planning. Relationship building should form an important part of any planning initiative.

In forest management as well as other areas of planning, these relationships should be based on the idea that there are mutual goals to be achieved in a broad range of areas. In this and other research (e.g., McGregor 2008; Ransom and Ettenger 2001; Sable *et al.* 2006), it has been found that to date those undertakings which have been *most*—but not entirely, at least as yet—successful have tended toward the following characteristics:

1. among non-Aboriginal participants, a high degree of awareness of Aboriginal issues and significant experience working with Aboriginal peoples;
2. a willingness among all those involved to accept both western and Aboriginal ideas and knowledge as valid;
3. a high degree of Aboriginal involvement throughout *all* stages of the initiative;
4. an established level of regular communication between Aboriginal and non-Aboriginal parties, even before the commencement of a particular undertaking;
5. a high degree of existing Aboriginal capacity in the area(s) of undertaking,;
6. a strong, stable commitment to supporting further Aboriginal capacity development in these areas; and,
7. a strong, sincere commitment on the behalf of all parties to work respectfully together to achieve agreed-upon goals.

With these in mind as basic principles, Aboriginal and non-Aboriginal parties involved in forest management and planning can work toward achieving relationships based on the idea of co-existence. Under the co-existence model, both Aboriginal and western world views and knowledge would be respected and encouraged to flourish, while at the same time the two would work together to achieve mutual goals such as the sustainable management of forests in Ontario. It is hoped that discussion fuelled by ongoing research and experimentation in this area will facilitate the achievement of co-existence relationships throughout this province and across Canada.

References Cited

Berger, T. (1991). *A Long and Terrible Shadow: White Values, Native Rights in the Americas 1492-1992.* Vancouver, BC: Douglas & McIntyre Ltd, 183p.

Berkes, F. (2008). *Sacred Ecology,* 2nd ed. Philadelphia: PA: Taylor and Francis, 336p.

Berneshawi, S. (1997). Resource management and the Mi'kmaq Nation. *Canadian Journal of Native Studies* 1: 115-148.

Berries, C., C. Higgins, and M. Retasket (2008). 'Ts'enwecw Te TmiCW: Our Sacred Land,' pp. 46-48 in *Plan Canada.* Canadian Institute of Planners. Summer 2008.

Bombay, H. (1995). *An Assessment of the Potential for Aboriginal Business Development in the Ontario Forest Sector.* Ottawa, ON: National Aboriginal Forestry Association, 67p.

Borrows. J. (2005). Crown *and Aboriginal Occupations of Land: A History & Comparison.* Research Paper prepared for the Ipperwash Inquiry. The Honourable Sidney B. Linden. Commissioner. Toronto, ON: Government of Ontario, 85p. Available online http://www.attorneygeneral.jus.gov.on.ca/inquiries/ipperwash/policy_part/index.html [Accessed April 2009]

Borrows, J. (1997). 'Wampum at Niagara: The Royal Proclamation, Canadian legal history, and self-government,' pp. 155-172 in M. Asch, ed., *Aboriginal and Treaty Rights in Canada: Essays on Law, Equality, and Respect for Difference.* Vancouver, BC: UBC Press.

Brubacher, D and D. McGregor (1998). *Aboriginal Forest-related Traditional Ecological Knowledge in Canada*. Contribution for the 19th session of the North American forest commission, Villahermosa, Mexico, November 16-20, 1998. Ottawa, ON: National Aboriginal Forestry Association, for the Canadian Forest Service, 21p.

Bryan, J. (2009). Where would be without them? Knowledge, space and power in Indigenous politics. *Futures* 41: 24-32.

Canadian Institute of Planning (2008). 'Celebrating best practices of Indigenous planning,' in *Plan Canada*. Canadian Institute of Planners. Summer 2008.

Chapeskie. A. (1995). *Land, Landscape, Culturescape: Aboriginal Relationships to Land and the Co-management of Natural Resources*. A report for the Royal Commission on Aboriginal Peoples. Ottawa, ON: Land, Resource and Environmental Regimes Project.

Cheveau, M., L. Imbeau, P. Drapeau, and L. Belanger. (2008). Current Status and future directions of traditional ecological knowledge in forest management: a review. *The Forestry Chronicle* 84(2): 231-243.

Davidson-Hunt, I. and M. O'Flaherty (2007). Researchers, Indigenous Peoples, and Place-Based Learning Communities. *Society and Natural Resources* 20: 291-305

Dickason, O. (1997). 'Toward a larger view of Canada's History: The Native factor,' pp. 7-19 in D. Long and O. Dickason, eds., *Visions of the Heart: Canadian Aboriginal Issues*. Toronto, ON: Harcourt Brace & Company.

Drengson, A. and D. Taylor, eds. (1997). Ecoforestry: *The Art and Science of Sustainable Forest Use*. Gabriola Island, BC: New Society Publishers, 312 pp.

EAB—Environmental Assessment Board (1994). *Reasons for Decision and Decision: Class Environmental Assessment by the Ministry of Natural Resources for Timber Management on Crown Lands in Ontario*. Toronto, ON: Queen's Printer for Ontario, 561p.

Ellis, S. (2005). Meaningful Consideration? A Review of Traditional Knowledge in Environmental Decision Making. *Arctic* 58(1): 66-77.

Erasmus, G. (1989). 'A Native Viewpoint,' pp. 92-98 in M. Hummel, ed., *Endangered Spaces: The future for Canada's Wilderness*. Toronto, ON: Key Porter Books.

Feit, H. (1998). 'Reflections on local knowledge and wildlife resource management: Differences, dominance and decentralization,' pp. 123-148 in L. Dorais, L. Muller-Wille, and M. Nagy, eds., *Aboriginal Environmental Knowledge in the North: Definitions and Dimensions*. Quebec, PQ: Groupe d'Etudes Inuit et Circumpolaires, Universite Laval.

Furniss, E. (1999). *The Burden of History: Colonialism and the Frontier Myth in a Rural Canadian Community*. Vancouver, BC: UBC Press, 237p.

Getty, I. and A. Lussier (1983). *As Long as the Sun Shines and Water Flows: A Reader in Canadian Native Studies*. Vancouver, BC: UBC Press, 303p.

Glaser, B., ed. (1994). *More Grounded Theory Methodology: A Reader*. Mill Valley, CA: Sociology Press, 388p.

Government of Canada (2008). *Aboriginal Consultation and Accommodation: Interim Guidelines for Federal Officials to Fulfill Legal Duty to Consult*. Ottawa: Indian and Northern Affairs Canada, 58p. www.anic-inac.gc.ca

Hammond, H. (2008). *Maintaining Whole Systems on Earth's Crown: Ecosystem-based Conservation Planning for the Boreal Forest*. Winlaw, BC: Silva Forest Foundation, 389p.

Hickey, C., and M. Nelson (2005). *Partnerships Between First Nations and the Forest Sector: A National Survey*. Edmonton, Alberta: Sustainable Forest Management Network, 30p.

Hill, R. (1990). Oral memory of the Haudenosaunee: Views of the Two Row Wampum. *Northeast Indian Quarterly* 11(1): 21-30.

Houde, N. (2007). The Six Faces of Traditional Ecological Knowledge: Challenges and Opportunities for Canadian Co-Management Arrangements. *Ecology and Society* 12(2): 34.

Johnson, R., R. Louis, and A. Pramano (2006). Facing the Future: Encouraging Critical Literacies in Indigenous Communities. *ACME. An International E-Journal for Critical Geographie*s. P. 81-98 www.Acme-journal.org/Volume 4-1.htm

LaDuke, W. (1997). 'Voices from white Earth: Gaa-waabaabiganikaag,' pp. 22-37 in H. Hannum, ed., *People, Land and Community. Collected E.F. Schumacher Society Lectures*. New Haven, CT: Yale University Press.

LaRose, D. (2009). *Adapting to Change and the Perceptions and Knowledge in the Involvement of Aboriginal Peoples in Forest Management: A Case Study with Lac Seul First Nation*. Masters of Science in Forestry. Unpublished manuscript. Faculty of Forestry. Thunder Bay, ON.: Lakehead University, 179p.

Linden, S. (2007). *Report of the Ipperwash Inquiry: Volume 4: Executive Summary*. Toronto, ON: Office of the Attorney General. Queen's Printer Press for Ontario. 115 pp. http://www.attorneygeneral.jus.gov.on.ca/inquiries/ipperwash/index.html

Lyons, O. (1988).' Land of the free/Home of the brave,' pp. 18-20 in J. Barreiro, ed., *Indian Roots of American Democracy: Cultural Encounter* 1. Special Constitution Bicentennial Edition of the Northeast Indian Quarterly, covering Vol. 4, No. 4, and Vol. 5, No. 1. Ithaca, NY: Cornell University.

MAA—Ministry of Aboriginal Affairs (2009). *Strengthening Relationships*. Government of Ontario. Toronto, ON., 9pp. Accessed at www.aboriginalaffairs.gov.on.ca (October 23, 2009).

McGregor, D.(2008). Linking Traditional Ecological Knowledge and Western Science: Aboriginal Perspectives on SOLEC. *Canadian Journal of Native Studies*. 28(1): 139-158.

McGregor, D. (2006). Aboriginal Involvement in Ontario Sustainable Forest Management: Moving Toward Collaboration. *Recherches Amerindiennes au Quebec* 36(2/3): 61-70.

McGregor, D. (2004). Coming Full Circle: Indigenous Knowledge, Environment and Our Future. *American Indian Quarterly* 28(3/4): 385-410.

McGregor, D. (2000). *From Exclusion to Co-existence: Aboriginal Participation in Ontario's Forest Management Planning*. Ph.D. dissertation. Faculty of Forestry. Toronto, On: University of Toronto, 254p.

McKibbon, G. (1999). *Term and Condition 77: Origins, Implementation and Future Prospects*. Prepared for Economic Renewal Workshop "Partnership trends: Aboriginal business in the forest sector," December 8-9,1999. Hamilton, ON: McKibbon Wakefield Inc., 18p.

Menzies, C. (2006). *Traditional Ecological Knowledge and Natural Resource Management*. Lincoln, NE: University of Nebraska Press.

Merkel, G. (2007). We Are all Connected: Globalization and Community Sustainability in the Boreal Forest, an Aboriginal Perspective. *Forestry Chronicle* 83(3): 362-366.

Miller, J., ed. (1991). *Sweet Promises: A Reader on Indian–White Relations in Canada*. Toronto, ON: University of Toronto Press, 468p.

ME—Ministry of the Environment (2003). *Declaration Order regarding MNR's Class Environmental Assessment Approval for Forest Management on Crown Lands in Ontario*. Available at www.ene.gov.on.ca/envision/env_reg/er/documents. (October 15, 2009).

MNR—Ministry of Natural Resources (2007). *Forest Management Guide for Cultural Values*. Toronto, ON: Queen's Printer Press for Ontario, 75p.

MNR—Ministry of Natural Resources (2004). *Forest Management Planning Manual for Ontario's Crown Forests*. Toronto, ON: Queen's Printer for Ontario, 440p.

MNR—Ministry of Natural Resources (1998). *Ontario's Forests: Management for Today and Tomorrow*. Peterborough, ON: Ministry of Natural Resources, 16p.

MNR—Ministry of Natural Resources (1996). *Forest Management Planning Manual for Ontario's Crown Forests*. Queen's Printer for Ontario, Toronto, ON, 452p.

MNR—Ministry of Natural Resources (1991). *Timber Management Guidelines for the Protection of Cultural Heritage Resources*. Toronto, ON: Queen's Printer for Ontario, 16p.

MOE—Ministry of the Environment (2003). *Declaration Order regarding MNR's Class Environmental Assessment Approval for Forest Management on Crown Lands in Ontario*. Environmental Assessment Board Toronto, ON: Queen's Printer for Ontario. http://ontariosforests.mnr.gov.on.ca

Mohawk, J. (1988). 'A symbol more powerful than paper,' pp. xiv-xvii in J. Barreiro, ed., *Indian Roots of American Democracy: Cultural Encounter* 1. Special constitution bicentennial edition of the Northeast Indian Quarterly, covering Vol. 4, No. 4, and Vol. 5, No. 1. Ithaca, NY: Cornell University.

NAFA—National Aboriginal Forestry Association (2007). *Looking Back, Looking Forward: RCAP in Review: Assessing Progress on Land and Resources Recommendations Made by the Royal Commission on Aboriginal Peoples as they Pertain to First Nations*. Ottawa, ON: National Aboriginal Forestry Association, 68p. (www.nafa.org).

NAFA—National Aboriginal Forestry Association (1993). *Forest Lands and Resources for Aboriginal People: An Intervention Submitted to the Royal Commission on Aboriginal Peoples*. Ottawa, ON: National Aboriginal Forestry Association, 39p.

Natcher, D. and S. Davis (2007). Rethinking Devolution: Challenges for Aboriginal Resource Management in the Yukon Territory. *Society and Natural Resources* 20(3):271-279.

Naysmith, J. (1996). *The Ontario First Nation Forestry Program: Issues and Opportunities*. A report to the Ontario First Nation Forestry Program Management Committee, 28p.

O'Flaherty, R., I. Davidson-Hunt and M. Manseau (2008). Indigenous knowledge and values in planning for sustainable forestry: Pikangikum First Nation and the Whitefeather Forest Initiative. *Ecology and Society* 13(1): 6-16.

Pearce, M. and R. Louis (2008). Mapping Indigenous depth of place. *American Indian Culture and Research Journal* 23(3): 107-126.

Poole, P. (1995a). Geomatics: Who needs it? *Cultural Survival Quarterly* 18(4): 1.

Poole, P. (1995b). Land-based communities, geomatics and biodiversity conservation: A survey of current activities. *Cultural Survival Quarterly* 18(4): 74-76.

Ransom, R. and K. Ettenger (2001). Polishing the Kaswentha: A Haudenosaunee View of Environmental Cooperation. *Environmental Science and Policy* 4:219-228.

Ray, R. and D. Harper (2008). 'Ts'enwecw Te TmiCW: Our sacred land,' pp. 43-45 in *Plan Canada*. Canadian Institute of Planners. Summer 2008.

Rayner, J. and M. Howlett (2007). The National Forest Strategy in Comparative Perspective. *Forestry Chronicle* 83(5): 651-657.

Richardson, B. (1993). *People of Terra Nullius: Betrayal and Rebirth in Aboriginal Canada*. Vancouver, BC: Douglas and McIntyre, 393p.

Robinson, M. and M. Ross (1997). Traditional land use and occupancy studies and their impact on forest planning and management in Alberta. *Forestry Chronicle* 73(5): 596-605.

RCAP—Royal Commission on Aboriginal Peoples (1996a). *People to People, Nation to Nation: Highlights from the Report of the Royal Commission on Aboriginal Peoples*. Ottawa, ON: Minister of Supply and Services, 149p.

RCAP—Royal Commission on Aboriginal Peoples (1996b). Lands and Resources,' pp. 421-685 in *Report of the Royal Commission on Aboriginal People, Volume 2:*

Restructuring the Relationship. Ottawa, ON: Canada Communication Group Publishing.

RCAP—Royal Commission on Aboriginal People (1993). *Integrated Research Plan*. Ottawa, ON: Minister of Supply and Services Canada, 64 pp.

Rude, D. and C. Deiter (2004). *From the Fur Trade to Free Trade: Forestry and First Nations Women in Canada. Status of Women in Canada*. Ottawa, ON.: Policy Research, 51p.

Sable, T., G. Howell, D. Wilson, and P. Penashue (2006). 'The Ashkui Project: Linking Western Science and Innu Environmental Knowledge in Creating a Sustainable Environment,' pp. 109-127 in P. Sillitoe, ed., *Local Science vs Global Science: Approaches to Indigenous Knowledge in International Development*. New York, NY : Berghahn Books.

Sapic, T., U. Runesson, and P. Smith (2009). Views of Aboriginal people in northern Ontario on Ontario's approach to Aboriginal values in forest management planning. *The Forestry Chronicle* 95(5): 789-801.

Saul, J. R. (2009). *A Fair Country: Telling the Truths About Canada*. Penguin Books.

Smith, P. (2007). *Creating a New Stage for sustainable forest management through co-management with Aboriginal People in Ontario*. Ph.D thesis. Faculty of Forestry. University of Toronto. Toronto, ON. 358p.

Stevenson, M. (2006). The possibility of difference: Rethinking co-management. *Human Organization* 65(2): 167-180.

Stevenson, M. (2005). *Traditional Knowledge and Sustainable Forest Management*. Edmonton, AB: Sustainable Forest Management Network, 16 pp.

Stevenson, M. (1999). 'What are we managing? Traditional systems of management and knowledge in cooperative and joint management,' pp. 161-169 in T. Veeman, D. Smith, B. Purdy, F. Salkie, and G. Larkin, eds., *Science and Practice: Sustaining the Boreal Forest*. Proceedings of the 1999 Sustainable Forest Management Network conference, Edmonton, AB: Sustainable Forest Management Network.

Strauss, A. (1987). *Qualitative Analysis for Social Scientists*. New York, NY: Cambridge University Press, 319p.

Telliet, J. (2005) *The Role of the Natural Resources Regulatory Regime in Aboriginal Rights Disputes in Ontario*. Office of the Attorney General, Government of Ontario. Prepared for the Ipperwash Inquiry, 78p. http://www.attorneygeneral.jus.gov.on.ca/inquiries/ipperwash/policy_part/research/index.htm

Tobias, T. (2000). *Chief Kerry's Moose: A Guidebook to Land Use and Occupancy Mapping, Research Design and Data Collection*. Union of BC Indian Chiefs & EcoTrust Canada. Vancouver BC.

UNDSD—United Nations Division for Sustainable Development (2004). *Agenda 21*. http://www.un.org/esa/sustdev/documents/agenda21/index.htm.

UNGA—United Nations General Assembly (2007). *United Nations Declaration on the Rights of Indigenous Peoples*. http://www.dd-rd.ca/site/_PDF/un/A_61_L67eng.pdf (accessed June 28, 2008

Wilson, J. and j. Graham (2005). *Relationships Between First Nations and Forest Industry: The Legal and Policy Context*. A report for the National Aboriginal Forestry Association, Forest Products Association of Canada, First Nations Forestry Program. Ottawa, ON: Institute on Governance, 87p.

White, G. (2006). Cultures in collision: Traditional knowledge and Euro-Canadian governance processes in northern land-claim boards. *Arctic* 59(4): 401-414.

WCED—World Commission on Environment and Development (1987). *Our Common Future*. Oxford, UK: Oxford University Press, 454p.

WIPO—World Intellectual Property Organization (2000). *Roundtable on Intellectual Property and Traditional Knowledge.* www.wipo.int/eng/meetings/1999/folklore/index_rt.htm

Wright, R. (1992). *Stolen Continents: The 'New World;' through Indian Eyes.* Toronto, ON: Penguin Books, 424p.

Wyatt, S. (2008). First Nations, Forest Lands, and 'Aboriginal' Forestry in Canada; From Exclusion to Co-management and Beyond. *Canadian Journal of Forest Research* 38: 171-1.

Chapter Eleven

First Nations' Criteria and Indicators of Sustainable Forest Management: A Review

M.A. (Peggy) Smith, Erin Symington and Sarah Allen

Introduction

In 1992, at the United Nations (UN0 Conference on Environment and Development (UNCED), the UN adopted 'sustainable forest management' (SFM) as an evolving international standard. In the Statement of Forest Principles (UN 1992), nation-states agreed to work toward a consensus on the meaning of SFM. To do this, national governments committed to 'environmentally sound national guidelines,' taking into account 'relevant internationally agreed methodologies and criteria.' Since 1992, national governments, individually and in groups, have undertaken the development of criteria and indicators (C&I) incorporating social, environmental and economic elements of SFM. The United Nations Food and Agriculture Organization (FAO 2001) described C&I as follows:

> Criteria define the range of forest values to be addressed and the essential elements or principles of forest management against which the sustainability of forests may be assessed. Each criterion relates to a key element of sustainability, and may be described by one or more indicators. Indicators measure specific quantitative and qualitative attributes (and reflect forest values as seen by the interest group defining each criterion) and help monitor trends in the sustainability of forest management over time. Changes in the indicators between periods indicate whether a country is moving towards, or away from, sustainability. Criteria and indicators, at both national and forest management unit level, are tools for monitoring trends and effects of forest management interventions. The ultimate aim of these tools is to promote improved forest management practices over time, and to further the development of a healthier and more productive forest estate.

In Canada, the Canadian Council of Forest Ministers (CCFM), comprising both provincial and federal ministers, developed its first set of C&I in 1995, revising them in 2003. Canada also joined the *Montreal Process* to work with several other boreal and temperate forest countries to develop a common set of C&I. Other C&I processes took place in different parts of the world (FAO 2001; McDonald and Lane 2004). With the development of these national sets, C&I became part of the lexicon of forest management. C&I have since become the formal method and tool for determining sustainability and have contributed to the adoption by the UN in 2007 of a 'Non-Legally Binding Instrument on All Types of Forests' that has set a global standard for SFM (UN 2007).

The C&I sets developed by nation-states apply at the national level and are being used in the development of international standards, but the usefulness of C&I has also been recognized at the local level. Provincial forest managers across Canada, as members of the CCFM, ratified Canada's C&I and are applying them at both the regional and forest management unit level. Non-governmental organizations have also explored C&I at both international and local levels. Internationally, C&I sets have been developed by the Center for International Forestry Research (CIFOR), initiated in 1998 (FAO 2001). Certification systems have been created by the Forest Stewardship Council and, in Canada, by the Canadian Standards Association (Lapointe 1998). In 2000 the Canadian Model Forest Network published a *User's Guide to Local Level Indicators of Sustainable Forest Management* which summarized the Network's experience developing and applying local C&I based on the CCFM's Criteria and Indicators (CMFN 2000).

It is not surprising, therefore, that a number of First Nation communities across Canada have been exploring the development of local-level criteria and indicators. In addition to having specific rights to forested lands, Indigenous peoples bring unique institutional, social, cultural and economic values to the forest management process.[1] In Canada, despite the constitutional recognition of Aboriginal and treaty rights since 1982, Aboriginal peoples have historically been excluded from forest development and planning. As a result of not sharing in the economic benefits from resource development in their territories, many Indigenous communities are economically disadvantaged. At the same time, most of the 80% of Indigenous communities located within Canada's commercial forest zone are forest-dependent, relying on forests for subsistence, culture and spirituality (Smith 2005), and maintaining their stewardship of their traditional territories. Increasingly, Indigenous peoples in Canada are influencing forest management and planning and getting involved in forest development, both to secure economic benefits and to ensure that economic goals are balanced by other values, especially social, cultural and ecological.

If there are existing frameworks of C&I of SFM, then why would First Nation communities develop their own instead of using those sets which already

---

[1] All of the communities surveyed in this study are First Nation communities. 'First Nations' refer to those Indigenous communities recognized as 'Bands' and 'status Indians' governed under the *Indian Act* in a special trust relationship with the federal government. These First Nations have limited autonomy under the Act, but many are negotiating various forms of 'self-government' with the Government of Canada. We also refer to 'Indigenous peoples,' the commonly used term at the international level. 'Aboriginal peoples' is used when referring to Canada's Indigenous peoples with constitutionally recognized Aboriginal rights, and includes 'Indian, Inuit and Métis.'

exist, and what are the processes underlying their development? Are First Nations C&I different than other sets? How have First Nations C&I been applied? Is C&I an effective tool for First Nations to protect their values in forest development and land use planning? This chapter seeks to address these questions by examining the efforts of six First Nations, described in Appendix A, to develop and use C&I:

- the Nuu-chah-nulth Nation's Iisaak Forest Resources Ltd. (IFRL) (British Columbia),
- Tl'azt'en Nation (British Columbia),
- Little Red River Cree Nation (LRRCN) (Alberta),
- the Algonquins of Barriere Lake (ABL) (Quebec),
- Waswanipi Cree Nation (Quebec), and
- the Innu Nation (Labrador).

Why these communities made these efforts, how they did so, and how C&I sets are being used are examined. A framework for developing, understanding and analyzing Aboriginal C&I is then proposed. Although the focus of the chapter is on the application of Aboriginal C&I for SFM, the methods and findings described below are equally germane to Aboriginal engagement and C&I development in regional land use planning. After all, what is effective SFM in many Aboriginal contexts and First Nation communities if not land use planning that addresses all dimensions of sustainability within a forested landscape?

What Have We Learned about Criteria and Indicators?

The broad literature on C&I provides background on its origins at UNCED and subsequent development through the United Nations Commission on Sustainable Development. The C&I literature that is most helpful for understanding the development of Aboriginal C&I are those studies that point to the importance of a range of factors:

- the measurability of indicators,
- building a baseline on which trends can be measured,
- community participation,
- a full suite of values (including environmental, social, economic and cultural), and
- monitoring

All of these elements support the use of C&I for adaptive management in forest development and land use planning. There are also several lessons to be learned, specifically from Aboriginal organizations and communities that have been involved in C&I development.

The measurability and monitoring of C&I are essential contributors to the sustainability of resource projects. Duinker (2001) points to the importance of having indicators meet certain requirements in order to ensure they can be measured and produce accurate information. He defines measurability in two ways: first, they need to be able to move in a direction or show a trend that can be measured over time, and, second, indicators need to have quantifiable units by which they can be measured. Duinker (2001) discusses eight ways that indicators can be deemed useful, including being relevant, practical, predictable and valid. The quality of the indicators themselves can help create C&I frameworks that are concise, manageable and successful. However, some indicators are difficult to measure, particularly those in social and cultural criteria where measures may be qualitative rather than quantitative and where the outcome of community well-being is linked with ecosystem health (Bombay *et al.* 1995). Not only can qualitative measurements be more challenging, but for Indigenous communities, different ways of knowing such as traditional knowledge and different cultural contexts demand different kinds of measures. The development of qualitative indicators demands more than technical expertise, requiring community involvement (Daniel *et al.* 2009; Adam and Kneeshaw 2009; O'Flaherty *et al.* 2009; Saint-Arnaud *et al.* 2009; Natcher and Hickey 2002). In fact, as Sherry *et al.* (2005) point out: "critical local values are defined as the spectrum of values and priorities *community members associate with the forest*" (emphasis added).

Garcia and Lescuyer (2008) show differences in how resource managers and local communities weigh indicators that identify what values they consider most important. Resource managers tend to focus on the resource while communities tend to focus on potential social gains. Berninger *et al.* (2009) support this finding with their comparison of economic, environmental and social indicators across three different boreal regions. The authors found that foresters placed more value on economic benefits and that the Innu of Labrador, an Indigenous group in one of the regions studied, considered social issues of higher value than other interest groups. O'Flaherty *et al.* (2009) demonstrate the different understandings held by Pikangikum elders and forest managers around the concept of 'tools.' Pikangikum elders see tools as mechanistic, preferring to focus on relationships to the land, 'using all of yourself to work the land, to be with the land.'

Burford de Oliveira (1999) discusses the results of tests on participation and knowledge of forest management in the development of C&I for CIFOR. Community participation was examined in two ways: first, by the extent to which community members participated in the development of C&I and second, by the emphasis placed on participation in C&I development. The results showed that community members who participated were generally wealthier, had political influence and were more outgoing. Although not always articulated by the word 'participation,' all three tests included some language around the importance of community involvement in C&I, particularly in areas of policy and law-making, conflict management and monitoring. The study showed that participatory processes for developing C&I had the potential to improve communication and the flow of information between participants, increase a sense of ownership and pride in the community, avoid conflict, and improve processes for decision-making and monitoring of those decisions. Participatory C&I development can incorporate multiple points of view that inform a more holistic, detailed inclusion of values.

The CIFOR findings complement those of Natcher and Hickey (2002) who point out that Indigenous communities are not of a single mind, but 'socially heterogeneous.' As a result, participatory processes at the local level are just as much about managing differences within communities as managing differences between local communities and outside agencies.

Garcia and Lescuyer (2008) discuss monitoring indicators, demonstrating that in a number of cases, the monitoring of C&I fails when the financial backers are no longer present. The authors also point to the importance of developing local monitoring systems. Many local communities have their own traditional systems of monitoring (i.e., what is important for them to monitor, why and what to look for). Outlining how monitoring will be carried out at the outset of C&I development would help to ensure more successful use of C&I in SFM and other land use planning initiatives.

The literature on Aboriginal involvement in C&I frameworks began in Canada with the National Aboriginal Forestry Association (NAFA). In 1995, as a result of NAFA's participation in the CCFM C&I process, and in keeping with its goal to increase Aboriginal participation in forest management in Canada, NAFA published *An Aboriginal Criterion for Sustainable Forest Management* (Bombay *et al.* 1995). In that paper, NAFA argued that the CCFM should consider a stand-alone criterion on Aboriginal rights and participation in forest management. NAFA took a rights-based approach arguing that "forest management is not sustainable if it does not respect and provide for Aboriginal and treaty rights." It continues to promote this position despite the failure of the CCFM to accept NAFA's recommendation. The CCFM opted instead to subsume Aboriginal indicators under Criterion 6, 'Accepting Society's Responsibility.'

It could be that the CCFM's reluctance to treat Aboriginal rights and participation as a separate criterion has contributed to gaps in indicator frameworks specifically in areas of social and Aboriginal criteria development. Hickey and Innes (2005) and Gough *et al.* (2008) conducted indicator reviews of C&I of SFM used in British Columbia, gathering thousands of indicators from initiatives used globally and condensed them into a shortlist. The authors examined the shortlist of indicators to determine why they were used, what the indicators measured, methods for monitoring, who used the indicators, how they were reported and issues associated with indicators. Their results suggest the strengthening of C&I processes through an inclusion of more indicators in social capital and Aboriginal areas. The authors were surprised by the lack of indicators in these areas because a considerable body of literature supports societal benefits and Aboriginal participation as components of sustainability.

While NAFA was challenging the CCFM approach at a national level, some Aboriginal communities began to explore C&I at the local level. Researchers working with those communities and others began to explore the challenges of measuring forest management that addressed Aboriginal values. Most recently, Adam and Kneeshaw (2009) provided a synthesis on *Formulating Aboriginal Criteria and Indicator Frameworks* for the Sustainable Forest Management Network, based on a literature review and comparison of C&I frameworks developed by Kitcisakik, Waswanipi, Tl'azt'en and Little Red River Cree. In reviewing the studies of Aboriginal C&I to date, including comparisons with national and international frameworks (Sherry *et al.* 2005; Gladu and Watkinson 2005; Ettenger *et al.* 2002; Beckley 1999), Adam and Kneeshaw (2009) concluded that:

1) environmental C&I are the most compatible across all frameworks (Sherry *et al.* 2005);
2) there is a need to further develop Aboriginal C&I in relation to economic opportunities, diversity and cultural areas, including traditional land use patterns and youth issues, as well as process (Sherry *et al.* 2005; Ettenger *et al.* 2002; Beckley 1999); and
3) using 'bottom-up' processes increases the relevance of C&I for Aboriginal communities (Sherry *et al.* 2005).

Gladu and Watkinson (2005), in their comparison of C&I by Aboriginal communities across Canada (Waswanipi, Algonquins of Barriere Lake, Little Red River Cree Nation and Innu Nation) and internationally, found the following common themes for indicators:

*   respect for legal and customary rights,
*   protection of cultural, social and spiritual values,
*   conservation of Indigenous and local knowledge,
*   identification and protection of archaeological and cultural sites,
*   participation in forestry activities and fair distribution of benefits, and
*   opportunities for employment and training.

In addition to these comparative and synthesis studies, new approaches have been developed by Pikangikum First Nation for their Whitefeather Forest Initiative (Shearer *et al.* 2009) and by Kitcisakik First Nation (Saint-Arnaud *et al.* 2009). Both of these initiatives bring in the new element of relationship, reflective of the worldview of the cultures of these communities. Kitcisakik's framework is still based broadly on the sustainable development approach but includes a new principle—the 'ethical' principle. *Inakonigewîn* includes: 'respect' for Aboriginal rights; for Anicinapek values of sharing, respect, mutual aid, equity and responsibility; for indigenous beliefs and spiritual systems; and for Aboriginal land tenure. It also includes forest management that avoids waste and pollution, and embraces consultation and local decision-making control. Pikangikum developed the *Whitefeather Forest Cultural Landscape Monitoring Framework* to help the community "understand whether forest conditions are perceived to be moving toward a desirable or undesirable state" (Shearer *et al.* 2009). Their framework focuses on relationships, dealing with the cycle of changes that occur through the seasons of life and land. Their indicators, or 'signs,' emerge from a worldview of 'keeping the land' and range from 'healing mind, body and spirit' to 'respectful relationships with people,' 'strong leadership,' and 'land-based livelihoods.' While maintaining elements of other

C&I frameworks, these two communities have broken out of the standardized sustainable development model with its discrete categories.

What makes our comparative study of Aboriginal C&I unique is its expansion to six different Aboriginal communities across Canada, its focus on the context and drivers for the development of C&I by each community, and the use of a framework that is inclusive of a full range of Aboriginal values within a broad sustainable development approach. We believe that Aboriginal communities can benefit from the work of other Aboriginal communities, as much as they can learn through internal discussions and analysis of national and international C&I, and therefore think it important to analyze and share information across communities.

Methods

Six First Nation communities (Fig. 1), known to have developed local criteria and indicators, were chosen for comparison:

- the Nuu-chah-nulth Nation's Iisaak Forest Resources Ltd.,
- the Tl'azt'en Nation,
- the Little Red River Cree Nation,
- the Algonquins of Barriere Lake,
- the Waswanipi Cree Nation, and
- the Innu Nation.

C&I were gathered from both academic, peer-reviewed sources and grey literature for each of the six communities. An Excel spreadsheet was used to organize the data. Each community's C&I were sorted under the criterion headings of environmental, social, economic, cultural, rights and institutions. These categories are based on the commonly accepted pillars of SFM. The addition of culture, rights and institutions addresses the unique position of Aboriginal communities and is incorporated into a framework developed by Smith (2005) that will be discussed later. A more detailed description of the context of each First Nation when developing its own C&I framework can be found in Appendix A.

The next step was to group all of the six communities' indicators under the appropriate criteria, using colour-coding to identify each individual community's indicators. The rights and institutions criteria, as per the Smith framework, were considered separately from the environmental, social, economic and cultural criteria. Definitions for these criteria, based on commonly accepted definitions and the community indicators review, were determined to establish a guideline for re-grouping the indicators:

- *Environmental*: Maintain forest health and integrity to ensure sustainable use.
- *Economic*: Increase economic benefits from forests for local First Nations.

- *Social*: Improve equity, health and stability for individuals and the community.
- *Cultural*: Protect Aboriginal cultural values in forest management planning.

Figure 1. *Locations of selected First Nation communities with local C&I sets for forest management.*

An example of the re-grouping according to these clearer definitions was the indicator 'use of Aboriginal language.' As a community indicator, it was originally placed under the social criterion, but was moved to the cultural criterion because it deals specifically with an Aboriginal cultural value.

Within each criterion, headings were chosen to group similar indicators together. For example, in the environmental criterion, the heading 'natural forest emulation' was used to group all indicators mentioning fire, blow-down, silvicultural prescriptions and native species as important to emulate natural patterns. Duplicates across communities were combined or 'rolled-up' to produce the final condensed charts. If a number of different communities had differently-worded indicators for the same criterion, one indicator was used, but each community's wording was tagged in the spreadsheet to maintain a record of the original wording. For example, in the social criterion, under the heading 'fair distribution of opportunities in the forest sector,' three communities had six differently worded indicators concerned with sharing economic benefits from forest development.

Following this grouping exercise, representatives from each community who had been involved in the development of the community's C&I were interviewed about the process, including what had prompted the development of C&I, how the development was done, and how the community planned to implement, monitor and adapt their C&I.

Methodological Challenges

Several methodological issues, some which challenged our analyses have also been identified by other authors attempting C&I comparisons. Adam and Kneeshaw (2009) point to limitations because of different hierarchies and goals of the purpose for which indicators are developed may lead to some aspects of sustainability being left out, while some sets may be more developed than others. Saint-Arnaud *et al.* (2009) point out that each C&I set is particular to the context of the First Nation developing it, including cultural, environmental and socio-economic variability. The authors also point to the confounding roles of the frameworks and methodologies used by researchers. Generally, comparisons are difficult given different forests types, cultures and drivers, including certification standards, provincial regulations, internal, expert-driven processes vs. community-developed and the nature of agreements with the state.

For us, it was difficult to group some indicators because an indicator might deal with more than one criterion—for example, monitoring environmental conditions at culturally sensitive location. In such cases, it was an arbitrary decision to place an indicator under the environmental or cultural criterion, although our preference was that if there was an explicit cultural component, we placed it under the cultural criterion. Communities also used varying approaches, making it difficult to compare with others using different methodologies, although we did find that all communities used some form of a sustainable development approach, considering ecological, economic and social indicators with the addition of cultural, rights and institutional aspects. Natcher and Hickey (2002) reflect in their work with the Little Red River Cree Nation what First Nations want in C&I development:

> By eliminating largely irrelevant criteria and indicators developed at the national level, and extending our attention beyond provisions of sustained timber yield, we initiated research that would identify environmental, social, cultural and economic criteria and indicators that could then be used as a guide for local forest management. This initiative was therefore designed to facilitate a system of adaptive community-based management that is responsive to the values, expectations, and changing needs of community members.

Table 1. Drivers and context for Aboriginal C&I development for SFM in Canada.

	Nuu-chah-nulth (BC)	Tl'azt'en Nation (BC)	Little Red River Cree Nation (AB)	Algonquins of Barriere Lake (QC)	Waswanipi Cree (QC)	Innu Nation (Labrador)
Policy and/or arrangement with Crown	Clayoquot Sound Science Panel, Interim Measures Agreement, Comprehensive Planning Board, Tree Farm & other timber licences, FSC certification	Tree Farm Licence, co-management with Univ. of Northern BC of John Prince Research Forest with Special Use Permit from BC	Treaty 8, 1899, Co-operative Management Planning Agreement, 1999-2002, Alberta Forest Planning Manual	Trilateral Agreement with Quebec & Canada for Integrated Resource Management Planning	James Bay & Northern Quebec Agreement 1975, Canadian Model Forest Network, 1997, Quebec timber licences, *Paix des Braves*	Comprehensive land claim, Forest Process Agreement with Government of Newfoundland and Labrador, 2001, for District 19
Ecosystem type	Coastal temperate rainforest	Boreal	Boreal	Boreal/Great Lakes-St. Lawrence	Boreal	Boreal
C&I development method	Forest Stewardship Council BC standard	Led by researchers, developed with community	Community-based, research-driven	Expert-driven using outside consultants, reviewed by community	James Bay Advisory Committee on the Environment, Cree Trappers Assoc., Model Forest Network Local C&I program	Expert-driven based on ecosystem planning framework, reviewed by community
Key documents	Sustainable Forest Management Plan (2006-2011), Mabee & Hoberg (2004)	Wilkerson and Baruah (2000), Karjala et al. (2003) on Aboriginal Forest Planning Process, Karjala et al. (2004)	Natcher & Hickey (2002), Natcher (2008)	Elias (n.d.), Trilateral Agreement	Ndoho Istchee: An Innovative Approach to Aboriginal Participation in Forest Management Planning.(2007)	Forest Ecosystem Strategy Plan for Forest Management, District 19, Labrador/ Nitassinan 2003-2023

Community involvement	Public fora within First Nations and with external stakeholders	Interviews with community members, focus groups, field trips, archival research	Interviews with community members, field trips, focus groups	Map biographies, field trips	Focus on input from trappers & families still practicing traditional activities	Public stakeholder meetings, Innu Forest Guardians
Use in forest management planning	Used in development & implementation of management plan & to support FSC certification	Use in management of John Prince Research Forest	Used in discussions with industry to modify forestry practices	Used in negotiations with industry & Quebec to modify forestry practices, but little implementation to date	Integration Round Table in which Cree trappers & forest managers share knowledge & address Cree values	Incorporated into forest management plan
Outside partnerships	ENGOs, forest industry joint venture	University of Northern BC	Sustainable Forest Management Network, Tolko Industries Ltd.	Governments of Canada, Quebec	Canadian Model Forest Network, Quebec, industry	Government of NF & Labrador, Canadian Model Forest Network, Canadian Boreal Initiative
Monitoring	Implementation & effectiveness monitoring included in FMP as part of adaptive management approach	Tl'azt'en Nation Community-Based Environmental Monitoring developed in 2009 with 252 environmental indicators (Yin 2009)	No specific mention in LRRCN C&I process, but monitoring a component of provincial forest planning	Little mention of monitoring except for social indicators which are 'dynamic,' to be refined over time by monitoring, data collection & analysis	Cree-Quebec Forestry Board with link to local communities (WCMF 2007)	Province & Innu as laid out in chapter on ecological, cultural & economic research & monitoring in FMP; Forest Guardian Program

Context and Drivers for Aboriginal Criteria and Indicators Development Processes

The six First Nations who developed C&I each have their own particular context and drivers that shaped the development of their C&I. Table 1 summarizes the context for the development of each First Nation's C&I under the headings of policy and arrangements with the Crown, ecosystem type, C&I development method, key documents, community involvement, outside partnerships and monitoring. Two First Nations (Iisaak and Tl'azt'en) were located in BC and another two in Quebec (Algonquins of Barrier Lake and Waswanipi), and thus share the same provincial contexts; the others are from Alberta and Labrador. All but Iisaak are located within the boreal ecosystem. Agreements with the Crown differ in each case, as do the methods for developing the C&I sets. The extent of community and/or expert involvement differs in all cases, as do outside partnerships developed to aid in C&I development. Monitoring mechanisms differ for each community, although all have some form of monitoring that will allow adaptation over time.

Condensing and Summarizing First Nations Criteria and Indicators

The literature review produced a total of 587 indicators from the six First Nations.[2] A breakdown of the indicators for each of the communities is shown in Table 2. The environmental criterion had the most numerous indicators at 163. Iisaak had the greatest number of indicators under this category with 96, while the least number of environmental indicators was developed by Little Red River Cree with only five, partially because of its shorter list of indicators. The rights criterion had the least amount of indicators (n=30), with Waswanipi having over half or 18 indicators related to Aboriginal rights. Iisaak and LRRCN also included rights in their indicators, while the Innu, Algonquins of Barrier Lake (ABL) and Tl'azt'en made no mention of rights. When the indicators were condensed and rolled up, their number was reduced to 169, with 74 under the environmental criterion, 36 under the cultural, 31 under the economic and 28 under the social criterion.

Figure 2 shows the 'rolled up' or combined indicators in each criterion, with the highest-numbered indicator placed at the top of the diagram, and lower numbered indicators following from right to left in order of importance. Under the environmental criterion, the highest number of indicators fell under forest operations with the lowest being climate change and site rehabilitation. Under

---

[2] Natcher and Hickey (2002), in their paper in *Human Organization* took a sample of 30 indicators from the total of 62 developed for the LRRCN. Since our methods are based on a review of the existing literature, the full LRRCN C&I set was not included. However, Natcher (pers. comm. 2009) indicated that the set of 30 indicators were representative of the larger set and criteria. While it would have been preferable to include the full set of indicators in our analysis, their exclusion in no way alters the conclusions forwarded in this chapter.

the cultural criterion, protection of traditional land use values ranked highest with cross-cultural learning ranked lowest. Under the economic criterion, the highest number of indicators was employment and business opportunities while the lowest was cost-benefit analysis. In the social category, community health and well-being ranked highest, while workers' rights and safety ranked lowest.

Table 2. A Summary of condensed criteria and indicators for six First Nations.

Community Name	Environmental	Economic	Social	Cultural	Institutional	Rights	TOTAL
Nuu-chah-nulth (Iisaak)	85	7	15	11	66	7	**191**
Innu Nation	17	46	6	16	35	0	**120**
Tl'azt'en Nation	22	13	22	3	30	0	**90**
Waswanipi First Nation	16	6	26	5	8	18	**79**
Algonquins of Barriere Lake	7	12	15	8	0	0	**42**
Little Red River Cree	5	5	10	5	0	5	**30***
TOTAL	**152**	**89**	**94**	**48**	**139**	**30**	**587**

* This total represents a preliminary set of indicators developed by the LRRCN; the final sample included 62 indicators (*see* footnote 2).

Common themes among all six First Nations under the environmental criterion were: timber harvesting and road building regulations; silvicultural direction; fish and wildlife habitat requirements; forest health and ecosystem biodiversity, and watershed and sensitive areas protection. Iisaak was unique with respect to indicators addressing plantations, the disposal of contaminants, and site rehabilitation in part because they adopted the FSC framework and developed an intensive set of regulations with the results from the Clayoquot Sound Scientific Panel Reports. Indicators for species protection were often unique to a First Nation's ecosystem location. For example, LRRCN addressed bison habitat, while the Innu Nation had a specific indicator for the protection of red wine caribou.

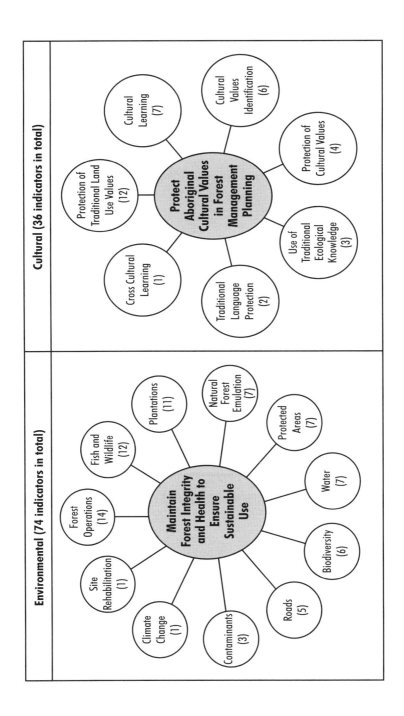

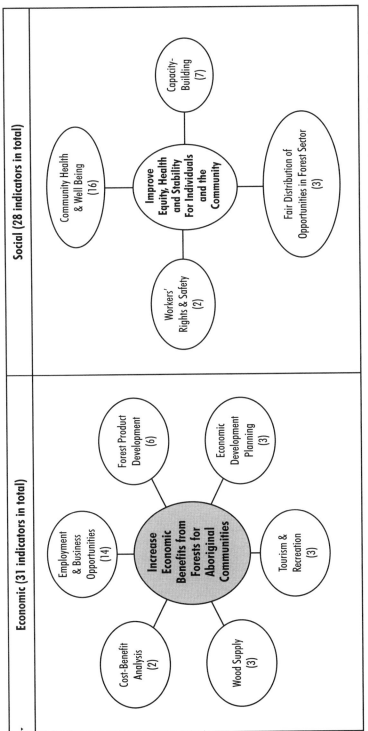

Figure 2. *Condensed community indicators for forest management grouped under environmental, economic, social & cultural criteria.*

Common themes under the cultural criterion were the protection of culturally significant areas using traditional ecological knowledge in management planning, and identifying and recording cultural values. Other themes were unique to certain First Nations such as areas identified for protection, e.g., LRRCN identified the Harper Creek caves. Tl'azt'en and the ABL identified the use of traditional language as an important indicator under culture. Waswanipi focused on the amount of available land for hunting, trapping, fishing and gathering, while Iisaak and Tl'azt'en identified cross-cultural learning as important.

The common theme to all six First Nations under the economic criterion was creating jobs for community members. Innu and Iisaak listed several indicators stressing the protection of tourism and recreational areas. The Innu Nation identified non-timber forest product marketing. LRRCN had an indicator on increasing the number of individually-owned businesses. Tl'azt'en included a holistic approach to economic planning that incorporates the social impacts of development. Interestingly, not one community mentioned profitability as a key consideration.

Under the social criterion, the common themes were improving education and health, and ensuring a fair distribution of wealth within communities. LRRCN addressed implementing forestry education at the elementary school level. Tl'azt'en developed indicators on personal health and participation in community activities, with measures such as the number of diabetics and the number of people volunteering in the community. The Innu Nation and the ABL addressed the participation of women in the forest sector.

Lessons from Aboriginal Criteria and Indicators Comparisons

C&I have been helpful for the First Nations involved in this comparison, although, because of the unique circumstances of each community, their utility varies from one to the next. All communities developed their sets in response to external forces and as a way of finding a language through which to convey their values to forest managers, both government and industry. The C&I sets articulated a range of values that could be considered as part of SFM, promoting cultural diversity and different ways of knowing. SFM can be enriched as a result of this process (Lertzman and Vredenburg 2005; Mabee and Hoberg 2006).

Interestingly, Aboriginal and treaty rights, or institutions, were addressed by only half the sample— Nuu-chah-nulth, LRRCN and Waswanipi. The centrality of constitutionally-recognized rights in Canada and the importance of institutional mechanisms that address decision-making raise the question of why these issues were not dealt with by the Innu, ABL and Tl'azt'en. Most of the C&I sets were developed as a way of protecting Aboriginal values in relation to the immediate pressures of forest management. Aboriginal and treaty rights may not have been considered a 'value,' or communities may have considered those rights inherent in everything they identified as a value. The lack of rights as a value in many C&I sets may also be a reflection of the widespread resistance of government and industry forest managers to recognize those rights and to accommodate them into current management practices. The Aboriginal communities may simply be acting in what Shatz (1987) termed an 'assertively pragmatic' fashion.

Community participation was key to these indicator development projects through focus groups, interviews, community meetings and joint committees between/among First Nations, government, industry and consultants. However, a common theme for all the communities was the limited understanding of the importance of C&I and how they can be used to protect their values and advance their rights and interests.

Despite the importance of community participation, none of the processes could be considered truly 'bottom up.' In all cases, there was a mix of 'top down' and 'bottom up' approaches (Adam and Kneeshaw 2009; Sherry *et al.* 2005). No community developed their indicators without the help of outside experts; all were 'co-produced' (Shearer *et al.* 2009). Tl'azt'en and LRRCN's processes were informed by university researchers, the ABL relied on expert social scientists, forest managers and GIS technicians. The Innu worked closely with government forest managers and hired outside consultants to inform their ecosystem-based management planning process. Waswanipi also relied on university researchers and the Canadian Model Forest Network. Iisaak was very much influenced and guided by the FSC's regional standards and took into account CCFM C&I and the Clayoquot Sound Science Panel recommendations.

For most of the communities, the development process was triggered by a response to existing forestry programs such as the Canadian Model Forest program or as a reaction to unsatisfactory resource development practices by governments and industry. Some communities took a broader land use planning approach before focusing on forest management planning. For example, processes such as the Clayoquot Science Panel and the Innu's ecosystem-based planning approach started with higher level land use planning decisions. This enabled these communities to look at a broader range of values and uses before focusing on forest management, development and planning.

Each initiative was aware of C&I sets being used by governments in Canada and internationally, although most did not implement them because they were considered too broad to capture local concerns and the unique circumstances of most communities. Most of the communities were not aware of the C&I sets being developed by other Aboriginal communities, probably because local processes occurred over roughly the same time period (late 1990s through to mid 2000s). It is only recently that a literature has developed around Aboriginal C&I beginning with the Tl'azt'en/UNBC Aboriginal Forest Planning Process in 2003 (Karjala *et al.* 2004).

All C&I sets are thought of as 'living documents' by the communities that developed them. The indicators are flexible in order to adapt to changing needs in the community and the environment. The usefulness of Aboriginal C&I sets may extend to a number of areas, including increased community awareness, but the true test for C&I sets is their ability to modify forestry operations. All communities indicated some level of operational or planning influence. Iisaak's C&I are being measured through FSC audits. Tl'azt'en's C&I are being used in the management of the John Prince Research Forest. The LRRCN has used C&I in its negotiations with the forest industry to modify operations (Webb *et al.* 2009). The Algonquins of Barrier Lake have used their C&I in negotiations with the province and industry. Waswanipi has developed an Integration Round Table, where Cree trappers and forest managers share knowledge and address Cree values, that oversees use of their C&I by the Cree-Quebec Forest Management Board established under the *Paix des Braves* agreement. The Innu have incorporated their C&I into their ecosystem-based plan.

However, monitoring and adaptation processes are not yet well developed by all communities. Few communities gave any indication of their level of confidence in the nature and quality of baseline data, or their capacity to measure indicators over time, calling into question the long-term value of C&I sets. Ultimately, this will be the litmus test of all Aboriginal C&I sets: Will they lead to real changes in forest management practices or land use planning initiatives? And will those changes result in the protection of Aboriginal social, cultural, economic and environmental values?

A Framework for Assessing Indigenous Participation in Forest Management

Two First Nations, Pikangikum and Kitcisakik, as described in Volume 1 of this series (Shearer *et al.* 2009; Saint-Arnaud *et al.* 2009), are moving beyond a sustainable development framework and attempting to develop C&I based on the worldviews of Aboriginal peoples. However, the First Nations compared in this chapter have adopted the sustainable development paradigm in spite of its limitations, with all of them introducing a cultural component. Some authors contend that Indigenous peoples, if they are able to assert their social and cultural values in the midst of the ongoing friction between development and the environment, may be able to strengthen the sustainable development paradigm (Loomis 2000; Groenfeldt 2003). Berkes and Folke (1998) have demonstrated the importance of social-ecological linkages, adding cultural capital to the human-made (economic), natural (environment) and social capitals that are critical elements in the ability of communities to modify and adapt to environmental changes. As long as the sustainable development paradigm is flexible enough to incorporate different values, then it remains a useful framework because of its global reach and acceptance.

The following framework thus adopts the sustainable development framework with the addition of a cultural criterion. Figure 3 modifies the classic Venn diagram for sustainable development to incorporate the cultural component. The 'ideal' centre portion of the diagram is where values are balanced, integrated and interconnected. The periphery of the circles is the region where imbalance and trade-offs occur in which some values are sacrificed for the protection of others.

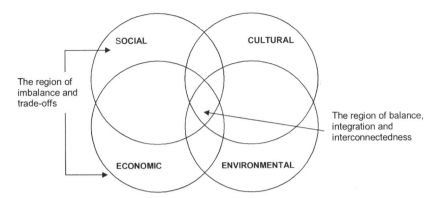

Figure 3. *A modified sustainable development diagram incorporating a cultural criterion.*

Table 3 expands on this sustainable development framework incorporating Aboriginal peoples' values in forest management. Community well-being and satisfaction are shown as the outcomes of an approach that combines environmental, social, cultural and economic criteria, founded on the recognition of Aboriginal rights with the appropriate institutional development to support community well-being.

Community well-being and satisfaction as outcomes generally mean that the basic human needs of food, clothing and shelter have been met. But well-being is much more than that, reflecting the aspirations of Aboriginal communities for self-governance and self-sufficiency. The core of the framework is the three pillars of sustainable forest management—economic, environmental and social criteria—with the addition of a cultural component that recognizes the importance of the maintenance of an Aboriginal way of life that maintains a strong connection to the land.

The foundation of the framework is the recognition, respect and accommodation of Aboriginal rights. In Canada, Aboriginal peoples still assert their inherent rights and sovereignty within the Canadian nation-state. This assertion of sovereignty is the basis for self-determination and self-sufficiency. There have been a number of different formal agreements between the Canadian state and Aboriginal peoples in which the state purports that Aboriginal rights have been limited or extinguished. However, Aboriginal peoples assert that their rights cannot be extinguished and therefore these rights continue despite agreements that have more clearly defined the nature of the relationship between Aboriginal peoples and the Canadian state. These agreements include historic treaties and modern land claims. In some areas of Canada, such agreements are still being negotiated. In the absence of any negotiated agreement between Aboriginal peoples and the Crown, Canadian courts have ruled that Aboriginal title, a unique or *sui generis* form of occupation and use of the land based on the prior occupation by Aboriginal peoples and their historic land practices, still exists.

Table 3. A sustainable development framework incorporating Aboriginal peoples' values in forest management (Smith 2005).

OUTCOME: COMMUNITY WELL-BEING/SATISFACTION
Basic needs (food, shelter, clothing), self-sufficiency, self-governance, adequate standard of living, high employment, fair distribution of economic benefits across households, cultural integrity, human health, education, political stability, access to lands & resources, sustainable forest management

SUSTAINABLE FOREST MANAGEMENT CRITERIA & INDICATORS

ENVIRONMENTAL Maintain forest integrity & health to ensure sustainable use	ECONOMIC Increase economic benefits from forests for local Aboriginal communities	SOCIAL Improve equity, health & stability for individuals & the community	CULTURAL Protect Aboriginal cultural values in forest management planning
• Condition of forest • Forest health • Sustainable extraction of renewable resource (amount & rate of logging) • Maintenance of habitat for all users, including wildlife • Environmental services (air, water, carbon storage, etc.)	• Jobs & income • Revenue-sharing • Business opportunities • Forest product development • Development planning • Taxation • Profitable businesses	• Capacity-building (education & skills, ability to hire resource managers long-term) • Increased community stability • Improved health • Equity among community members in decision-making & distribution of wealth • Workers rights	• Maintenance of way of life • Language retention • Cultural learning (knowledge) • Traditional land use • Identification & protection of cultural values

THE FOUNDATION	
ABORIGINAL RIGHTS & RESPONSIBILITIES	**INSTITUTIONS**
<ul><li>Recognition, protection and accommodation of Aboriginal & treaty rights = sovereignty and/or self-government</li><li>Rights are inherent, cannot be extinguished and so apply to all areas: historic treaties, modern land claims, areas without agreements to which Aboriginal title or ownership/use apply</li><li>Responsibility for forest stewardship is considered an integral part of rights.</li></ul>	<ul><li>Shared decision-making (ranging from advisory to consent & control, including joint decision-making or 'co-management')</li><li>Conflict resolution</li><li>Consultation and participation</li><li>Communication</li><li>Mutual learning/respect for knowledge systems</li><li>Monitoring and adaptation mechanisms</li></ul>

The recognition of Aboriginal rights necessitates the joint development by Aboriginal peoples and the state of new institutions for the management and use of forest resources. Such institutional arrangements might include shared decision-making, ranging from advisory to consent and control, including joint decision-making or 'co-management.' Mechanisms for conflict resolution, assessment and monitoring would be important for adaptation over time.

For the environmental criteria, the condition of the forest (Gibson *et al.* 2005) as a result of the exploitation of resources—either industrial extraction or traditional use—can be measured by the amount and rate of extraction. This is important for many Aboriginal communities where traditional use is part of their economies, maintenance and use of habitat, especially for wildlife.

For the economic criteria, classical measures of jobs, income and profit (Trosper *et al.* 2008) are important. However, given the recognition of Aboriginal rights and shared control with the state, sharing of resource revenues would also be a measure of success. In order to maintain traditional land use, and the cultural benefits and values that are derived from such use, the maintenance of a mixed economy with both traditional land uses and a cash economy based on industrial extraction should be considered.

In the social arena, one of the impacts of the exclusion of Aboriginal communities from the industrial forest-based sector has been the lack of skills to participate. An increase in forest business and management skills, gained both practically and through formal education, and the ability of Aboriginal communities to hire resource and business managers are criteria that lend themselves to measuring change over time. It has been posited that the well-being of the community and individuals within it are integrally tied to the health of the land (Bombay *et al.* 1995). Although difficult to prove a causal relationship, there have been some studies in Canada that explored the link between environmental and human health. In one study by Usher (2003), violence was explored as a social indicator of the impact of environmental degradation. In the First Nation communities of Grassy Narrows and Whitedog, which lost access to a traditional fishery because of mercury pollution from a nearby pulp mill, the study found that:

> …a sharp spike occurred in violent deaths at both Grassy Narrows and Whitedog, precisely as the crisis of harvest disruption deepened, and which did not occur on a nearby reserve that we used as a control. Think of what 17 violent deaths in one year means to a community of a few hundred people (Usher 2003).

Criteria such as community stability and improved health need to be further explored as indicators of sustainable development.

Finally, in the interests of encouraging development that maintains the cultural diversity of Aboriginal communities, cultural criteria appropriate to the way of life of the community are important. Language retention is a prime indicator of cultural integrity. In Canada Aboriginal languages are vibrant in some areas and threatened in others, with many communities attempting to restore almost extinct languages. In forest-based economic development, historically, Aboriginal knowledge has been integral to the success of the larger economy, especially during the early period of colonization and the fur trade. Today, with scarce and diminishing forest resources, Aboriginal knowledge may once again play an important role in new types of forest-based development. The criterion of the role of Aboriginal knowledge or cultural learning in developing and conserving forests is another key aspect of cultural diversity.

Conclusion

Imperfect as it is, Aboriginal C&I development has been an important process for formalizing and introducing Aboriginal values into the forest management process. Local level C&I development initiatives may also be a useful and practical tool for those First Nations who are engaged in regional land use planning processes. Communities respond to their unique conditions and context to develop C&I that reflect community values and respond to forest managers' demands for measurable indicators of SFM. Locally developed Aboriginal C&I sets are introducing Aboriginal cultural values and worldviews into a forest management system that is only beginning to understand and implement the sustainable development paradigm. Aboriginal communities are taking a leadership role in demonstrating how to achieve a balance among environmental, social, economic and cultural values in a way that truly integrates and accommodates all of these aspects.

References Cited

Adam, M-C. and D. Kneeshaw (2009). *Formulating Aboriginal criteria and indicators frameworks.* Edmonton AB: Sustainable Forest Management Network. http://www.sfmnetwork.ca/html/report_synthesis_listall_page_1_e.html. Nov. 9/09.

Beckley, T.M. (1999). *Sustainability for whom: Social indicators for forest-dependent communities in Canada.* Edmonton AB: Sustainable Forest Management Network. http://www.sfmnetwork.ca/docs/e/PR_1999-24.pdf. Nov. 9/09.

Berkes, F. and C. Folke (1998). 'Linking social and ecological systems for resilience and Sustainability,' pp. 1-25 in F. Berkes, and C. Folke, eds., *Linking social and ecological systems: Management practices and social mechanisms for building resilience.* UK: Cambridge University Press.

Berninger, K., D. Kneeshaw, and C. Messier (2009). The role of cultural models in local perceptions of SFM: Differences and similarities of interest groups from three boreal regions. *Journal of Environmental Management* 90(2):740-751.

Bombay, H., P. Smith, and D. Wright (1995). *An Aboriginal criterion for sustainable forest management.* Ottawa: National Aboriginal Forestry Association.

Burford de Oliveira, N. (1999). Community participation in developing and applying criteria and indicators of sustainable and equitable forest management. CIFOR Project Report: Testing Criteria and Indicators for the Sustainable Management of Forests.http://www.cifor.cgiar.org/publications/pdf_files/ComPar.pdf. Nov. 6/09.

CMFN—Canadian Model Forest Network (2000). *A user's guide to local level indicators of sustainable forest management: Experiences from the Canadian model forest network.* Ottawa: Canadian Forest Service, Natural Resources Canada.

Daniel, M., M. Cargo, E. Marks, C. Paquet, D. Simmons, M. Williams, K. Rowley and K. O'Dea (2009). Rating health and social indicators for use with Indigenous communities: A tool for balancing cultural and scientific utility. *Social Indicators Research* 94(2):241-256.

Duinker, P.N. (2001). Criteria and indicators of sustainable forest management in Canada: Progress and problems in integrating science and politics at the local level. In *Criteria and indicators for sustainable forest management at the forest management unit level.* ed.

Ettenger, K., in collaboration with Local Elders, Hunters and Tallymen (2002). Muskuuchii (Bear Mountain): Protecting a Traditional Iyiyuuch Wildlife Preserve and Sacred Site: Based on the Oral History, Knowledge and Values of the Waskaganish Cree First Nation. Forestry Working Group of the Grand Council of the Crees of Quebec. 24 pp. http://www.gcc.ca/pdf/ENV000000006.pdf. Nov. 6/09.

FAO—Food and Agriculture Organization (2001). Criteria and indicators for sustainable forest management: A compendium. Forest Management Working Paper 005. http://www.fao.org/DOCREP/004/AC135E/AC135E00.HTM. Jul. 14/08

Garcia, C.A., and G. Lescuyer (2008). Monitoring, indicators and community based forest management in the tropics: pretexts or red herrings? *Biodiversity Conservation* 17:1303-1317.

Gibson, C., E. Ostrom, and M. McKean (2005). *Forests, people, and governance: Some initial theoretical lessons* pp. 227-242 *in* Gibson, C., M.A. McKean, and E. Ostrom (eds.) People and Forests: Communities, Institutions, and Governance. Cambridge, MA: Massachusetts Institute of Technology, 298p.

Gladu, J.P., and C. Watkinson (2005). *Measuring sustainable forest management: A compilation of Aboriginal indicators.* A report prepared for the Canadian Model Forest Network-Aboriginal Strategic Initiative. Model Forest Network, Ottawa, ON. 56 pp. http://www.modelforest.net/cmfn/en/find_out_more/aboriginal/publications_record.aspx?title_id=4665. Nov. 9/09.

Gough, A.D., J.L. Innes, and S.D. Allen (2008). Development of common indicators of sustainable forest management. *Ecological Indicators* 8:425-430.

Groenfeldt, D. (2003). The future of Indigenous values: Cultural relativism in the face of economic development. *Futures* 35(9):917-929.

Hickey, G.M., and J.L. Innes (2005). Scientific review and gap analysis of sustainable forest management criteria and indicators initiatives. FORREX Series 17. Kamloops BC:

Innes, L., and L. Moores (2003). The ecosystem approach in practice: Developing sustainable forestry in central Labrador, Canada. Paper 0717-C1 presented to the XIII World Forestry Congress. Quebec City, Canada. September 2003. http://www.fao.org/DOCREP/ARTICLE/WFC/XII/0717-C1.HTM. Nov. 6/09.

Innu Nation (n.d.a) Forestry in Nitassinan. http://www.innu.ca/. Apr. 27/08.

Innu Nation (n.d.b) Home. http://www.innu.ca. Nov. 6/09.

Lapointe, G. (1998). Sustainable forest management certification: The Canadian programme. *Forestry Chronicle* 74(2):227-230.

Lertzman, D., and H. Vredenburg (2005). Indigenous peoples, resource extraction and sustainable development: An ethical approach. *Journal of Business Ethics* 56(3):239-254.

Loomis, T.M. (2000). Indigenous populations and sustainable development: Building on Indigenous approaches to holistic, self-determined development. *World Development* 28(5):893-910.

Mabee, H.S., and G. Hoberg (2006). Equal partners? Assessing comanagement of forest resources in Clayoquot Sound. *Society and Natural Resources* 19(10):875-888.

McDonald, G.T., and M.B. Lane (2004). Converging global indicators for sustainable forest management. *Forest Policy and Economics* 6(1):63-70.

Natcher, D.C., and C.G. Hickey (2002). Putting the community back into community-based resource management: A criteria and indicators approach to sustainability. *Human Organization* 61 (4): 350-363.

O'Flaherty, R.M., I.J. Davidson-Hunt, and A.M. Miller (2009). 'Anishinaabe stewardship values for sustainable forest management of the Whitefeather Forest, Pikangikum First Nation, Ontario,' pp. 19-34 in M. Stevenson, and D. Natche, eds., *Changing the culture of forestry in Canada: Building effective institutions for Aboriginal engagement in sustainable forest management.* Vol. 1. Special edition advanced copy. Edmonton, AB: CCI Press and Sustainable Forest Management Network.

Saint-Arnaud, M., H. Asselin, C. Dubé, Y. Croteau, and C. Papatie (2009). 'Developing criteria and indicators for Aboriginal forestry: Mutual learning through collaborative research,' pp. 85-105 in M. Stevenson, and D. Natcher, eds., *Changing the culture of forestry in Canada: Building effective institutions for Aboriginal engagement in sustainable forest management.* Vol. 1. Special edition advanced copy. Edmonton, AB: CCI Press and Sustainable Forest Management Network.

Shatz, S. (1987). 'Assertive pragmatism and the multinational enterprise,' pp. 93-105 in D.G. Becker, and R. Sklar, eds., *International capitalism and development in the late twentieth century* Boulder, CO: Lynne Rienner. Cited in Anderson 1997.

Shearer, J., P. Peters, and I.J. Davidson-Hunt (2009). 'Co-producing a Whitefeather Forest cultural landscape monitoring framework,' pp 63-84 in M. Stevenson, and D. Natcher, eds., *Changing the culture of forestry in Canada: Building effective institutions for Aboriginal engagement in sustainable forest management.* Vol. 1. Special edition advanced copy. Edmonton, AB: CCI Press and Sustainable Forest Management Network.

Sherry, E., R. Halseth, G. Fondahl, M. Karjala, and B. Leon (2005). Local-level criteria and indicators: An Aboriginal perspective on sustainable forest management. *Forestry* 78(5):513-539.

Smith, P. (2005). 'Managing the commons: Indigenous rights, economic development & identity. Emerging issues, conclusions and recommendations,' pp. 66-70 in L. Merino, and J. Robson, eds. *Managing the commons: Indigenous rights, economic development and identity.*. Consejo Civil Mexicano para la Silvicultura Sostenible A.C. Mexico City: The Christensen Fund, Ford Foundation, Secretaria de Medio Ambiente y Recursos Naturales and Instituto Nacional de Ecologia (INE).

Trosper, R.L., H. Nelson, G. Hoberg, P. Smith, and W. Nikolakis (2008). Institutional Determinants of Profitable Commercial Forestry Enterprises among First Nations in Canada. *Canadian Journal of Forest Research* 38(2):226-238.

UN—United Nations (1992). Non-legally binding authoritative statement of principles for a global consensus on the management, conservation and sustainable development of all types of forests. http://www.un.org/documents/ga/conf151/aconf151263annex3.htm. Jul. 14/08.

UN—United Nations (2007). United Nations forum on forests report of the seventh session (24 February 2006 to 16 to 27 April 2007). E/2007/42 E/CN.18/2007/8. http://daccessdds.un.org/doc/UNDOC/GEN/N07/349/31/PDF/N0734931.pdf ?OpenElement. Jul. 14/08.

Usher, P.J. (2003). Environment, race and nation reconsidered: Reflections on Aboriginal land claims in Canada. The Canadian Geographer 474: 365-382.

Webb, J., H. Sewepagaham, and C. Sewepagaham (2009). 'Negotiating cultural sustainability: Deep consultation with the Little Red River Cree Nation in the Wabasca-Mikkwa Lowlands, Alberta,' pp. 107-126 in M. Stevenson, and D. Natcher, eds., *Changing the culture of forestry in Canada: Building effective institutions for Aboriginal engagement in sustainable forest management.* Vol. 1. Special edition advanced copy. Edmonton, AB: CCI Press and Sustainable Forest Management Network.

Appendix A:
Six First Nations Approaches to C&I Development for SFM

Iisaak, Nuu-chah-nulth

Clayoquot Sound is situated on Vancouver Island in a landscape characterized by mountains, heavy precipitation and a dense, temperate rainforest. The area is located within the coastal western hemlock biogeoclimatic zone as well as the mountain hemlock and alpine tundra zones (CSSP 1995) where western hemlock (*Tsuga heterophylla*, (Raf.) Sarg.), western red cedar (*Thuja plicata* Don ex D. Don), yellow cedar (*Chamaecyparis nootkatensis* D. Don), Douglas fir (*Pseudotsuga menziesii* Mirb.) and lodgepole pine (*Pinus contorta* Dougl. Ex. Loud.) predominate. The Nuu-chah-nulth use the forest for hunting, fishing and gathering, and for cultural, social and economic reasons. In it, they practice selective burning to create animal habitat and improve berry production. They cherish the scenic landscape because it helps define the people of the area (CSSP 1995). From their ties to the land has arisen a Nuu-chah-nulth worldview, captured in their language and stories (Umeek 2004).

The Nuu-chah-nulth are one of the many First Nations in British Columbia that have not entered into any agreement with the Crown. In 1993 the five First Nations of the Nuu-chah-nulth Central Region—Ahousat (population 1,905), Hesquiaht (population 678), Tla-o-qui-aht (population 960), Toquaht (population 137) and Ucluelet (population 618) (INAC 2009)—signed an *Interim Measures Agreement* with the Province of BC, providing for joint management of Nuu-chah-nulth lands during completion of treaty negotiations which began in BC in 1992. The mechanism for this joint management was a Central Region Board with a mandate to advise the province on all matters relating to development in the Clayoquot Sound region. The Central Regional Board mandate includes:

- promoting sustainability, economic development and diversification of local communities;
- providing a viable, sustainable forest industry;
- respecting the *Interim Measures Agreement* and its implementation;
- reducing the 70% unemployment levels within Nuu-chah-nulth communities;
- enhancing fish and wildlife, and restoring damaged streams and forest areas;
- assessing compliance with world-class forest standards, such as the Scientific Panel recommendations and Forest Practices Code;
- ensuring that visual values and ecological integrity are maintained and given high priority in any proposed resource extraction or development;
- increasing local ownership within the forest industry;
- working toward reconciling the interests the interests of environmentalists, labour, industry, First Nations, recreational users,

governments, and all others with interests and concerns in Clayoquot Sound; and

- encouraging respect for Aboriginal heritage (Iisaak n.d.).

After a decade of conflict and failed initiatives to improve the forestry situation in Clayoquot Sound, the Government of British Columbia created the Clayoquot Sound Scientific Panel (CSSP) to review forestry operations and make recommendations for improvements. The panel comprised representatives from the Nuu-chah-nulth, scientists, foresters and engineers who made recommendations in seven key areas: silviculture, harvesting, transportation, scenic and tourism values, planning for sustainable management, monitoring, and inclusive of First Nation perspectives (CSSP 1995). The Panel also took a tremendous step forward in recognizing the Nuu-chah-nulth worldview and knowledge which informed the Panel's recommendations (Lertzman and Vredenburg 2005). The CSSP made recommendations to protect Nuu-chah-nulth cultural sites from logging, stipulating that the identification of such sites and measures for their protection be identified by the Nuu-chah-nulth (Mabee and Hoberg 2004).

In 1997, the Ma-Mook Development Corporation was established to represent the collective economic interests of the Nuu-chah-nulth. Iisaak Forest Resources Ltd. started as a joint venture between the Nuu-Chah-Nulth First Nation and Macmillan Bloedel in 1998. The joint venture was transferred to Weyerhaeuser when Macmillan Bloedel went out of business in 1999. The company became fully owned by the Nuu-Chah-Nulth when they bought the remaining Weyerhaeuser shares in 2005. In 1999 Iisaak signed a Memorandum of Understanding with environmental non-governmental organizations (ENGOs) Greenpeace, the Natural Resources Defense Council, the Sierra Club of BC and the Western Canada Wilderness Committee. On their side, the ENGOs agreed to support Iisaak, promote the marketing of the company's products and find mechanisms for ongoing co-operation. Iisaak agreed to seek Forest Stewardship Council (FSC) certification and manage *eehmiis* or 'precious' areas emphasizing non-timber values (Iisaak n.d.).

Iisaak holds several provincial licenses—a Tree Farm License (#57) and a number of timber licenses. Iisaak's most recent Sustainable Forest Management Plan (SFMP) 2006-2011 articulates with the requirements of the *BC Forest Act*, the *Forest Range and Practices Act*, the *Forest Practices Code of BC Act*, the *Clayoquot Sound Scientific Panel Recommendations*, the *Clayoquot Sound Watershed Planning Process* and other relevant government legislation and associated regulations. The plan includes criteria and indicators that meet both provincial standards and the Forest Stewardship Council (FSC) BC Regional Standard. The plan includes a monitoring program based on the principles of adaptive management (IFRL 2006).

The development of criteria and indicators was necessary for Iisaak as requirements of the CSSP recommendations, provincial forest planning regulations, and the FSC BC Regional Standard. The criteria and indicator data were collected through public forums which included participants from the Nuu-chah-nulth and other stakeholders with vested interests.

The criteria outlined in Iisaak's SFMP are organized in four categories: cultural, economic, ecological, and social. The SFMP cross-references Iisaak's 14 criteria with both FSC principles and CCFM criteria as shown in Table 4.

Table 4. Iisaak's Criteria for SFM referenced to FSC & CCFM C&I (IFRL 2006).

IISAAK CRITERIA OF C&I	FSC PRINCIPLE	CCFM CRITERIA
CULTURAL VALUES		
Criterion 1: First Nations' customary and legal rights are recognized & respected.	**Principle 3:** Indigenous Peoples' Rights	**Criterion 6:** Society's Responsibility
Criterion 2: Free and informed consent is obtained from First Nations shareholders directly impacted by forest management activities.	**Principle 3:** Indigenous Peoples' Rights	**Criterion 6:** Society's Responsibility
Criterion 3: First Nations training, employment and capacity building.	**Principle 4:** Community Relations and Worker's Rights	**Criterion 6:** Society's Responsibility
ECONOMIC VALUES		
Criterion 4: Sustained economic benefits are generated by forest operations.	**Principle 5:** Benefits from the Forest	**Criterion 5:** Economic & Social Benefits
Criterion 5: Economic opportunities are sustained for non-timber forest products.	**Principle 5:** Benefits from the Forest	**Criterion 5:** Economic & Social Benefits
ECOLOGICAL VALUES		
Criterion 6: Soil condition & productivity of forests are sustained.	**Principle 5:** Benefits from the Forest **Principle 6:** Environmental Impacts **Principle 7:** Management Plan **Principle 9:** Maintenance of High Conservation Value Forests	**Criterion 2:** Ecosystem Condition & Productivity **Criterion 3:** Soil & Water **Criterion 4:** Role in Global Ecological Cycles
Criterion 7: Genetic & species diversity are sustained.	**Principle 6:** Environmental Impacts **Principle 7:** Management Plan **Principle 9:** Maintenance of High Conservation Value Forests	**Criterion 1:** Biological Diversity
Criterion 8: Forest ecosystem diversity is sustained.	**Principle 6:** Environmental Impacts **Principle 7:** Management Plan	**Criterion 2:** Ecosystem Condition & Productivity **Criterion 4:** Role in Global Ecological Cycles

	Principle 9: Maintenance of High Conservation Value Forests	
Criterion 9: Water quality & flow are sustained.	**Principle 6:** Environmental Impacts **Principle 9:** Maintenance of High Conservation Value Forests	**Criterion 2:** Ecosystem Condition & Productivity **Criterion 3:** Soil & Water
Criterion 10: High Conservation Value Forests attributes are maintained.	**Principle 5:** Benefits from the Forest. **Principle 6:** Environmental Impacts. **Principle 9:** Maintenance of High Conservation Value Forests	**Criterion 1:** Biological Diversity **Criterion 2:** Ecosystem Condition & Productivity **Criterion 3:** Soil & Water **Criterion 4:** Role in Global Ecological Cycles
SOCIAL VALUES		
Criterion 11: Forest management shall respect all national & local laws & administrative requirements.	**Principle 1:** Compliance with Laws and FSC Principles	**Criterion 5:** Economic & Social Benefits
Criterion 12: Economic benefits & contributions to the local communities are sustained.	**Principle 1:** Compliance with Laws & FSC Principles **Principle 5:** Benefits from the Forest	**Criterion 5:** Eonomic & Social Benefits
Criterion 13: The diversity of public interests are sustained in forest management planning processes.	**Principle 4:** Community Relations & Worker's Rights	**Criterion 5:** Economic & Social Benefits **Criterion 6:** Society's Responsibility
Criterion 14: Tourism, recreation and scenic values are sustained through forest management activities.	**Principle 4:** Community Relations & Worker's Rights	**Criterion 5:** Economic & Social Benefits

Tl'azt'en

The John Prince Research Forest (JPRF) is situated in the northern interior of British Columbia, 250 kilometers northwest of Prince George. The forest is managed under a partnership between the Tl'azt'en First Nation and the University of Northern British Columbia for the purposes of research, teaching and training (Karjala and Dewhurst 2003). The Tl'azt'en First Nation has a population of 1,595 (INAC 2009) distributed between four communities: Tache, Binche, Dzitl'ainli and K'uzche. Tache is the main community which hosts the majority of the community's services.

The Tl'azt'en use *keyohs* (family territories) to denote traditional resource use areas where community members hunt, fish, trap, and gather traditional plants. The main wildlife species used by this First Nation include bear, moose, deer, caribou, salmon, whitefish and trout (Tl'azt'en 2009). The community uses family campgrounds in the summer and collects food for the winter months, a tradition that has always been a part of their culture.

The forest region in which the JPRF is located is the sub-boreal spruce biogeoclimatic zone of British Columbia. White spruce (*Picea glauca* (Moench) Voss) and Douglas fir (*Pseudotsuga. menziesii*) dominate the forest, with sub-alpine fir (*Abies lasiocarpa* (Hook.) Nutt.), black spruce (*Picea mariana* (P. Mill.) B.S.P.), lodgepole pine (ditto), cottonwood (*Populus deltoids* Bartr. ex Marsh.), trembling aspen (*Populus tremuloides* Michx.) and paper birch (*Betula papyrifera* Marsh.) also present. The region is characterized by cold winters with a lot of snow and warm summers (JPRF n.d.). Current activities occurring on the land are tourism, hunting, fishing, and forestry.

In 2001, graduate researchers identified C&I as a means of including the Tl'azt'en more meaningfully in forest management to meet community needs. National C&I sets were found unsuitable at the local level. The national sets had a lot of information dealing with ecologic and economic indicators, but there was a lack of indicators related to social, cultural and non-timber values. It was decided that local C&I would be incorporated into a computer model developed by UNBC that could track changes in community values in relation to the landscape, and assist in the co-management of the JPRF. Community values were gathered using interviews, archived information, focus groups and field trips with community members through the Aboriginal Forest Planning Process (AFPP) (Karjala *et al.* 2003). The AFPP was designed to simplify and encourage First Nation participation in forest management, involving community members to ensure that local values are documented and respected, facilitating communication between stakeholders and gaining education about forestry planning and Aboriginal values (Karjala *et al.* 2003).

Three broad questions directed the research: What is important to people in this community? What are their concerns? and, What ideas emerge as solutions to some of their resource and social problems? Four 'criteria themes' were used to organize the data:

1) human factors with the sub-themes 'education,' 'community' and 'employment;'

2) economy, with 'economic development' and 'the bush/subsistence economy' as sub-themes;

3) land management with the sub-themes 'current approaches,' 'alternative approaches,' 'traditional approaches and philosophies,' 'knowledge,' 'research and communication;' and,

4) resource/environmental concerns with 'wildlife,' 'fish,' 'trees and plants,' 'water quality and access' as sub-themes (Karjala *et al.* 2004).

This framework was modeled after Kearney's approach developed through stakeholder perceptions of SFM in the U.S. Pacific Northwest (Kearney *et al.* 1998).

The AFPP process suggests dividing C&I into three categories: 1) spatial—to determine if forest management practices are being applied in the right places; 2) quantitative—to determine how much of a value should be represented; and 3) qualitative—to determine if community members are satisfied with the planning process. Under the spatial category, the two 'criteria themes' given as examples are 'resource/environmental concerns' and 'human factors and indicators.' These themes indicate the amount of buffer placed around spatial sites, such as streams or spiritual or archaeological sites. Quantitative criteria include themes of human factors, economy and resource/environmental concerns addressing community/employment, the bush economy, economic development and wildlife. Indicators point to measuring numbers, whether it be person days/year for employment or yearly profits from value-added businesses (Karjala *et al.* 2003).

In comparing Tl'azt'en's C&I with national frameworks, Sherry *et al.* (2005) concluded that:

> Tl'azt'en C&I go beyond jobs and income to address other supportive roles forests can play in the achievement of community sustainability, such as cultural revitalization, capacity building, intergenerational equity, amenity values, and ownership of forestland. Tl'azt'en C&I call for identification of ways to address and resolve social problems, enhance community cohesiveness and resilience, and build relationships.

In the AFPP, monitoring and feedback is identified as an important step in adaptive management, allowing periodic review of C&I to meet community needs over time. In 2009, a more detailed examination of monitoring was undertaken with the Tl'azt'en Nation Community-Based Environmental Monitoring project with an additional 252 environmental indicators developed (Yin 2009).

Little Red River Cree Nation

Three communities in Northern Alberta make up the LRRCN: Fox Lake, Garden River and John d'or Prairie. They have a combined registered population of 4,454 (INAC 2009), and are located in two boreal eco-regions: the mixed-wood and the sub-arctic (Natcher and Hickey 2002). The boreal mixed-wood forest is characterized by balsam poplar (*Abies balsamea* (L.) P. Mill.), trembling aspen, jack pine (*Pinus banksiana* Lamb.) black spruce (*Picea mariana* (P. Mill.) and white spruce (*Picea glauca* (Voss). The sub-arctic region is characterized by black spruce and permafrost soils due to its proximity of Caribou Mountains and higher elevations.

The LRRCN has a staggering unemployment rate of over 85 percent. The few existing jobs held by the community are with government services or seasonal work (Natcher and Hickey 2002). These two factors make the community critically dependent on natural resources to provide basic necessities such as food. Many households, not just a few active ones, continue to rely on the bush economy. In the past, caribou and bison were the preferred food species; however, a decline in their population has made the community turn to moose (Nelson et al. 2008).

In Alberta, Forest Management Agreements are awarded to companies for a 20-year term. Extractive industries such as forestry and oil and gas have a huge impact on the environment with road building, laying pipeline, and timber harvesting. This impact is affecting the Aboriginal and Treaty rights of the LRRCN, rights that were secured with the signing of Treaty 8 in 1899 (Natcher and Hickey 2002).

With large tracts of forest going to industry and no profits being shared with LRRCN, the communities became concerned and sought to engage government at both levels in order to protect lands and resources vital to them. As a result, an agreement was reached in 1999, and the LRRCN and Tallcree First Nations were awarded a Special Management Area (SMA)—a territory of 35,000 square kilometers (Webb *et al.* 2009). A decision-making board was established composed of representatives from First Nations, government, and the forestry and petroleum industries.

The SMA included a commitment to undertake joint research to be used in planning. LRRCN became a member of the Sustainable Forest Management Network (SFMN), engaging in 20 research projects after joining the Network in 1996. C&I were examined as a means of managing SMA sustainably and was one area of SFMN research. The LRRCN began work to develop a local set of indicators as the national and international sets did not address local level concerns. Using interviews with community members, local values were documented. It is from these values that the C&I were created (Natcher and Hickey 2002). The process resulted in six criteria that were action-based:

- modify forestry operations to reduce negative impacts on wildlife species;
- ensure continued community access to lands and resources;
- provide protection to sites identified by community members as having biological, cultural and/or historical significance;

- recognize and protect Aboriginal and treaty rights to hunt, fish, trap and gather;
- increase forest-based economic opportunities for community members; and
- increase the involvement of community members in decision-making.

Criteria were broken down to critical elements, local values, goals, indicators and actions. The initial C&I set (n=30) was seen as dynamic and part of an ongoing community-based assessment process. The sharing of knowledge required to develop the criteria and subsequent actions to address them were seen as an important piece in promoting both community participation and the discussion of tradeoffs with industry. As Natcher and Hickey concluded:

> ... by making local ecological knowledge available to the management process, the nested relationship between community members, wildlife habitat, and industrial development is further clarified. As a result, decisions and/or trade-offs can now be made between habitat enhancement and economic development objectives through a framework that is transparent, accessible, and inclusive to all community members.

Although the Province of Alberta refused to renew their co-operative management agreement with LRRCN in 2002, LRRCN used their C&I as the basis for discussions with Tolko, the main Forest Management Agreement licence holder in northwestern Alberta, about the development of their *Detailed Forest Management Plan* (DFMP). LRRCN translated their C&I into 'Values, Objectives, Indicators and Targets' (VOITs) as laid out within the Alberta Sustainable Resource Development forest management planning regulations. Following a 'deep consultation' approach, LRRCN convinced Tolko to modify their logging operations in critical community use areas, including reducing the amount of old growth harvested, adopting a modified three-pass logging system, increasing buffer areas and deferring logging in cut blocks with a significant number of traditional use sites (Webb *et al.* 2009).

Waswanipi Cree

The Waswanipi Cree Model Forest (WCMF) in Quebec is the only Aboriginal Forest in the Canadian Model Forest Network (CMFN 2009). It is located approximately 800 kilometers north of Montreal in the boreal forest region with the current townsite established in 1976 (Pelletier 2002). The registered population is 1,856 members (INAC 2009) and the community still depends largely on natural resources to maintain its culture and to provide livelihoods based on fishing, hunting, and trapping. The community is focused on maintaining traditional Cree culture with an emphasis on "..Cree language and the relationship to the land of those who speak it" (WCMF 2007), but is also involved in contemporary forestry operations.

In 1983 Mishtuk Corporation was established to harvest timber from lands controlled by the community under the *James Bay and Northern Quebec Agreement* (JBNQA) signed in 1975. The company entered into a joint venture with Domtar in 1997 building Nabaktuk sawmill in the community. Waswanipi was also part of a larger movement of JBNQA Cree communities that undertook legal action against the Quebec government and the forest industry for forestry practices that threatened the Cree traditional way of life. This legal threat eventually led to the *Paix des Braves* agreement with Quebec in 2002 that outlines revenue-sharing from resource development, adopts Cree traplines as forest management units and establishes joint decision-making arrangements, including the Cree–Québec Forestry Board (WCMF 2007).

As a condition of Model Forest Network funding, Waswanipi had to develop a set of C&I for sustainable forest management. However, WCMF found C&I difficult to develop because the concept was too new and difficult to grasp. To begin the work, Waswanipi built on what had been done by the James Bay Advisory Committee on the Environment (JBACE) and the Cree Trappers Association. The forestry sub-committee of the JBACE started work on C&I as a response to the environmental concerns of the Cree. With forestry operations in their traditional territory, the Cree felt that their participation was limited and that their Aboriginal rights were being ignored by industry as were their concerns about land use, biodiversity and ecology (JBACE 1998). The indicator development was supported by legislation in Quebec; both the *James Bay and Northern Quebec Agreement* and the *Quebec Forest Act* made commitments to sustainable forest management, the basis for criteria and indicators. The JBNQA addressed actions to maintain or improve the Cree environment, society, culture and rights (JBACE 1998), supporting in law the inclusion of First Nations in forestry planning.

The main purpose of the C&I developed by the JBACE was to measure forestry impacts on the community, and to establish ground rules to improve the consultation process and participation level of the Cree in management planning. The set of indicators that was developed was preliminary, as on-the-ground testing was required to adapt the indicators if they were not found to be satisfactory. Baseline data also had to be collected as a starting point for monitoring. A second set of consultations with the Cree community and relevant stakeholders was required to determine the effectiveness of the indicators. The JBACE felt that the WCMF would be the ideal location for data collection as the people involved had the means to carry out the necessary work and the facilities to analyze data and store records. The WCMF also benefitted from the work already undertaken by the JBACE so that duplication would not occur (JBACE 1998).

Waswanipi also built on their own earlier efforts with the Trapline/Forestry Project initiated in 1997 to replace consultation efforts on forest management impacts by the forest industry with Cree trappers. On the basis of the trappers' dissatisfaction, Waswinipi entered into discussions with the forest industry to develop a collaborative learning process. This approach evolved into the Ndoho Istchee Integration Vision which combined all WCMF research in their efforts to work with the community and 'in the bush with trappers' to gain a deeper understanding of what the Cree needed to protect their way of life. This effort resulted in *Ndoho Istchee* (WCMF 2007).

The first phase of *Ndoho Istchee* involved documenting Cree conservation values with a second phase focused on Cree interactions with forest managers. The documentation of Cree values resulted in a vision statement and guiding principles (or criteria). The vision statement, "All Eenou are to maintain and enhance the Eenou way of life and the land that supports it" is supported by the following principles:

- protect and preserve the land and wildlife,
- ensure that all stakeholders can co-exist within our traditional territory,
- value, support and promote Eenou stewardship of the land,
- leave a living legacy for future generations,
- maintain control of Cree traditional lands and strengthen Cree cultural identity,
- learn all heritage waterway systems, and
- acknowledge the sacredness of trees (WCMF 2007).

Waswanipi continues to look to the Cree–Quebec Forestry Board under the *Paix des Braves Agreement* with the Province of Quebec to protect their values.

Algonquins of Barriere Lake

The Algonquins of Barriere Lake (ABL) have a registered population of 664 members (INAC 2009). The community lives on the Rapid Lake Reserve, north of Montreal. The local environment is a mix of boreal and Great Lakes–St. Lawrence ecoregions with diverse tree species such as black spruce (*Picea mariana* (P. Mill.) and white spruce (*Picea glauca* (Voss), white and yellow birch (*Betula alleghaniensis* Britt.), red pine (*Pinus resinosa* Soland.) and maples (*Acer spp.*). The Algonquins are tied to the natural environment as a way of life. The community did not have a reserve until the 1960s, nor modern housing until the 1970s. Prior to this, the community lived throughout the traditional territory, a connection that remains strong with seasonal fluctuations of population as members return to family territories for hunting and trapping. Some of the key species are bear, moose and deer. In the spring and fall the community has 'beaver breaks' to accommodate the hunt with the school term incorporating the breaks for children to accompany their families.

In the 1980s the Algonquins saw many negative impacts of development on their traditional territory. Clearcutting was affecting the land, dams were being built that affected their waterways and logging roads increased competition for hunting and trapping with non-Aboriginal people from outside the territory. The community struggles with several social problems, including an unemployment rate of between 80 and 90 percent, overcrowding in homes, low levels of education and a high dependence on government transfer payments. Millions of dollars are generated from forestry, fishing, hunting, and tourism in the Algonquin traditional territory. The community could see many benefits if they had access to some of the profits generated from these industries.

In 1991 the *Barriere Lake Trilateral Agreement* was signed with the Federal Government and the Government of Quebec. This agreement was created for the purpose of developing an Integrated Resource Management Plan (IRMP) to manage an area of approximately 1 million hectares. The management plan was to be based on the principles of the *Brundtland Report* to support sustainable development while maintaining the Algonquin traditional way and listening to their environmental concerns (Notzke 1993).

Work began on the IRMP with data collection and analysis, inventory and research on local natural resources and their uses by the community. A draft plan was created and recommendations for implementation of the plan were presented. As a means of measuring the effects of the plan on the community, criteria and indicators were identified. The guiding principles for the C&I development were the continuation of the traditional way-of-life, conservation, versatile use and adaptive ecosystem-based management. For place-based, environmental impacts, the ABL took an areas-of-concern approach, outlining prescriptions for each area. For example, for medicinal plant collection, the maintenance of so many hectares of Algonquin-identified collection areas throughout the 20-year forest management planning was prescribed.

The ABL decided to focus on social indicators as a key area for measuring ecological and cultural sustainability. Other indicators included economic health, community statistics and other non-forestry related indicators that can help to determine community health.

The ABL have struggled with highly publicized internal divisions and conflict, exacerbated by federal government interference in local governance and the withdrawal of support at various times from the Federal or Quebec governments. Despite the acrimony, the community continues to pursue the vision outlined in the *Trilateral Agreement*, using the indicators they have developed to negotiate modifications to forestry operations in their territory.

Innu Nation

Nitassinan is the Innu word meaning 'our land' and refers to the traditional lands occupied by the Innu in parts of eastern Quebec and Labrador. Sheshatshiu and Natuashish are the prominent communities of the Innu Nation in Labrador with a total population of approximately 2,200 (Innu Nation n.d. a,b), with 1,279 registered in Sheshatshiu (INAC 2009).

The cultural environment in Nitassinan is one with ties to the past and looking toward the future. Traditionally, Innu have depended on the boreal forest to provide the necessary materials for tools, food, shelter and medicine. Canoes paddles, snowshoes and fishing equipment all came from the forest. Even today there is a reliance on the environment to maintain the Innu ways of living.

Current economic development in the community include hydro, mining, hunting and fishing, forestry and tourism. Development projects must protect the needs of the land, animals and people of Nitassinan using an ecosystem-based approach. The Innu believe that economic stability will provide a future for the community, but they also want to maintain their subsistence lifestyle. They believe that providing for the future while protecting the past can strengthen the community.

Large-scale development projects have not always taken Innu issues or concerns into consideration, and have occurred without consultation or consideration of the community and its needs. The 1970s saw the beginning of forestry operations in Nitassinan. Due to several constraints, forestry operations failed, leaving behind a poor image of forestry in Labrador.

In the 1980s and 1990s the Innu Nation began taking a proactive approach to protecting their lands. Road blockades were set up to limit clear cutting in sensitive areas and science was beginning to play a role in documenting the harmful environmental impacts of harvesting on the unique landscape in Labrador.

The Innu Nation and the Government of Newfoundland and Labrador signed a historic agreement in 2001 called *The Forest Process Agreement* (Higgins 2008, that entailed full participation of the Innu Nation in forestry operations using an ecosystem-based approach. The agreement paved the way for the Ecosystem-Based Forest Management Plan for District 19 in central Labrador within the boreal forest ecosystem. District 19 is divided into three units totaling 7.1 million hectares. The management plan focuses on unit 19A which has an area of 2.1 million hectares.

This forest is characterized by large quantities of timber, mainly spruce and fir. Due to the cold climate in Labrador, the majority of the land is characterized by shallow soils which are nutrient-poor supporting little vegetation. There are a number of lakes and rivers which help define the landscape. Labrador also has populations of wildlife species including the woodland caribou, harlequin ducks and marten. Unit 19A hosts the majority of the forest in Labrador, making it an excellent location for forestry development.

As part of the *Forest Process Agreement*, the Innu became co-authors with the Government of Newfoundland and Labrador of the forest management plan (Forsyth *et al.* 2003). The process was also facilitated by the Model Forest of Newfoundland and Labrador (MFNL n.d.). It is within this plan that a set of objectives and actions are defined and the strategies to be implemented to meet community objectives are outlined. Between 1996 and 1998, planners held several public stakeholder meetings about the development of C&I and, in the end, decided to follow the CCFM C&I framework. Because the CCFM C&I are weak in the cultural area, the planners drew from the CIFOR C&I which focused on rural and Indigenous communities and FSC principles and criteria (Forsyth *et al.* 2003). The data used to create the objectives and actions was gathered by public consultation and through the work of the Forest Guardian Program that also has responsibility for monitoring the implementation of the plan (Courtois n.d.). As Table 5 shows, the Innu put the ecosystem first with social, cultural and economic factors dependent on the maintenance of ecosystem health.

Table 5. Ecosystem-Based Planning Objectives for District 19, Labrador (Innes and Moores 2003).

Theme	*Ecosystem-Based Planning Objectives*
Ecological Integrity	Respect the ecological limits of various ecosystems to human disturbance. Natural biological diversity and natural disturbance regimes will be protected and maintained through historic range and variability in order to maintain natural forest functioning.
Hierarchical Context	Ensure that all plans and activities protect, maintain and, where necessary, restore forest functioning at the landscape, watershed and stand level scales.
Ecological Boundaries	Focus on the ecological features to retain and utilize ecological boundaries at all levels of planning.
Values	Aboriginal and non-aboriginal cultural values will be respected and protected.
Humans Embedded in Nature	Plan and carry out diverse, balanced activities to encourage ecological, social, and economic well-being and stability. The maintenance of ecosystem health is recognized as the basis for sustaining cultures and economies.
Adaptive Management	Apply the precautionary principle to all plans and activities utilizing monitoring, assessment and adaptive management.
Data Collection	Undertake research on ecosystem structure, function, sensitive habitats, disturbance regime dynamics and impacts of timber harvesting.
Interagency Cooperation	Ensure effective communication and cooperation channels are created between management organizations. Ensure all management organizations accept and support the listed guiding principles.
Organizational Change	Encourage management organizations to adapt past practices and operating structures in order to facilitate an EBP approach and build trust between other management organizations.
Monitoring	Review and evaluate the success of all forest activities in meeting the previous nine principles.

References Cited in Appendices

CMFN—Canadian Model Forest Network (2009). *Cree Research and Development Institute*. http://www.modelforest.ca/cmfn/en/forests/waswanipi/default.aspx. Nov. 6/09.

CSSP—Clayoquot Sound Scientific Panel (1995). *Sustainable Ecosystem Management in Clayoquot Sound: Planning and Practices*. Cortex http://www.cortex.org/Rep5c1-2.pdf. .

Courtois, V. (n.d.) *Sustaining Nitassinan Forests: The Innu Nation Forest Guardian and Ecosystem-based Forest Management in Labrador*. Ottawa: Canadian Boreal Initiative. http://www.borealcanada.ca/documents/innu-forest-guardians.pdf. Nov. 9/09.

Forsyth, J., L. Innes, K. Deering, and L. Moores (2003). *Forest Ecosystem Strategy Plan for Forest Management District 19, Labrador/Nitassinan 2003-2023*. Innu Nation and Dept. of Forest Resources and Agrifoods. Government of Newfoundland and Labrador. http://www.env.gov.nl.ca/env/env/ea%202001/Archival%20EA%20Documents/pdf%20files/1062%20-%20Crown%20District%2019A%20Forest%20Plan%20-%20Goose%20Bay/Strategy%20Plan/Text.pdf. Nov. 6/09.

Higgins, J. (2008). *Innu Organizations and Land Claims*. Newfoundland and Labrador heritage website. http://www.innu.ca/index.php?option=com_content&view=article&id=10&Itemid=7&langn. Nov. 9/09

Iisaak. (n.d). *Clayoquot Sound Background*. http://www.iisaak.com/history.html. Nov. 6/09.

Indian and Northern Affairs Canada (2009). *First Nation Profiles*. http://pse5esd5.aincinac.gc.ca/fnp/Main/Index.aspx?lang=eng. Nov. 5/09.

JBACE—James Bay Advisory Committee on the Environment (1998). *A proposal for the first approximation of criteria and indicators of sustainable forest management for Eeyou Istchee*.

JPRF—John Prince Research Forest (n.d.). *Background*. http://researchforest.unbc.ca/jprf/jprf.htm. Nov. 6/09.

Karjala, M.K. and S.M. Dewhurst (2003). Including Aboriginal issues in forest planning: A case study in central interior British Columbia, Canada. *Landscape and Urban Planning* 64(1-2): 1-17.

Karjala, M., E. Sherry, and S. Dewhurst (2003). *The Aboriginal Forest Planning Process: A Guidebook for Identifying Community-level Criteria and Indicators*. Ecosystem Science and Management Program. Prince George, BC: University of Northern British Columbia.

Karjala, M., E. Sherry, and S. Dewhurst (2004). Criteria and indicators for sustainable forest planning: A framework for recording Aboriginal resource and social values. *Forest Policy and Economics* 6(2): 95-110.

Kearney, A.R., G. Bradley, R. Kaplan, and S. Kaplan (1998). Stakeholder perspectives on appropriate forest management in the Pacific Northwest. *Forest Science* 45: 62-73.

Lertzman, D. and H. Vredenburg (2005). Indigenous peoples, resource extraction and sustainable development: An ethical approach. *Journal of Business Ethics* 56(3): 239-254.

Mabee, H.S. and G. Hoberg (2004). Protecting culturally significant areas through watershed planning in Clayoquot Sound. *Forestry Chronicle* 80(2): 229-240.

MFNL—Model Forest of Newfoundland and Labrador (n.d.) *Canadian Model Forest Network: Case Study*. Futures from Forests series. Ottawa: Canadian Forest Service, Natural Resources Canada http://www.wnmf.com/files/futuresFiles/Canadian%20Model%20Forest%20Netw ork.pdf. Nov. 06/09.

Natcher, D.C. and C.G. Hickey (2002). Putting the community back into community-based resource management: A criteria and indicators approach to sustainability. *Human Organization* 61(4): 350-363.

Nelson, M., D.C. Natcher, and C.G. Hickey (2008). 'Subsistence harvesting and the cultural sustainability of the Little Red River Cree Nation,' pp. 29-40 in D.C. Natcher, ed., *Seeing Beyond the Trees: The Social Dimensions of Aboriginal Forest Management*. Concord, ON: Captus Press.

Notzke, C. (1993). *The Barriere Lake Trilateral Agreement*. Barriere Lake Indian Government.

Pelletier, M. (2002). *Enhancing Cree Participation by Improving the Forest Management Planning Process*. Quebec: Waswanipi Cree Model Forest. http://www.modelforest.net/cmfn/en/publications/publications/publications_ record.aspx?title_id=3418. Apr. 27/08.

Sherry, E., R. Halseth, G. Fondahl, M. Karjala, and B. Leon (2005). Local-level criteria and indicators: An Aboriginal perspective on sustainable forest management. *Forestry* 78(5): 513-539.

Tl'azt'en Nation (2009). *About Us*. http://pse5-esd5.ainc inac.gc.ca/fnp/Main/Search/FNRegPopulation.aspx?BAND_NUMBER=617 &lang=eng. Nov. 5/09.

Umeek [Atleo, R.E.] (2004). *Tsawalk: A Nuu-chah-nulth Worldview*. Vancouver: UBC Press.

WCMF—Waswinipi Cree Model Forest (2007). Ndoho Istchee: *An innovative Approach to Aboriginal Participation in Forest Management Planning*.

Yin, D.K.Y. (2009). *Evolving Co-management Practice: Developing a Community-based Environmental Monitoring Framework with Tl'azt'en Nation on the John Prince Research Forest*. Master's thesis, University of Northern British Columbia.

SECTION 4:
CAPACITIES FOR CO-EXISTENCE

Chapter Twelve

Capacity for What?
Aboriginal Capacity and Canada's Natural Resource Development and Management Sectors[1]

Marc G. Stevenson and Pamela Perreault

Introduction

In response to the settlement of land claims, court decisions on Aboriginal and treaty rights, social and economic problems in many Aboriginal communities and other drivers, capacity building has emerged as a key priority for many Aboriginal communities.[2] Federal/provincial/territorial governments have also responded to the issue with whatever resources and political will they can muster *vis-à-vis* shrinking budgets and other challenges. With natural resource extraction taking place on traditional Aboriginal territories, many Aboriginal communities look to natural resource development as a way to break the shackles of poverty and dependency, and embark on a path to economic self-sufficiency. However, a lack of 'capacity' among Aboriginal peoples is most often cited by all parties as the key barrier to realizing these goals. Economic remedies to the many social problems plaguing Aboriginal forest-dependent communities (poverty, high unemployment, high welfare dependency, etc.) are sought and seen as the most proximate solution.

A focus has thus emerged in most provincial and territorial jurisdictions on providing Aboriginal peoples with the necessary education, training and skills to capture employment and business opportunities in the resource development sector. This approach, which we refer to as the 'capacity deficit model,' continues to drive nearly all government and industry, and even some Aboriginal, capacity building initiatives.

However, these initiatives have generally failed to improve the economic, social and cultural well-being of forest-dependent Aboriginal peoples and

---

[1] This chapter is based on the Sustainable Forest Management Network's publication, *Capacity for What? Capacity for Whom? Aboriginal Capacity and Canada's Forest Sector* by Stevenson and Perreault (2008). The authors would like to thank the Network for permission to rework and present various sections of that report in this chapter.

[2] Aboriginal community is used here to include forest-dependent First Nations and Métis communities. Community does not necessarily denote a group of people living at the same location, but refers to people bound together by a number of common features (e.g., history, culture, values, kinship, etc.) that define them as a community.

communities. Indeed, many capacity building initiatives for Aboriginal peoples in the natural resource development sector build skills and promote values incongruous with traditional Aboriginal worldviews and values. Given the current state of affairs in Canada's forestry, mining and oil/gas sectors, we must ask whether Aboriginal engagement in these sectors provides a solid foundation upon which to build sustainable Aboriginal communities and preserve their cultural values, knowledge and traditions. Capacity building initiatives aimed exclusively at increasing Aboriginal participation in the resource development sector may, in fact, be setting up Aboriginal peoples for failure, disappointment, cultural loss and ultimately greater dependency.

Frequently, non-Aboriginal interests and funding sponsors view Aboriginal capacity needs within a constellation of existing economic, technical, social and political relationships, institutions and systems. These views underpin and give momentum to the 'capacity deficit model.' While such approaches may suffice to meet mutual short-term economic interests of Aboriginal peoples, they fail to prepare all parties with the necessary skill sets and professional competency to address effectively existing and emerging social, economic, cultural and political realities in Indian Country, or to develop sustainable economic and ecological relationships with forested lands and resources.

A more thorough, reflective and grounded answer to the question 'Capacity for What?' necessitates a review of Aboriginal capacity building programs in the natural resource development sector, and their success in Aboriginal communities. While we previously conducted such a review for the forest sector (Stevenson and Perreault 2008), Aboriginal capacity building initiatives targeted at the broader resource development sector also need to be analyzed from both a 'full cost accounting' and an Aboriginal perspective. Although this examination is beyond the scope of this chapter, it is worthwhile noting that our investigation of four First Nations—Tla'zt'en Nation, Little Red River Cree Nation, Waswanipi Cree First Nation and Innu Nation—that attempted to create meaningful employment for their members in the forest sector through a variety of capacity building initiatives reveal that the 'operation was a success, but the patient died.' Our analysis (Stevenson and Perreault 2008) indicated that these First Nations derived very few benefits from their participation in the forest sector. Common themes that emerged from these case studies relate to issues of 'cultural fit' and the uncertainty of political and economic forces with respect to the security and sustainability of Aboriginal communities and economies. The most notable successes of government-sponsored Aboriginal capacity building programs appear to be experienced as gains in individual skill sets that further personal aspirations of economic self-sufficiency. Whether or not these successes contribute positively to community needs, goals and aspirations is a matter of debate.

These case studies and a review of existing programs revealed the need to develop a more nuanced approach to building capacity in forest/natural resource-dependent Aboriginal communities (Stevenson and Perreault 2008). Such approaches must make room for Aboriginal peoples, not just as 'hewers of wood,' but in all aspects of forest land use decision-making, planning, management and resource extraction. 'Capacity deficit' and 'top down' approaches to Aboriginal capacity building are only part of the solution, and must be met in equal or greater measure with 'bottom-up' approaches, which put Aboriginal peoples at the centre of determining and realizing their capacity needs.

In this chapter, we explore Aboriginal capacity building *vis-à-vis* the natural resource development and management sectors from multiple perspectives, scales and dimensions. Theoretically grounded in a growing body of literature on Aboriginal and indigenous empowerment where local communities drive the design and delivery of capacity building programs, this approach builds on and situates existing Aboriginal capacity building initiatives within a conceptual framework that, if implemented, should facilitate Aboriginal peoples' aspirations to become true architects of their future.

We would also like to acknowledge the work of our associates on the *National Forest Strategy Theme Three Capacity Working Group*, and their involvement in the formulation of the ideas expressed in this chapter. Composed of representatives from Aboriginal organizations (First Nations and Métis), non-government organizations, academic institutions and provincial and federal government departments, the Aboriginal Capacity Working Group (ACWG) was formed to address the National Forest Strategy Coalition's (2003) commitments to develop the institutional arrangements and support needed to accommodate Aboriginal rights and participation in the sustainable use of Canada's forests. Both of us were active members of the ACWG, and in many respects, the thinking that evolved during the course of our work and the work of the ACWG complement and mirror each other. For a summary of what we perceive to be the key findings and conclusions of the ACWG see Appendix A.

Framing the Issue

Various governments, national Aboriginal organizations, federal commissions and senate committees acknowledge that Canada's Aboriginal peoples have not received their fair share of benefits from natural resource extractions on their traditional lands (e.g., AFN 2005; British Columbia 2005; Government of Canada 2007; NRTEE 2005; RCAP 1996). At the same time, most federal, provincial and territorial agencies now recognize that major natural resource developments and management decisions cannot proceed without significant Aboriginal support (NTREE 2005:44). Yet, as currently conceived, regulated and practiced, forestry, mining, oil/gas and other industrial activities have simply not improved the lives of Aboriginal peoples and communities to the extent envisioned by all or most parties. In fact, natural resource extraction is beginning to be viewed by Aboriginal peoples as more a curse than an opportunity (Gibson and Klinck 2004).

The promise of jobs in the natural resource development sector is often used by government and industry to gain access to lands and resources where Aboriginal peoples assert rights protected by the *Canadian Constitution Act, 1982*. As the prevailing dogma goes, if only Aboriginal peoples had the education, training and skills (i.e., 'capacity') to participate in natural resource extraction, planning and management, they could rise above their social and economic problems and become productive members of Canadian society.[3] This thinking is perpetuated and reinforced, not only by existing government-sponsored capacity building initiatives, but by some Aboriginal governments and organizations who look to natural resource development as a way to improve economic and social conditions within their communities, but only if carried out in a manner that is ethical and where the land is properly cared for and respected. But rarely is development ethical, as corporate social responsibility takes a backseat to the 'bottom-line.' The focus is on capturing employment and business opportunities in the natural resource development sector. As a result, education, training and skills development in a variety of fields ranging from resource extraction to planning, are seen as the solution.

This 'capacity deficit model' continues to dominate nearly all government and industry, and even some Aboriginal, approaches to capacity building (NAFA 1993). While some of these initiatives have achieved a modicum of success, noticeable improvements to the economic and social well-being of forest-dependent Aboriginal peoples and communities are rare. Unemployment, poverty and welfare dependency rates in most forest dependent Aboriginal communities remain unacceptably high and are many times the national average. Indeed, recent analyses reveal that Aboriginal communities in forested environments remain significantly poorer than those in non-forested environments (Gysbers and Lee 2003).

The current state of affairs in Canada's forestry sector is tenuous, while its energy, particularly oil/gas, and mining sectors are only marginally better off. Fluctuations in commodity prices, parity in the Canadian/US dollar, increasing international competition, the world-wide economic downturn, slowing foreign demand for natural gas, minerals and lumber, and other factors force us to ask whether Canada's natural resource development sectors provide adequate foundations upon which to build sustainable Aboriginal communities, when experience suggests that they may have little to contribute to their long-term economic, social and cultural stability.

Compounding matters is the all too common practice of natural resource development proceeding in the absence of a thorough understanding of the impacts on Aboriginal peoples and communities, who are left to face the consequences of development before, during and long after resource extraction ends. For thousands of Aboriginal peoples, the promise of 'development' through resource extraction is weighed daily against the understanding that to act on this opportunity is to compromise their long-standing relationship with, and responsibilities to, lands and resources that have sustained them for generations. Nevertheless, the mobilization of Aboriginal peoples in natural resource development has become the mantra of many stakeholders (governments, academia, industries and Aboriginal).

At the same time, many Aboriginal peoples view the issue of 'capacity' as a two-way street. Industry and government need capacity too! Industry,

---

[3] A more pessimistic view holds that the promise of 'capacity development' is often used a carrot by government and industry to gain access to Aboriginal lands and resources, particularly those that are highly contested.

provincial governments, and other non-Aboriginal interests rarely acknowledge that they need the capacity to address the three pillars of sustainability— environmental, social and economic. Such attitudes ill prepare policy and decision-makers with the necessary skill sets and professional competency to effectively address existing and emerging realities in Indian Country. In particular, there is a lack of existing capacity within government and industry circles to manage effectively for the broad range and complex articulation of issues, values, needs, rights and interests that need to be considered in the quest for sustainability. Thus, capacity to develop and sustain viable economic and ecological relationships with forested lands and resources is not specific to Aboriginal peoples. It is also a requirement for non-Aboriginal governments and industries, especially with respect to the design and implementation of institutions that recognize and accommodate Aboriginal needs, rights and interests and create space for their knowledge, values and management systems in land use planning and decision-making.

Building Aboriginal Capacity from the Ground Up

> *Capacity building refers to the need for First Nations People and First Nations organizations to gain the competence and ability to do various things. In Burnt Church it was a term used by the government to say that the Burnt Church people were not ready to fish for lobster, nor ready to manage the fishery in a responsible way, or to engage in business and economic development. Capacity building has become a polite and politically correct way for governments and others to say to the First Nations: 'You are not ready to do this yet. But if you wait; if you are patient; if you get more training; if you make the arrangements we suggest; if you just do this our way, sooner or later you will have the capacity to do what we do. And when you accomplish this: when you have qualified for our programmes, when you have slowly managed to gain the qualifications we require, then we will consider some kind of partnership with you.* (Matthew Coon Come 2001).

The words of the former Grand Chief of the Assembly of First Nations speak eloquently to the thinking behind, and shortcomings of, most government- and industry-sponsored Aboriginal capacity building programs in the natural resource development and management sectors. In order to participate in Canadian society and economy, Aboriginal individuals must develop skills and knowledge (i.e., capacities) that are valued and rewarded by this hegemony. In many cases, these programs have benefited Aboriginal peoples, economically, personally, psychologically and in other ways. Because of this, funding support for current programs and initiatives that target the capacities of Aboriginal individuals should be increased substantially and made more user-friendly to those peoples they are intended to serve.

271

Current capacity building initiatives to engage forest-dependent Aboriginal peoples in the natural resource development management sectors alone, however, are not sufficient, nor should they be the first priority. Forest-dependent communities generally are inherently unstable compared, for example, to agricultural- or tourism-dependent communities. Using a broad range of indicators, Drielsma (1984) found that forest-dependent communities are among the least stable and prosperous because they tend to have high population turnover and more social problems (e.g., divorce, suicide, low social cohesion) than other communities. For a variety of reasons, these problems are even more acute and compounded in forest-dependent Aboriginal communities. Moreover, recent analyses reveal that Aboriginal communities in forested environments have significantly lower average incomes and employment rates than those in non-forested environments (Gysbers and Lee 2003). Aboriginal communities and their governments require access to resources, financial and otherwise, to:

1) construct their vision of the future and their desired relationships with their lands and resources, and
2) build and walk down the path(s) to get there.

In the meantime, Aboriginal peoples and communities ought not to ignore existing opportunities and challenges. Short-term goals and objectives to address current challenges and to take advantage of employment/business opportunities need to be set. Existing 'capacity strengths,' not just capacity needs, of the community to achieve these targets also need to be identified and supported. Too often, capacity building initiatives for Aboriginal peoples in the natural resource development sectors are framed by the needs and aspirations of the dominant society. Such a notion obfuscates the fact that such programs are created by political agendas, institutional arrangements and economic processes that Aboriginal peoples had little hand in creating. Collectively, we need to ask: *What are the existing capacity strengths and capacity needs of Aboriginal peoples and communities to construct, implement and realize their vision of the future with respect to their own culture, identity, traditional lands and resources?* The real capacity needs of forest-dependent Aboriginal peoples and communities have a much better chance of being properly identified and ultimately accommodated by framing the capacity issue in this way.

While it is important to address local capacity needs and capacity strengths over the immediate/short-term, this cannot be accomplished at the expense of the long-term objectives and aspirations of Aboriginal communities. The capacity strengths and needs of the community and its members to achieve their ultimate goals need to be identified and accommodated. It is also critical that the capacity issue be addressed in a framework that includes not only the natural resource development and management sectors, but all sectors (health, education, local government, traditional economy, etc.) relevant to building and sustaining the human, natural, social and cultural capitals or assets of Aboriginal peoples and communities. In other words, the capacity to develop and implement economically and ecologically sustainable relationships with traditional lands and resources cannot be considered in isolation of other capacity needs. By addressing the capacity question within an integrated and comprehensive community-driven strategy, the chances of success over both the short-term and long-term will be increased substantially.

This approach would target not just a sustainable economic future, but a future that sustains and enhances the well-being and ecological, social, cultural and other values of the community and its members; all of which are interrelated and cannot be separated without undermining Aboriginal institutions and cultures. All members of the community (elders, youth, women, men, traditional land users, health care workers, educational workers, chief and council, etc.) should be involved in this visioning exercise. Fundamental to the success of this project is the realization that Aboriginal communities need the time, opportunity and resources (financial and otherwise) to:

1) document, assess, prioritize and develop consensus about their uses, values and needs with respect to their lands and natural resources;

2) undertake the planning and other related research (trade-off analyses, market assessments, traditional land use studies, resource inventories, full costs accounting, etc.) to produce economic development, land use management and other plans (health, education/training, etc.) commensurate with the community's needs and desired relationship with their lands and resources; and,

3) negotiate, with government and industry, and develop appropriate institutional and other arrangements to effect these plans and create win-win situations for all involved.

In order to undertake and implement these steps in a comprehensive and effective manner, local capacity needs and strengths will need to be identified, addressed and accommodated. It is within this scoping exercise that assessments of, and engagements with, current government-sponsored capacity building programs should take place. At the same time, Aboriginal peoples and communities may, in fact, be found to already possess many strengths required for their empowerment, long-term sustainability and for implementing the above initiatives.

The capacity to take advantage of, and engage in, current employment and business development opportunities in the natural resource development sectors is one of the key capacities that forest-dependent Aboriginal peoples and communities may want to consider over the short-term. However, focusing on the development of short-term capacity needs at the expense of capacities required to realize long-term individual and community goals and aspirations needs to be avoided. The capacity issue challenges Aboriginal peoples and communities at both scales:

- short-term: to capitalize on existing employment/economic development opportunities and address current challenges, and,

- long-term: to realize their desired relationships with their lands and resources.

In order to achieve personal and community goals and aspirations, Aboriginal peoples must also have the capacity to represent both themselves and their communities effectively in their engagements with government and industry. The ability to 'represent' requires a culturally grounded understanding of community values, goals, needs, rights, interests and their articulation—a tall order without the necessary financial support and resourcing to undertake the type of community visioning exercises advocated in this chapter. By way of example, elders with a lifetime worth of experience on the land may be well positioned to represent the needs, values and aspirations of their communities, but they might have little capacity to participate in existing engagements with government and industry. Conversely, Aboriginal youth who leave their communities to obtain education and training in fields that capitalize on existing employment opportunities in natural resource extraction, planning and management, may lack the capacity to effectively represent community interests, rights and values. 'Engagement' capacities cannot be built in the absence of 'representative' capacities, and can be avoided by financing and resourcing the type of processes considered below.

Capacity-building for forest-dependent Aboriginal communities can be conceptualized as a three-dimensional box or contingency table structured along three axes (Fig. 1):

1) engagement/representation on the 'X' axis,
2) short-term/long-term on the 'Y' axis, and
3) individual/community along the 'Z' axis.

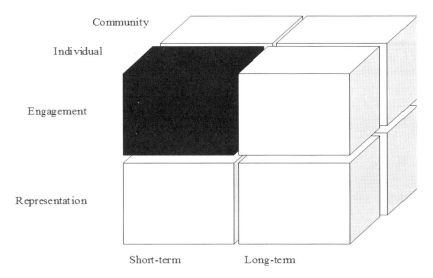

Figure 1. *Relationship of individual/community, engagement/representation, and short-term/long-term capacities.*

To date, most government-sponsored Aboriginal capacity building efforts
in the natural resource development sector have focused on <u>individual</u> capacities
to <u>engage</u> in available employment/business development opportunities over the
<u>short term</u> (black box in Fig. 1). Any capacities that are built in areas related to
community, representation and/or long-term needs are a collateral benefit of this
focus. Another way of looking at the issue is to conceptualize the black box in
Figure 1 as forming the inner core of the larger circle (Fig. 2). Operating under
the notion that we should 'begin where we are at,' individual/engagement/short-
term capacities, i.e., the capacities targeted by existing programs and initiatives,
would provide the building blocks or core capacities needed to achieve
community/representation/long-term capacities.

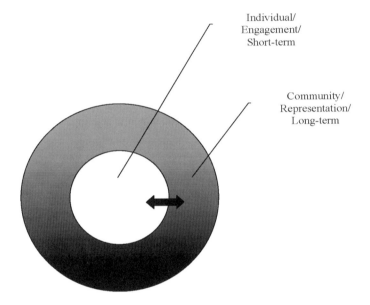

Figure 2. *Relationship of core and broader capacities required for Aboriginal
empowerment and sustainability.*

The ACWG (Kepkay 2007) puts this relationship nicely in the context of forest
tenure arrangements:

> …pragmatic opportunities for incremental progress within existing
> frameworks also need to be engaged. For example, although a new
> type of forest tenure specifically tailored to the traditions of
> Aboriginal Peoples may require many years to develop, in the
> meantime it is possible to take smaller steps towards the same goal
> by adapting the terms of an existing tenure. The lessons being
> learned and the skills being developed by communities and
> businesses holding conventional tenures today will build their
> capacity to represent and pursue their interests in the innovative
> arrangements being developed for tomorrow.

These figures illustrate two truths. First, current capacity building efforts, no matter how successful, target only a limited percentage of the capacity requirements of forest-dependent Aboriginal peoples and communities. Second, the issue of capacity in this context, and our collective response to it, needs to consider all these dimensions and their articulation.

In Consideration of Aboriginal Values and Assets

Another way of looking at the capacity issue, particularly with regard to social and cultural sustainability, is to consider the many benefits that Aboriginal forest-dependent peoples derive from their relationship to and dependency on their traditional lands and resources. A short list of these values include economic, nutritional, physical, spiritual, psychological, environmental, ecological, cultural and social values (*see* Stevenson and Perreault 2008:29-30 for a comprehensive definition of each value). The question that must be asked of all capacity building initiatives aimed at Aboriginal peoples is: Does the initiative enhance or erode the social, cultural, natural, human and other capitals or assets of Aboriginal communities? All too frequently, only the economic benefits that might accrue to individuals are taken into consideration. What is not considered is the impact of such initiatives on the social, cultural and other capitals (and their integration) necessary to sustain Aboriginal peoples and communities. In fact, these other considerations, and how they might be affected by such scenarios, are viewed as 'externalities' (i.e., outside the scope of the sponsoring agency's mandate), to be managed by Aboriginal beneficiaries. While most capacity building programs are well intended, and not necessarily antagonistic to many Aboriginal values, they become grand experiments in social and cultural engineering when they do not consider the fundamental linkages and relationships among these capitals, and how changes in one may affect changes in others.

As the case studies described in Stevenson and Perreault (2008) suggest, the social capital and assets necessary to sustain Aboriginal communities may not be supported by existing capacity building programs. Witness the fact that none of the 30 members of the Little Red Cree Nation that were trained as log haul truck drivers found employment in that profession. The nature of the job (i.e., long hours in an isolated job) likely eroded rather than enhanced social roles, relationships and responsibilities, while the economic incentives were simply not great enough to outweigh the losses (social, cultural, other) that would result from being gainfully employed in the forest sector.[4] The same could be said of the Waswanipi Cree First Nation where a high desertion rate was identified as the greatest problem by both Aboriginal and employer interviewees (Rousseau 2006).

---

[4] This is not to suggest that they did not receive any benefit from their training.

Grounding the Model: Considering Community Capacity

Capacity building is broader than human resource development of individuals....Individual needs should fit within an overall system in which the collective needs of the Aboriginal community are addressed (Bombay 2007).

Beckley *et al*. (2008) define community capacity as the 'collective ability of a group (the community) to combine various forms of capital within institutional and relational contexts to produce desired results or outcomes.' Capacity outcomes may be defined either narrowly or more broadly. In the model being advanced here, Aboriginal communities may want to consider both macro (long-term) and micro (short-term) approaches to capacity building. Because external capacity building initiatives tend to support more short-term, expedient capacity outcomes, communities should develop their own indicators of measures of success accordingly.

Nadeau *et al*. (2003) examine the notion of 'community' from multiple perspectives advancing three inter-related conceptual frameworks: community capacity, community well-being and community resiliency. Community capacity and community resiliency are closely related and, in turn, result in community well-being. In other words, the issue of community capacity cannot be considered in isolation from community well-being or resiliency. Community capacity in forestry has been used to estimate the collective ability of residents to respond to external and internal stress, to create and take advantage of opportunities, and to meet their diverse needs (Kusel 1996). The major challenge of community capacity assessments is to identify the specific attributes of a community that facilitate or impede its ability to respond to problems or external threats (Nadeau *et al*. 2003). According to these authors, the major attributes that determine community capacity are:

1. Social Capital,
2. Human Capital (skills and abilities of individuals),
3. Environmental Capital (natural resources, environmental integrity/biodiversity), and
4. Economic Capital (physical/financial infrastructure/resources).

On Social Capital
Social capital has been used to refer to features of social organization such as networks, norms, values and social trust that result in, and are the result of, collective and socially negotiated ties and relationships. Recent work by a number of authors (e.g., Edwards 2002; Flora and Flora 1993; Haley 2007; Naryan 1999; Putnam 2000) is beginning to converge on the notion that social capital operates on two different, but complementary, levels: 1) horizontally to facilitate the inclusion of individuals, groups, ideas and values into communities (bonding social capital), and 2) vertically to facilitate interaction between individual communities and external organizations, institutions and other communities (bridging or linking social capital) (Nayran 1999). Beckley *et al*.

277

(2008) further identify at least four basic types of social relations that constitute 'bonding' capital: market relations, bureaucratic relations, associative relations and communal relations. It is the latter two that perhaps are the most pervasive in structuring social relations and interaction in most forest-dependent Aboriginal communities.

Haley (2007) conceptualizes social capital in northern Aboriginal communities as being characterized by the interplay of bonding and linking social capitals (Fig. 3). Communities with low internal cohesion and high external linking capital risk a breakdown of internal social networks and losing people to out-migration. Alternatively, high internal social cohesion and few external linkages may lead to social, cultural and economic paralysis. Aboriginal communities that score high on both counts (internally cohesive with access to diverse resources and opportunities resulting from viable external relationships and connections) have the greatest prospects for sustainability.[5] In today's world, a strong measure of both may be needed to sustain most forest-dependent Aboriginal communities. Communities that are not well integrated internally and have few external linkages that provide access to resources, people, institutions and opportunities outside the community are likely not sustainable (Fig. 3).

Internal Cohesion

		Low	High
External Linkages	Low	**Poor**	**Moderate**
	High	**Moderate**	**Good**

Figure 3. *Interplay of internal cohesion and external linkages in Aboriginal communities, with projected outcomes (in boxes) for long-term sustainability.*

---

[5] In essence, the kinship systems of most Indigenous peoples may be viewed as an attempt to balance intra-group cohesiveness and intergroup cooperation. For example, those groups characterized by endogamous marriage rules and hostile relations with outsiders are at risk of extinction.

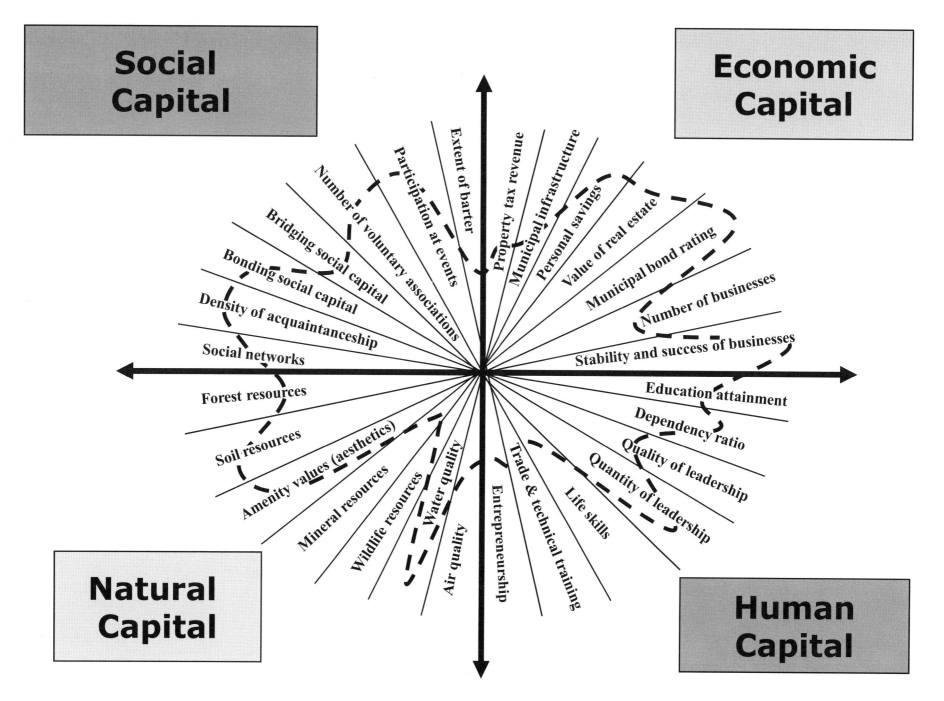

Figure 4. *Beckly* et al. *'s (2008) community asset amoeba model (reproduced with permission of the authors).*

Current Aboriginal capacity building programs most often aim to improve the 'linking' social capital of individuals to external economic development opportunities, without regard to supporting the 'bonding' social capital within communities. Framed in this way, the natural resource development and management sectors, and the capacity building programs that facilitate Aboriginal participation in these sectors, have a critical role to play. Alternatively, one can readily grasp how too great an emphasis on accessing and sustaining external linkages without careful consideration of the impacts on the internal or bonding social capital of the community may be detrimental to community interests.

Beckley *et al.* (2008) envision social, human, natural and economic capitals as providing the minimum asset base for any given community. Moreover, these can be combined and organized to produce a range of outcomes, including capacities to:

1) maintain or enhance economic vitality (economic capital),
2) access resources from the state, e.g., revenues, political will, infrastructure (financial capital?),
3) maintain a vital civic culture (social and cultural capital),
4) subsist and persist (human capital),
5) maintain ecological integrity (natural capital), and
6) maintain human health (human capital)

Capacity outcomes relating to employment and economic enhancement have traditionally been the major, if not exclusive, concern of many politicians, community developers and business leaders (Beckley *et al.* 2008). On another hand, the capacity to subsist or persist while sustaining desired social networks, human health, ecological integrity and cultural values is likely of greater importance to many forest-dependent Aboriginal communities. However, at any one time, a community may respond more effectively to existing challenges and opportunities by emphasizing capacity outcomes in one area over another. As an organizational device and planning tool for Aboriginal communities to assess and implement their capacity strengths and needs over both the long- and short-term, Beckley *et al.*'s (2008) hypothetical 'community asset-amoeba' model, with some modifications to reflect Aboriginal needs and interests, may have great utility (Fig. 4).

Capacities for Community Empowerment
The traditional human resource development model of Aboriginal capacity building fails to see that empowerment is multidimensional (occurring within sociological, psychological, economic, political and other dimensions), operative at a number of levels (individual, group, community), and is less an outcome than a process (Hur 2006). Personal or individual empowerment is not the same as community or collective empowerment. The former relates to the way the people think about themselves, as well as the knowledge, capacities, skills and mastery they actually possess (Staples 1990:32). The latter are the processes by which individuals join together to change their condition, assist one another, learn together and develop skills for collective action (Hur 2006:530).

Most Aboriginal capacity building initiatives implicitly assume that by increasing the skill sets of individuals to participate in existing economic opportunities, the whole community will ultimately benefit. This is a flawed assumption. Although individuals can become empowered through personal development, they do not always become effective in helping to build their community's collective empowerment (Hur 2006:530). It is not uncommon in many Aboriginal communities to witness personal empowerment occur at the expense of collective empowerment whereby new, externally oriented socioeconomic relationships replace older, internally generated ones.

Ideally, personal (individual) and collective empowerment should be complementary (Staples 1990). The goal of individual empowerment is to achieve a state of liberation strong enough to impact one's power in life, community and society, whereas the goal of collective empowerment is to achieve a sense of security, freedom, belonging and power that can lead to constructive change (Hur 2006:535). Existing approaches to Aboriginal capacity building are designed to empower individuals, not communities. No specific attention is given to the design or implementation of programs, processes and institutions that would empower the latter. No real consideration is given to the impacts that capacity building programs have on the social, cultural and natural assets of Aboriginal communities. Viewed in this way, Aboriginal communities are in the best position to determine their own capacity needs and strengths for individual and collective empowerment. What is missing is the political will to create and finance institutions and programs for this to proceed in an effective and appropriate manner.

Capacity for Social Entrepreneurship
Aboriginal capacity building programs that target and promote the entrepreneurial skills of individuals, while beneficial in many ways to individuals (e.g., increased income, standard of living, feelings of self-worth, etc.) have ignored community goals, needs and aspirations, leaving it up to individual beneficiaries to contribute to the greater good of the community. Experience shows that some do and some don't. A potential solution to this disconnect may be the consideration of capacity building initiatives in 'social entrepreneurship.' The concept of 'social entrepreneurship' has emerged to encourage entrepreneurship in support of social sustainability, whereby "social purpose is achieved primarily through entrepreneurship; there is little if any distribution of profit to individuals, as any surplus is reinvested for the long-term benefit of the community; constituents are democratically involved; and there is accountability" (Anderson *et al.* 2006). These authors examined a number of case studies in Canada that provided powerful evidence of the importance of 'social entrepreneurship' as a tool for community empowerment:

> Especially evident are the prevalence of community ownership and the acknowledgement of the importance of long-term profitability and growth of businesses created, not as an end but as a means to an end. And it is these ends that make their activities social entrepreneurship. Some of these ends included the creation of employment with characteristics that 'fit' the interest, capabilities,

and preferred lifestyles of community members; control of traditional lands and activities on these lands; and the creation of wealth to fund education, health and wellness, housing and other social programs… . While what these Aboriginal groups have done as they have identified opportunities and created business is clearly entrepreneurship, their reasons for doing so and the organizational forms they have adopted extend far beyond wealth creation for the entrepreneur(s)/owners involved. The wealth is generated to fund social objectives, broadly defined (Andersen *et al.* 2006:46,54).

Capacity building programs in support of social entrepreneurship and community empowerment have not received much attention from federal and provincial government agencies, and perhaps understandably so. Such programs, if they are to 'fit' socially, culturally and economically with short- and long-term community aspirations, cannot be directed by outside interests or agendas, but must be driven by the Aboriginal community itself. The approach advanced in this chapter provides one such model to sustain and empower in a complementary manner, both Aboriginal individuals and the communities of which they are a part. While funding for Aboriginal capacity building programs that target individuals should be increased substantially, commensurate with the need of Aboriginal communities, new institutional approaches and programmes that support the capacity of Aboriginal communities to design and realize a sustainable future are also needed.

Rationalizing the Need for New Institutions and Approaches to Building and Implementing Capacity in Aboriginal Communities

> If what Aboriginal peoples thought they had won had been delivered—a reasonable share of lands and resources for their exclusive use, protection for their traditional economic activities, resource revenues from shared lands, and support for their participation in the new economy being shaped by the settlers, the position of Aboriginal peoples in Canada today would be very different. They would be economically self-reliant. Some would be prosperous (RCAP 1996).

> If government promises capacity building, then I want real education and training. But I do not want a government to come and tell our First Nation that we are not ready to participate in economic development, or not ready to exploit our own natural resources, or that we do not know how to responsibly manage our own affairs. Because those things are not true. Those are myths, the lies, the misrepresentations. They are the excuses for keeping things as they are" (Matthew Coon Come 2001).

Current capacity building programs aimed at increasing Aboriginal participation in the natural resource development and management sectors represent only a partial and initial first step toward addressing the needs, rights and interests of forest-dependent Aboriginal peoples and communities. 'Bottoms-up' approaches, whereby Aboriginal communities assume ownership, control and responsibility for developing its members' capacities, are needed. The latter is not a replacement for, but is complementary to, existing 'top-down' approaches. Both approaches are of value to Aboriginal peoples and communities, and should not proceed independently of each other. By way of summarizing the 'bottom-up' approach and facilitating the participation of forest-dependent Aboriginal peoples and communities in Canadian society and economy on their terms, they must be supported to:

1) document, assess and prioritize Aboriginal uses, values and needs with respect to their lands and resources;
2) plan for sustainable economic development, land use management and other plans based upon the community's desired relationship with their lands and resources, and to integrate these plans in a coordinated manner so as to support and achieve the community's vision;
3) identify the capacity strengths and requirements of community members to implement this vision, its plans and its constituent components;
4) identify, develop and negotiate the appropriate institutional frameworks and processes required to implement these plans; and,
5) seek to achieve and implement these capacities so as to sustain and build upon the human, intellectual, economic, social, cultural and natural capitals and assets of Aboriginal communities.

The engagement of Aboriginal peoples in natural resource development and land-use planning and management is a central and complex challenge facing Canada and its Aboriginal peoples. Upwards of a million Aboriginal peoples may live in Canada's boreal forest, where "their identity and relationship to the land is both spiritual and material, not only one of livelihood, but of community and continuity of their cultures and societies" (NRTEE 2005:44). At the same time, unemployment, poverty and birth rates as well as other social and health problems in most forest-dependent Aboriginal communities remain many times the national average. If proactive and coordinated measures are not undertaken soon to complement current initiatives that work and to replace ones that don't, a social crisis of unimaginable proportions to Canadian society may be the result. The 'cost of dependency' is simply not sustainable, acceptable, moral or ethical. The future will likely rest on the ability of "Canadian governments and Aboriginal peoples to cooperatively address the need for significant institutional reform and focused capacity development" (NTREE 2005:44). Current institutional arrangements and capacity building initiatives alone are not getting the job done.

The lack of effective institutions to engage Aboriginal peoples and communities in natural resource development and land-use planning is part of the colonial legacy of all Canadians. While comprehensive land claims agreements (modern-day treaties) have begun to enable Aboriginal communities to participate in and benefit from natural resource development and land-use planning, these institutions are, for the most part, restricted to the northern parts of the country. Many forest-dependent First Nations communities are covered by historic treaties that provide little direction to their signatories regarding the participation of Aboriginal peoples in the sustainable development of Canada's natural resources. Against this background, governments continue to construe the rights of First Nation and Métis peoples narrowly, while granting rights to third party interests (forestry, mining and energy companies) to explore for and develop natural resources on Aboriginal territories. As Aboriginal peoples assert their rights and as conservation issues intensify, the potential for conflict is great. A key challenge for all governments, including Aboriginal governments, is the creation of institutions that will give Aboriginal and treaty rights substance and effect on a day-to-day basis, and provide Aboriginal communities with the opportunity and support to design and realize a sustainable future on terms and conditions acceptable to them.

Other than land claims settlements and the consultation policies of a few provinces, the direct involvement of Aboriginal peoples in land use planning and management is absent. Moreover, those few government-sponsored programs that are amenable to building Aboriginal peoples' capacities in land use planning and management (i.e., First Nation Forestry Program, Building Environmental Aboriginal Human Resources) require much greater resourcing to meet the short-term needs of Aboriginal communities to represent their rights and interests in these processes effectively. The importance of building Aboriginal capacities in land-use planning, management and governance has been highlighted by many sources, including the National Round Table on the Environment and Economy (NRTEE 2005, Recommendation #7):

> A major challenge to the involvement of Aboriginal peoples in the future of the boreal is the limited capacity at the community level to participate effectively in management and planning processes related to resource development and conservation. Currently, Aboriginal communities are characterized as having scarce technical, human and financial resources: low levels of educational attainment; and a small base of professional and technical expertise upon which to draw. These concerns about limited capacity are compounded by the increasing demands for consultation being placed on Aboriginal communities…Federal, provincial, territorial, and Aboriginal governments and society organizations should support capacity-building of Aboriginal communities, enabling them to effectively manage their interests in the boreal.

Increased support for existing capacity building initiatives that appear to be working alone is not sufficient. Nor is it likely to produce the conditions that will allow Aboriginal communities to find the right formula to sustain their cultures, societies, values and desired relationships with their lands and resources. Canada is an experiment in cultural diversity and sustainability that few other countries have attempted, and at which even fewer have succeeded. Current Aboriginal capacity building programs, however, contribute little to the success of this project. NAFA (Bombay 2007) advocates that Aboriginal forest management capacity is dependent on the implementation of institutional arrangements that entrench and support Aboriginal values and rights. Canada needs to create institutional arrangements with its Aboriginal peoples that enable them to participate in Canadian economy and society on their terms and conditions, or at least on terms and conditions that are negotiated, not unilaterally set by one party.

Aboriginal peoples and communities need, and in many cases already possess, the capacities to participate in the design of such institutions. However, the institutions and rules of engagement in which Aboriginal peoples must participate have already been created to serve the other political, economic and social agendas. It is in the context of creating new institutional initiatives that Aboriginal peoples and communities may have the greatest opportunity to plan their future, and to assess and implement their capacity needs and strengths to get there.

Yet, it would seem that existing institutional arrangements and Aboriginal capacity building programs are too constrained by current government policies and economic initiatives to achieve these objectives. Existing institutions and capacity programs can perhaps be tweaked or expanded to better address the immediate needs of Aboriginal peoples. However, the creation of new Aboriginal capacity building initiatives backed by the development of more effective institutional arrangements between Canada and its Aboriginal peoples would seem a more appropriate and effective strategy.[6] Before this can happen, Aboriginal and First Nations leaders need to initiate and participate in a dialogue to reach consensus on the most appropriate and effective course of action to empower and sustain their communities.

Building Aboriginal capacity and creating effective institutional arrangements must go hand in hand. It is not enough for Aboriginal peoples to develop capacities to participate in existing political and economic arrangements. Not only are such arrangements not particularly adaptive, they are less than shining examples of integrating multiple values and ways of knowing. Rather, they are part of the colonial legacy of all Canadians that began with the first comprehensive land claims agreement in the mid-1970s. The general terms and conditions of such institutional arrangements, including the rules, language and concepts of discourse were set by Canada, leaving Aboriginal parties to figure out how to best fit their values, understandings, rights and interests into this new currency.

Nation-to-nation relationships must be negotiated in an environment of mutual respect and equality. Ideally, negotiated institutions would integrate multiple ways of knowing, and both 'top-down' and 'bottom-up' approaches.

---

[6] In this scenario, Aboriginal communities would assume ownership, responsibility and control of their capacity development, and have access to financial and other resources to do so.

Creating integrated and adaptive institutions of 'know what' (knowledge) and 'know how' (practice), however, requires an examination of social, decision-making and learning processes at play, including the myths, values, beliefs and power systems at work, and how these interact with one another to produce a range of outcomes (Wilkinson *et al.* 2007).

Integration and adaptation will not happen by themselves; "they require motivated people, with awareness of their own standpoints and biases, a commitment to mutual respect and the skills to find common ground" (Wilkinson *et al.* 2007). If Aboriginal forest-dependent (and other) communities are going to be sustainable, these are capacities that both Aboriginal and non-Aboriginal participants involved in the discourse will need as they seek to create new institutions and programs. If developing new institutions that integrate diverse views and knowledge into democratic, adaptive learning processes is a means for Aboriginal communities to achieve sustainability, we must come to grips with the 'blind spots' and 'sound bites' that undermine this goal (Wilkinson *et al.* 2007).

Blind Spots and Sound Bites: Barriers to Reform
Behind every quest for sustainability, and every conservation effort, lies one fundamental problem: "How do we create and sustain a healthy relationship with our world" (Doremus 2000; Doob 1995; Wilkinson *et al.* 2007). Positivistic science and the view that it should have a privileged role in society, policy and practice because of its 'objectivity' have become 'institutionalized' in government practice. There is nothing inherently wrong with the scientific method; science is an insightful and self-correcting tool (Wilkinson *et al.* 2007). However, those who subscribe to it often dismiss the knowledge and views of others who do not endorse such a narrowing of vision or fragmentation of understanding. Often this is done unintentionally, even unconsciously, by well-meaning professionals. However, those who champion the scientific method in their interactions with Aboriginal peoples might do well to remember that all knowledge is culturally constructed and replete with biases and assumptions, even—and some would say, especially—western science. Science is just one of many ways to organize experience and create understanding, and no one knowledge system has a monopoly on the 'truth,' or the right way to achieve a sustainable relationship with the world (Wilkinson *et al.* 2007). All viewpoints and knowledge systems are needed, and should to be brought to bear, in a respectful and complementary manner, to address the challenges at hand. The supposed superiority of 'science' is reinforced by other 'blind spots' that:

- view humans as external (i.e., as a disturbance or outlier) to 'natural systems';
- believe humans can control and manage 'nature,' and that we have sufficient knowledge and means to do so; and
- perceive environmental change and ecological flux as disorder, i.e., something to be corrected and managed to achieve some level of ecological stability (Stevenson 2006).

Together with concepts and language embedded in the Canadian legal system, these 'blind spots' have strongly influenced the design of comprehensive land claim agreements and other arrangements with Canada's Aboriginal peoples involving their lands and resources, especially rules of engagement and operation. Together, they form a practically insurmountable barrier to effective Aboriginal participation in our common quest to develop a sustainable relationship with our world.

Aboriginal peoples involved in natural resource development and management discourses are not immune to 'blind spots.' Their portrayal as 'conservationists' and 'managers' of wildlife is somewhat misleading, even disempowering; neither are descriptions of Canada's Aboriginal peoples' relationships to animals historically accurate (*see* Stevenson 2006, 2009). Another blind spot that sometimes hinders Aboriginal peoples' advancement in natural resource development and management discourses is the promotion and advocacy of Aboriginal (and treaty) rights in the absence of any discussion about Aboriginal 'responsibilities.' Provincial and federal governments have been glacially slow in translating Aboriginal rights into the design of new policies and institutions that would facilitate the meaningful participation of Aboriginal peoples in addressing the sustainability challenge.[7] What is needed is a new discourse that clarifies the responsibilities that attend those rights. Aboriginal rights and responsibilities are two sides of the same coin. Yet, somewhere, somehow in the discussions leading to the drafting of sec. 35 of the *Canadian Constitution Act of 1982*, the concept of Aboriginal 'responsibilities' fell off the table. Continued assertion of Aboriginal rights in the absence of clarifying Aboriginal and government responsibilities, and designing institutions that accommodate them is a 'blind spot' that will continue to hinder Aboriginal peoples' quest for the recognition of their rights. Moreover, a focus on responsibilities has the potential to foster a mutually cooperative and respectful exploration of strategies and institutions that will allow the rights and responsibilities of both Aboriginal communities and individual Canadians to be exercised.

In existing institutions and approaches to Aboriginal capacity building, the problems identified and solutions sought are framed by those in/with power. In an effort to make problems and solutions tractable, well-intentioned professionals often oversimplify complex realities (Wilkinson *et al.* 2007), rendering initiatives far less effective than initially anticipated. By narrowing focus, decoupling relationships among otherwise interdependent variables, their

---

[7] British Columbia's 'New Relationship' policy appears to be an exception and appears to be a step in the right direction (Government of BC 2005). The real challenge in BC will be to translate this new policy into practice. While the New Relationship Trust sets aside $100 million for First Nations to pursue objectives in education, capacity, economic development, youth & Elders, and culture & language, only $6 million was spent on capacity building over the first three years (2007-10). Capacity is defined as "the tools, training, and resources First Nations need to participate in land and resource management and planning processes, and to develop social, cultural and economic programs in their communities" (Government of BC 2008).

efforts have a greater chance at being rewarded by the system that sets the rules. However, any time we reduce the complex world around us to a tractable problem or area of specific concern, we put 'blinders' on (Wilkinson *et al.* 2007). With its focus on enhancing the capacities of Aboriginal peoples to engage in existing employment opportunities in the economic development and management of forests, this is what has happened to Aboriginal capacity building agenda in the forestry sector.

The myths that underpin this approach are many. Myths are the stories we tell ourselves about how the world is or ought to be, and translate easily into metaphors. Botkin (1990) argues that it is not a shortage of technical or scientific knowledge that hinders our ability to perceive and constructively address key issues, it is the underlying myths and metaphors that shape our understandings. In the realm of Aboriginal capacity building, these metaphors can be translated into the 'sound bites' some of which are echoed in Grand Chief Matthew Coon Come's comments above:

- We are all the same!
- What's good for us, is good for you!
- To be successful, you need to become more like us!
- You are not yet ready to be like us, you need training, you need patience!
- Once you are like us, then we will be partners!
- If we build it, you will come, and if you don't, it's not our fault, it's yours!

Current government-sponsored approaches to Aboriginal capacity building in the natural resource development and management sectors run the risk of becoming stuck in their own unexamined myths, metaphors and sound bites, thus limiting their effectiveness and our ability to perceive, understand and engage in effective dialogue about the central challenges at hand. To maximize their effectiveness, existing approaches must be construed within, and become an integral component of a broader, more comprehensive Aboriginal capacity building project that possesses the types of characteristics identified in this chapter. But barriers must be overcome in order to achieve this objective.

The key to successful joint problem identification and solving is to recognize the strengths and limitations of the focus we each may have and to create processes and institutions wherein multiple voices, methods and streams of understanding are valued and considered (Wilkinson *et al.* 2007). Institutions that set the stage for alternative epistemologies and effective dialogue among different cultures and knowledge systems maximize our chances for a more holistic identification and understanding of the problems we face, and the solutions needed to resolve them. Such institutions can only be created out of negotiation whereby mutual respect, equality of voice and consensus are achieved in a common quest to develop a sustainable relationship with our world.

The benefits of having viable and effective institutions and capacity building programs that create the space for Aboriginal communities to plan their future and set the stage to get there are considerable. While each Aboriginal community will likely differ with respect to their formula for social, cultural, and economic sustainability, communities with a common vision and a sense of purpose know what they must do and are more likely to be proactive. Strong, vibrant Aboriginal communities, such as might be predicted by those with high 'bonding' and 'linking' capitals, are also capable of engaging industry, governments and others on an equal footing.[8] They make good partners, are resilient to change and are more likely to negotiate rules of engagement that nurture and sustain viable economic, political and other arrangements with non-Aboriginal interests so as to create win-win situations for both. They are also more apt to identify and address their capacity needs and strengths, and to achieve a sustainable future, independent of government relief and support. Multiplied by several hundred communities, representing a million or more people, the cost savings alone to Canadian taxpayers could be in the billions of dollars annually, freeing up monies for more proactive initiatives. Strong, viable Aboriginal communities in rural settings may also serve to ground Aboriginal peoples in urban environments to their history, culture, value systems and communities, contributing to their survival in the multicultural fabric of Canada's cities and building a strong bridge between on-reserve and off-reserve Aboriginal populations.

Substantial environmental benefits may also be an outcome of integral, vibrant, sustainable Aboriginal communities. Biological and cultural diversity are positively correlated (Stevenson and Webb 2004; Turner *et al*. 2003, among others). Where Indigenous peoples, communities and governments are strong, their relationships to their lands and resources, which depend on and indeed sustain biological diversity, are also strong. It has also been demonstrated that countries with marked economic disparity, such as highly industrialized nation-states, demonstrate a significantly greater loss of biodiversity (Mikkelson *et al*. 2007), and we would argue cultural diversity, than countries where wealth is more evenly distributed.

It is particularly important that all members of Aboriginal communities (women, men, elders, youth, etc.) have a strong voice in planning their collective and individual futures, including the identification of their capacity strengths, and needs to get there. All voices must be represented and appropriately accommodated within community visioning/planning/capacity, identification exercises, lest one or a few agendas come to dominate others. Strategies for sustainability must embrace the complexity and articulation of ecological, economic, social and cultural and other needs of people within Aboriginal communities at all scales (individual, family, group, community). A focus on one dimension (e.g., economic vs ecological) or scale of sustainability (individual vs community, etc.) at the risk of not considering others will also not

---

[8] Aboriginal communities that score low in both types of social capital may be incapable of maintaining sustainable relationships with their lands, resources and with others. The future of such communities is strongly vested in the status quo and currently left in the hands of others.

likely be viable over the long term. Where they do not exist, skill sets needs to be developed within Aboriginal communities to ensure that all dimensions and scales of sustainability are appropriately considered, accommodated and integrated on the path to a viable future.

Even though conventional Aboriginal capacity building programs are based on the myths and metaphors of non-Aboriginal society, they do have something to offer; sometimes showing up in a community with an expedient answer to a problem provides some benefits. But again, this is an interim fix and not the ultimate solution to the many challenges faced by forest-dependent Aboriginal communities. There is a need for innovative approaches to Aboriginal capacity building that are driven from the 'ground-up,' that articulate with conventional 'top-down' programs, and that seek to accommodate all dimensions and scales of sustainability. Innovative Aboriginal capacity building programs, and the institutions that support them, may not always have the answers *prima facie*, but they will at least work with Aboriginal communities to identify the problems and design appropriate responses on their path to sustainability.

Conclusion

> *(That) is what I would call real 'capacity building, building a land and resource base that will create sustainable economies for First Nations"* (Matthew Coon Come 2001).

In the absence of recognizing and accommodating the rights and responsibilities of Aboriginal peoples to their lands and resources, it would seem doubtful, regardless of the capacity building initiative undertaken, whether First Nations communities will ever become self-sustaining, and thus true partners in confederation. In this regard, the RCAP report (1996) recommended that "federal, provincial and territorial governments, through negotiation... provide Aboriginal nations with lands that are sufficient in size and quality to foster Aboriginal economic self-reliance and cultural and political autonomy." While this has happened to some extent with the negotiation of comprehensive land claims in northern Canada, with few exceptions (e.g., *Nisga'a Final Agreement*), it has not happened in other parts of Canada where competition for resources and population densities are higher.

The creation of new Aboriginal capacity building initiatives must be tied to existing programs and the creation of new institutional relationships that provide Aboriginal peoples and communities with increased rights of access to resources and revenue-sharing. These are matters to be negotiated between the Aboriginal leaders of this country and the relevant federal/provincial/territorial authorities. In the interim, we offer them the following broad observations and recommendations to guide them in our mutual quest to create a Canada where Aboriginal communities are self-sustaining, true partners in confederation and engaged participants in a process that will lead us collectively to develop a sustainable relationship with our natural resources:

1) Aboriginal peoples, communities and governments alone must own the processes of determining their capacity needs and requirements, and of implementing their existing capacity strengths, and this must be done in such a way as to enhance community rights, interests, responsibilities, goals and aspirations.

2) Resourcing for existing government- and industry-sponsored Aboriginal capacity building programs in the natural resource development and management sectors that appear to be working should be increased substantially, commensurate with the needs of Canada's forest-dependent Aboriginal communities.

3) A new government-funded Aboriginal capacity building initiative should be created to provide funding and administrative support to forest-dependent Aboriginal communities and governments to undertake the community-driven research, planning and visioning exercises necessary to achieve sustainability.

4) In the spirit of reconciliation, an Aboriginal Natural Resources Research and Policy Institute should be created with federal and provincial assistance to provide the balanced research and policy analyses needed to inform the development and implementation of new Aboriginal capacity initiatives, and of new institutional arrangements that provide Aboriginal peoples and communities greater access to their lands and resources.

5) Industry and government should assess their capacity strengths and weaknesses to accommodate Aboriginal needs, rights and interests in the context of natural resource development and sustainable land-use planning and management, and, where appropriate, undertake measures to address capacity deficiencies.

6) Relevant post-secondary educational institutions need to re-design and develop programs and courses that create the space for the equitable and meaningful participation of Aboriginal peoples and communities in the natural resource development and management sectors.

In the final analysis, 'top down' approaches to Aboriginal capacity building must be met with 'bottom up' approaches to achieve synergies and mutual aspirations for improving the lives of Aboriginal peoples in forest-dependent communities, and conserving the cultural and biological diversity of our forests. Greater support of existing Aboriginal capacity building programs must be met with the creation of new institutional approaches to building capacity in Aboriginal communities that support their efforts to plan and realize a sustainable future from their lands and resources based on their goals and priorities. No longer can the issue of Aboriginal capacity be approached in a narrow and piecemeal manner, and, no longer is it appropriate to speak of 'capacity'—a term rejected by a growing number of Aboriginal and non-Aboriginal peoples—without asking:

Capacity for What? And for Whom?

References Cited

Anderson, R.B., L.P. Leo, and T.E. Dana (2006). Indigenous land rights, entrepreneurship, and economic development in Canada: Opting-in to the global economy. *Journal of World Business* 41: 45-55.

AFN—Assembly of First Nations (2005). *A First Nations-federal Crown political accord on the recognition and implementation of First Nations governments.* Ottawa.

Beckley, T.D., D. Martz, S. Nadeau, E. Wall, and W.C. Reimer (2008). Multiple capacities, multiple outcomes: Delving deeper into the meaning of community capacity. *Journal of Rural and Community Development* 3(3): 56-75.

Bombay, H. (2007). *NAFA's Aboriginal capacity building initiatives.* Ottawa: National Aboriginal Forestry Association.

Botkin, D. (1990). *Discordant harmonies: A new ecology for the 21st century.* New York: Oxford University Press.

Government of British Columbia (2005). The new relationship. http://www.newrelationship.gov.bc.ca/shared/downloads/new_relationship.pdf

Government of British Columbia (2008). New relationship trust: Building strong First Nations. http://www.newrelationshiptrust.ca/faq

Government of Canada (2007). *Sharing Canada's prosperity: A hand up, not a hand out*: Final Report, Special Study on the Involvement of Aboriginal Communities and Businesses in Economic Development Opportunities in Canada. Standing Senate Committee on Aboriginal Peoples, Ottawa. March 2007.

Coon Come, M. (2001). 'Capacity building in First Nations,' in *Aboriginal forestry 2001: Capacity building, partnerships, business development, and opportunities for Aboriginal youth.* Proceedings of the Conference of the First Nations Forest Program. Saskatoon. 21-24 January, 2001.

Doob, L.W. (1995). *Sustainers and sustainability: Attitudes, attributes and actions.* Westport, CT: Pralleger.

Doremus, H. (2000). The rhetoric and reality of nature protection: Toward a new discourse. *Washington and Lee Law Review* 57 (1): 11-73.

Drielsma, J.H. (1984). *The influence of forest-based industries on rural communities.* PhD diss., Yale University.

Edwards, R. (2002). *Social capital, a Sloan work and family encyclopedia.* Chestnut Hill, MA: The Sloan Work and Family Research Network.

Flora, C.B. and J.L. Flora (1993). Entrepreneurial social infrastructures: A necessary ingredient. *The Annals of American Academy of Political and Social Science* 529: 48-58.

Gibson, G. and J. Klinck (2004). Canada's resilient north: The impact of mining on Aboriginal communities. *Pimatisiwin: A Journal of Aboriginal and Indigenous Community Health* 3(1): 116-140.

Gysbers, J.D. and P. Lee (2003). *Aboriginal communities in forest regions in Canada: Disparities in Socio-economic conditions.* Edmonton: Global Forest Watch.

Haley, S. (2007). *The impact of resource development on social ties. Theory and methods for assessment.* Paper presented at the Society for Applied Anthropology Meeting. Tampa, FL.

Hur, M.H. (2006). Empowerment in terms of theoretical perspectives: Exploring a typology of the process and components across disciplines. *Journal of Community Psychology* 34(5): 523-540.

Kepkay, M. (2007). *Building capacity of Aboriginal peoples in Canada's forest sector: Rationale, model and needs.* In association with Thematic Team 3 of the National Forest Strategy Coalition, 31 Ottawa: National Aboriginal Forestry Association. March 2007.

Kusel, J. (1996). Well-being in forest-dependent communities, part 1: A new approach. In *Sierra. Nevada ecosystem project. Final report to Congress, Vol. II, assessments and scientific basis for management options,* 361-374. University of California, Davis: Centers for Water and Wildland Resources.

Mikkelson, G.M., A. Gonzalez, and G. D. Peterson (2007). Economic inequality predicts biodiversity loss. *PLoS One,* wwwplosone.org. May/Issue 5/e444: 1-5.

Nadeau, S., B. A. Shindler, and C. Kayoynnis (2003). Beyond the economic model: Assessing sustainability in forest communities. In *Two paths to sustainable forests: Public values in Canada and the United States.* ed. B.A. Shindler, T. Beckley, and C. Finley. OSU Press.

NAFA—National Aboriginal Forestry Association (1993). Aboriginal forestry training and employment review: Final report, executive summary phase I. Ottawa. National Forest Strategy Coalition. 2003. *National Forest Strategy 2003-2008: A sustainable forest, the Canadian commitment.* Ottawa.

Nayran, D. (1999). *Bonds and bridges: Social capital and poverty, policy research working paper* No. 2167. Washington, DC: The World Bank.

NFSC-National Forest Strategy Coalition (2003) *National Forest Strategy (2003-2008), A Sustainable Forest: The Canadian Commitment.* http://www.ccfm.org/english/coreproducts-nscf.asp

NRTEE—National Round Table on the Environment and Economy (2005). *Boreal futures: governance, conservation and development in Canada's boreal.* Ottawa: Queen's Printer.

Putnam, R.D. (2000). *Bowling alone: The collapse and revival of American community.* New York: Simon and Shuster.

RCAP—Royal Commission on Aboriginal Peoples (1996). *Report of the Royal Commission on Aboriginal Peoples.* Ottawa: Queen's Printer.

Rousseau, M.H. (2006). Economic development project: Final report. Waswanipi Cree Model Forest. 31 March 2006.

Staples, L.H. (1990). Powerful ideas about empowerment. *Administration in Social Work* 14 (2): 29-42.

Stevenson, M.G. (2006). The possibility of difference: Rethinking co-management. *Human Organization* 65(2):167-180.

Stevenson, M.G. (2009). 'Negotiating research relationships with Aboriginal communities: Ethical considerations and principles,' pp. 197-210 in M.G. Stevenson and D.C. Natcher, eds., *Changing the culture of forestry in Canada: Building effective institutions for Aboriginal engagement in sustainable forest management.* Research and Insights form the Aboriginal Research Program of the Sustainable Forest Management Network. Edmonton: CCI Press.

Stevenson, M.G., and P. Perreault (2008). *Capacity for what? Capacity for whom? Aboriginal capacity and Canada's forest sector.* Edmonton: Sustainable Forest Management Network.

Stevenson, M.G., and J. Webb (2004). First Nations: Measures and monitors of boreal forest biodiversity. *Ecological Bulletins* 51:83-92.

Turner, N.J., I.J. Davidson-Hunt, and M. O'Flaherty (2003). Living on the edge: Ecological and cultural edges of diversity for socio-ecological resilience. *Human Ecology* 31 (3): 439-461.

Wilkinson, K.M., S.G. Clark, and W.R. Burch (2007). *Other voices, other ways, better practices: Bridging local and professional environmental knowledge.* Yale University of Forestry and Environmental Studies Bulletin 14.

Appendix A:
Aboriginal Capacity Working Group

There are a variety of government-sponsored capacity building programs available to Aboriginal peoples and communities to increase their abilities to participate in and benefit from the forest sector.[9] In partial response to the inability of these programs to deliver on their stated objectives, a number of Aboriginal-led capacity building initiatives have emerged under the auspices of the National Aboriginal Forestry Association. Foremost among these was the *National Forest Strategy Theme Three Capacity Working Group.* Composed of representatives from Aboriginal organizations (First Nations and Métis), non-government organizations, academic institutions, and provincial and federal government departments, this working group was formed to address the National Forest Strategy Coalition's (2003) commitments to addressing the institutional arrangements and support needed to accommodate Aboriginal rights and participation in the sustainable use of Canada's forests.

The authors of this chapter were active members of the Aboriginal Capacity Working Group (ACWG), and in many respects, the thinking that evolved during the course of their work and that of the ACWG was complementary. The ACWG (Kepkay 2007) discussion paper stresses the need to build on the institutional and cultural resources already in existence in Aboriginal communities, including the traditional knowledge of Aboriginal peoples. Further, it considers the need for the cooperative development of new institutions as essential in building Aboriginal capacity to participate in and benefit from economic opportunities in the forest sector. The ACWG (Kepkay 2007) felt unequivocally that "Aboriginal peoples hold primary responsibility for building their own capacity." Because recent court decisions indicate that the provinces are also responsible for consulting with Aboriginal peoples regarding forest practices and policy, meaningful consultation must include support to develop capacities to effectively engage and represent their interests, rights and traditions in consultation processes. In all cases, Aboriginal peoples need to be informed, to represent their interests and to engage in consultations, all of which require capacity. Meanwhile, the provinces contend that, if there is any responsibility for supporting Aboriginal capacity building, it arises from the federal Crown's fiduciary obligations. But the latter has always justified its variable levels of support for Aboriginal capacity building as a humanitarian policy, rather than a legal responsibility. In the meantime, Aboriginal peoples maintain that both levels of government are responsible for providing meaningful support to Aboriginal peoples to develop their capacities to engage and represent their rights and interests in consultation with government and industry over resource development. As noted by the ACWG, because Crown government approaches to Aboriginal capacity are not grounded with clear reference to constitutional and statutory authorities, the result at the national level is a patchwork of arrangements that are neither coordinated nor robust

---

[9] For a more thorough review of these programs see Stevenson and Perreault (2008).

against political vagaries—a fact highlighted by the Standing Senate Committee on Aboriginal Peoples (Kepkay 2007):

> [T]he time has come for the federal government to stop treating Aboriginal economic development as "discretionary". The federal government must make meaningful investments in Aboriginal economic development, anchored by a newly formulated Canadian Aboriginal Economic Development Strategy designed to meet Aboriginal economic development aspirations and achieve measurable results. This strategy should take a coordinated and integrated approach across sectors, connecting to education, skills development and training, infrastructure development, institutional and governance capacity, capital development and access to lands and resources.

As pointed out by the ACWG (Kepkay 2007), the current complex set of institutions and programs, often lacking in even rudimentary strategic coordination among them, creates an enormous, often redundant and sometimes conflicting burden on the resources of Aboriginal communities. The Auditor General of Canada has identified several key guidelines for the development of new capacity building initiatives that may produce positive outcomes across a number of areas of concern simultaneously, while reducing cost:

> We identified seven factors that appear to have been critical in the successful implementation of our recommendations. These include the sustained attention of management, co-ordination of government programs, meaningful consultation with First Nations, development of First Nations capacity, establishment of First Nations institutions, development of an appropriate legislative base for programs, and consideration of the conflicting roles of Indian and Northern Affairs Canada. In our view, ensuring that these factors are fully considered when adjusting existing programs and implementing new ones will make a significant difference in the lives of Aboriginal people (Auditor General of Canada 2006, para. 5.64).

Key recommendations advanced and conclusions reached by the ACWG (Kepkay 2007) include:

- Aboriginal peoples hold the primary responsibility for building their own capacity, but they must be financed and resourced to take on this role, and institutional barriers to the exercise of their primary responsibility must be removed. New institutional arrangements are needed in order to provide assurances that constitutional, fiduciary and treaty obligations will be honoured.
- New capacity builds on the foundations of existing capacity. The first step is to recognize the existing capacity represented

by the distinct traditions and values of Aboriginal peoples, as well as other strengths. Cultural fit is key in any capacity building or institutional development initiative, and initiatives need to be flexible and adaptive to different community conditions.

- The fiduciary doctrine and the duty to uphold the Crown's honour suggest that the federal government has a role to play in building capacity for Aboriginal peoples to represent their interests and to engage in economic development opportunities on provincial/territorial Crown land.
- Fiduciary obligations and the imperative to uphold the honour of the Crown give rise to the duty to consult and accommodate Aboriginal peoples' interests when any decision or action by the Crown holds the potential to infringe their rights.
- A central component of meaningful consultation is the capacity of Aboriginal communities and organizations to take informed positions. Becoming informed about a proposed project or land use question and its potential impacts on one's interests involves drawing on a range of diverse resources. Accordingly, the idea is becoming more widely accepted that measures taken to ensure a meaningful consultation process logically should include the provision of support for Aboriginal capacity building.
- Apart from the legal accountabilities discussed above, Aboriginal peoples and the Crown share a common interest in the mutual benefits that will arise from capacity building for Aboriginal Peoples' rights and participation in the forest sector.

By way of moving forward, the ACWG (Kepkay 2007) envisions a 'two-pronged' approach to institutional development and capacity building:

On one hand, there is a need to address the fundamental lack of shared understandings and commitments regarding roles and responsibilities. The patchy, piecemeal progress on these issues to date needs to be tied together in proactive, overarching agreements and standards with regard to roles and responsibilities. Assigning clear jurisdiction and authority is a key issue. Gaps and failures at the legislative level need to be addressed, self-governance needs to be advanced, and the capacity for taking on these new authorities and new liabilities needs to be ensured. Modern-day treaties and land claims are a special sub-category of this type of work.

On the other hand, pragmatic opportunities for incremental progress within existing frameworks also need to be engaged. For example, although a new type of forest tenure specifically tailored to the traditions of Aboriginal peoples may require many years to develop, in the meantime it is possible to take smaller steps towards the same goal by adapting the terms of an existing tenure. The lessons being learned and the skills being developed by communities and businesses holding conventional tenures today will build their capacity to represent and pursue their interests in the innovative arrangements being developed for tomorrow.

There is little doubt that many government-sponsored Aboriginal capacity building initiatives have resulted in tangible benefits to many Aboriginal peoples in terms of increased individual well-being and standards of living. However, despite the fact that Aboriginal peoples are becoming more involved in the forest industry across all sectors "the lack of professional forestry and business training within Aboriginal communities remains the largest limiting factor for increased participation" (Aboriginal Strategy Group 2007). While there are surely other important and contributing factors, every program has experienced a number of recurring themes that have reduced their effectiveness.

All governments must begin to view the issue as a legal responsibility and begin to address it in a comprehensive and coordinated manner. There needs to be greater coordination in the delivery of regional and national Aboriginal capacity building initiatives. There also needs to be long-term institutional support and frameworks in place to allow First Nation and Métis communities to implement their capacity building strategies.

Government-sponsored Aboriginal capacity building programs, however, cannot address the consensus-building and integration exercises that are necessary to accommodate the multifaceted issues at play within a community. Only the community itself can do the work and identify the solutions that will best fit their unique circumstances. For example, communities with significant control over education delivery could consider designing a capacity building program that parallels the fundamental learning and life skills taught in elementary school and carries on through high school and higher levels of education to the job market. It is well demonstrated that 'capacity begets capacity' and human resources are needed not only to envision this type of program, but to also develop and implement projects that meet the goals and objectives guided by the community vision. Government needs to play an important role in facilitating these processes, including the willingness and ability to be flexible to community needs at the local or regional levels. By way of improving the delivery and efficacy of existing Aboriginal capacity building programs, there is a need to:

- Encourage concordance between/among various initiatives,
- Move away from proposal-driven processes,
- Develop appropriate evaluation and monitoring criteria,
- Institute flexible reporting requirements,
- Open funding windows to support long-term capacity development plans,
- Substantially increase program budgets, and
- Develop a community focus that integrates with an individual client focus.

References Cited in Appendix

Aboriginal Strategy Group (2007). *A Quantitative Assessment of Aboriginal Involvement in the Canadian Forestry Sector.* 30 April 2007. Ottawa: National Aboriginal Forestry Association.

Kepkay, M. (2007). *Building Capacity of Aboriginal Peoples in Canada's Forest Sector: Rationale, Model and Needs.* In association with Thematic Team 3 of the National Forest Strategy Coalition, 31 Ottawa: National Aboriginal Forestry Association. March 2007.

Stevenson, M.G., and P. Perreault (2008). *Capacity for What? Capacity for Whom? Aboriginal Capacity and Canada's Forest Sector.* Edmonton: Sustainable Forest Management Network.

Chapter Thirteen

Capacity-Building Moose Cree Style: Moose Cree Strategies for Becoming a Goose Hunter[1]

Brent Kuefler, Adrian Tanner and David C. Natcher

Introduction

Building the capacities of Aboriginal peoples to participate in natural resource development has been become a rallying call for government, industry as well as Aboriginal peoples, communities and nations. However, such 'capacity building' initiatives too often ignore the need for Aboriginal peoples to sustain and strengthen capacities rooted in their cultures, collective identities and fundamental connections and stewardship responsibilities to the land. As argued in the previous chapter by Stevenson and Perreault, Aboriginal capacity building projects must not only address capacities to engage in existing opportunities and challenges, but they must also ensure that those so engaged have the capacity to provide sustainable cultural and social futures for Aboriginal communities.

For the Moose Cree First Nation (MCFN), enhancing the traditional land-based skills and knowledge of Moose Cree youth, and the social values and responsibilities that attend them, remains a vitally important social and cultural priority. Building the skills of individuals to participate in what the outside world has brought to them remains an important activity to the survival of the Moose Cree people, but not to the detriment of cultural and social imperatives and obligations. Indeed, it might even be argued that without the latter, the abilities of Aboriginal peoples to successfully navigate the challenges and complexities of the modern world are diminished, set adrift, and left without a solid foundation.

In this chapter, we describe the process through which MCFN children learn to become goose hunters and the impediments that some of these children today face when wanting to learn about hunting geese. Goose hunting is an important activity for many members of the Moose Cree First Nation in northern Ontario. A successful hunt depends on the use of techniques that are based on an intricate body of local knowledge and values, which shape the ethical and environmental principles governing the hunt. Historically, both have been passed to younger generations through their interactions with experienced

---

[1] Ethnographic data for this chapter is drawn from fieldwork by B. Kuefler conducted in Moose Factory in the autumn of 2007. Semi-structured interviews were conducted with 21 members of the MCFN as well as informal conversations with others. In the spring of 2008 Kuefler visited a goose camp and observed the hunting practices of camp members.

hunters while on the land. In recent years, however, some children have not had the opportunity to learn from experienced hunters, and there is a growing concern amongst some members of the MCFN that some younger hunters no longer respect traditional practices, and the values and responsibilities that attend them. In response to these concerns, outdoor education programs have been implemented in Moose Factory schools to give children the opportunity to travel to bush camps to learn bush skills from elders and experienced hunters.

Learning to become a goose hunter in Moose Factory is more than just learning the activities needed to kill geese; it is a process by which children (and adults who did not have the chance to learn as children) acquire a body of knowledge about geese and the environment that they can apply to becoming good hunters. Goose hunting is a technique based on this knowledge, in which hunters are aware of their responsibilities toward geese and act to fulfill those social responsibilities. When referring to this learned body of knowledge, we will eschew the use of the term traditional ecological knowledge (TEK). TEK is problematic because of implicit assumptions that are present within the concepts of 'traditional' and 'ecological' (Nadasdy 1999:4-5). Rather, we utilize the term local knowledge, which refers to "tacit knowledge embodied in life experience and reproduced in everyday behaviour and speech" (Cruikshank 2005:10). In this way, Cree local knowledge is (re)produced through complex interactions with the land and with each other.

Background

Moose Factory is located in the Hudson Plains and the James Bay Lowland ecoregion of northern Ontario (Environment Canada 2005). The James Bay Lowland is a transitional area between the forested areas to the south and the tundra to the north, and covers an area that reaches from Quebec in the east to the Attawapiskat River in northern Ontario. Much of the region slopes gently toward James Bay and is poorly drained, with wetlands covering 75 percent of the ecoregion in the north and 50 percent of the ecoregion in the south (Environment Canada 2005).

The wetlands of the Hudson Plains are productive waterfowl areas and the area around Moose Factory is host to several species of geese, ducks and loons, in addition to other migratory birds (Berkes 1982:26). Canada geese (*Branta canadensis*) from the Southern James Bay population migrate past Moose Factory and the areas used by the MCFN membership in the spring as they head to nest on Akimiski Island or areas to the south and west of James Bay. These geese again pass through this region as they migrate to their wintering areas which extend from southern Ontario to Mississippi, Alabama, Georgia and South Carolina. Recent scientific surveys indicate that there are approximately 77,500 geese in this population with 69,200 breeding geese (USF&WS 2009:41).

The mid-continent population (geese) also migrates through the area used by the members of the MCFN. This population consists mostly of lesser snow geese (*Chen caerulescens caerulescens*) with growing numbers of Ross' geese (*Chen rossii*); and nests primarily on Baffin and Southampton Islands with some geese nesting to the west of Hudson Bay. These geese winter in eastern

Texas, Louisiana, and Arkansas (USF&WS 2009:49). Population estimates suggest that this population increased by 12% between 2008 and 2009 to 2,753,400 geese (USF&WS 2009:49). Although a survey undertaken by the U.S. Fish and Wildlife Service suggests that the population is increasing, goose hunters in Moose Factory are noticing that fewer snow geese are flying through their hunting areas. Many MCFN members who spoke of this did not think that there are fewer geese, rather they suggest that the geese have changed their migration path and are flying to the west or flying at night.

The Mississippi Flyway Giant Population is a third goose population that uses the James Bay lowlands. This population comprises Giant Canada geese (*Branta canadensis maxima*) and is numbered at around two million (USF&WS 2009:42). These geese migrate to the James Bay Lowlands to molt in the late spring (Prevett *et al.* 1983:190) and are often referred to by the members of the MCFN as popcorn geese.

Goose Hunting

In the past, geese served as a vital food source for members of the Moose Cree First Nation. Geese hunted in the fall were preserved and eaten during the winter. The spring migration of Canada geese was especially critical as the return of the geese provided many families the first opportunity to procure fresh meat after the winter and in some instances was critical in warding off starvation. Although some members of the MCFN may now kill fewer geese each year, goose hunting continues to be an important activity. The goose hunt, in addition to providing members of the MCFN with an important source of food, also provides non-material benefits such as relaxation and time for social interaction with family members away from the distractions and demands of settlement life.

Today, there continues to be two primary hunting seasons for MCFN goose hunters. Canada geese are hunted in the spring as they migrate north; snow geese are hunted in the autumn as they migrate south. The spring hunt is more productive than the fall hunt and attracts more participants. Giant Canada geese are not hunted extensively in the late spring and summer when they are around Moose Factory, although some people hunt them opportunistically. The spring goose hunt, which has been described as having "religious significance for the Cree" (Prevett *et al.* 1983:190), is the most significant for the MCFN membership and many people get 'goose fever' as they await the northward migration of the Canada geese.

The members of the MCFN who participated in Kuefler's research described their relationship with geese is premised on an understanding that geese are 'persons' who participate in the hunt. The attribution of intelligence and personhood to geese, and to other non-human species, is a noted feature of Cree culture (Brightman 1993:76; Feit 1994:185; Scott 1996:73, 76; Tanner 1979:114, 130). In this Cree worldview, it is assumed that there are common connections between humans and non-human persons—the supposition of "the unity of spirit and the diversity of bodies" (Viveiros de Castro 2005:37). Humans and non-humans, which include animals, spirits and geophysical features, all share the fundamental similarity of being alive.

Members of the MCFN often emphasized the role that geese have in the hunt. Geese are viewed as actively engaging with the hunters and playing a vital role in the success of the hunt and are said to sacrifice themselves to hunters in order for them to have food. During an interview a hunter told Kuefler:

> I think the way I look at it, is that the Canada goose has given its life to you ... I mean anybody that gives their life so another person can live ... that's one of the biggest sacrifices that can be made. It doesn't matter if it [is] a human or an animal [be]cause animals do that too.

Geese act with humans in mind when they sacrifice themselves to the goose hunter and provide the hunter with food. MCFN participants in the research believe that goose hunters must consider the needs of geese when hunting them by respecting the animal, which has or will sacrifice itself. This is done by following certain prescriptive hunting practices that limit the disturbance of the geese being hunted and minimize the wastage of the animals that are killed. By respecting geese for the sacrifice they make, goose hunters trust that geese will continue to sacrifice themselves to them.

If humans are disrespectful to geese they will become unsuccessful in goose hunting. A hunter told Kuefler about an exchange he once had with some unsuccessful hunters who were walking around looking for geese:

> They do that [scare away feeding geese], then you see them walking away disappointed. That's not the way to do that kind of thing. What do we do? I don't know, ...you get more geese... sitting in your blind. You know— get in your blind and don't wander around, and that type of thing. We bumped into a few guys [and said] ... "get in your blinds or we're going to get your geese." They [reply] "oh where did you get your geese?" Well we're sitting in our blind. ... That type of thing.

Another hunter remarked that:

> There's a term we use in Cree ... which translates as 'what goes around comes around.' One year, you know, you could be killing 30-40 geese; next year, nothing ... and it's because you were disrespectful.

The same hunter later explained that hunters who do not hunt in ways that are respectful will be unsuccessful in their hunting efforts and may also have undesirable things happen to them:

> Like ... I said before, bad things will come your way. Accidents might start happening to you, or bad luck. I believe ... one of the big things is that things might not always go your way because of disrespect.

Following the prescriptive hunting practices that show geese respect ensures that MCFN hunters and geese maintain a mutually sustainable relationship. To reciprocate for being treated respectfully the geese allow themselves to be hunted. If a hunter does not follow the proper etiquette for hunting geese, the hunter will begin to experience 'bad luck' while hunting, as the geese will not come to the hunter.

Many of the MCFN members to whom Kuefler talked feel that some members of the First Nation are now hunting in disrespectful ways toward geese. A hunter explained that:

> Hunting styles have changed. You know there are certain things that you are not supposed to do. Like when you're hunting geese you stay back away from them; you let them do what they would normally do. Like you know when they're eating and feeding and they fly and now a lot of people just walk up to them. Like our hunting practices have changed a lot and it's not good for the birds. You know the ones that we hunt.

Many of these people believe that the reason some people are beginning to be disrespectful to geese is because there is a growing number of MCFN members who have never had the proper instruction on how to hunt properly.

The Teachers

Children ideally learn the skills needed to participate in goose harvesting activities from the experts in goose hunting—usually older members of the Moose Cree First Nation. Cree children learn through what Preston (1982:300) terms action-models and word-models. Action-models refer to the observations children make of their experiences and the actions of others, which provide examples of proper or improper ways to act. Word-models are the shared experiences of others, which grant the child access to another's experiences and teach the child moral lessons, important cultural categorizations, and how to perceive and understand events that occur around them. Children develop the knowledge, skills and values required to be goose hunters as they "grow into apprenticeship relationships with older people, watching how they do something and then imitatively playing at it, or as they grow up, learning by watching, followed by more serious 'play' in repeated trials" (Preston 1982:301). For a child to become a goose hunter who hunts in accordance with the traditional values held by members of the MCFN, having a relationship with an experienced goose hunter is vital. A young man explained the reason why he and his friends are good hunters: "I guess all of my friends grew up with their parents and their grandparents so we've always been taught the right way."

Generally, or ideally, parents assume the role of teacher to their children and are responsible for passing on their knowledge of geese and related goose hunting activities to them. However, when parents are unable to take their children hunting, members of the child's extended family ideally take the place of parents as educators. A mother explained to me that while she and her husband took their children goose hunting in the spring, they did not participate in the fall hunt. In this case, her son would go with his aunt and uncle to hunt geese in the fall. In other instances, grandparents could take their grandchildren with them and teach them how to hunt.

Educating children to be goose hunters extends to non-family members as well. Goose camps may house several families in close proximity and thus people may teach children from other families. A man talking about teaching his son how to hunt geese explained that he and his wife would "try to teach him as much as we can and not only him, but all the kids that come out to the blinds." Similarly, an elderly woman talked about her role in teaching children at a camp:

> Oh ... we do teach them. There are also other families that camp with us and they have little children with them so these kids come and spend some time in my tent, you know, and we talk and do things together. There's a little girl ... that came around this spring, and the spring before there were two other little girls from another family. So... we all get together and talk, and do little things and teach them. They sit there and help me pluck, or try to pluck you know, ducks and geese—whatever we have there... You know, they're so willing to do things.

What Boys and Girls Learn

Members of the MCFN have generalized ideas of the labour that men and women are supposed to perform while goose hunting. These are summed up by an interview informant who told Kuefler that: "there are certain jobs that women do compared to men, ... like the women would clean the geese, the men would hunt only." Thus, men are responsible for killing geese, while women are responsible for cleaning and preparing geese. However, in reality there are no strict rules governing gendered division of labour, and the responsibility for undertaking each of the tasks required at a goose camp differs from family to family. A family from Moose Factory that Kuefler came to know provides an example of this. Plucking geese is generally considered to be 'women's' work and is done by women away from the blinds at the camp. However, the men in this family would pluck the geese in the blind shortly after being killed. Two reasons were given for this. Firstly, geese are easier to pluck when they are still warm. By plucking them at the blind, it saved their wives effort later on. Additionally, plucking the geese at the blind gave the men something to do while they were waiting for geese to fly by.

Many of the MCFN members with whom Kuefler had spoken were trained in many, if not all, of the jobs that needed to be done while hunting geese, allowing for greater self-reliance and independence in their activities. Knowing how to accomplish all tasks was an obvious source of pride for a young man who explained that:

> I was hunting for myself and my family so I was always taught to do things myself. Now when I do go out in the bush I don't have to

rely on my mother to do my own geese for me. I can do everything myself.

Learning to Hunt Geese

The education of MCFN children begins at a young age, when they are taken to goose camps in the spring or fall. Initial training includes learning proper behavior by spending time around people involved with goose hunting and observing them. This is also a time for children to be introduced to, and learn about, the non-human elements that are present in their environment. These elements in the education of children are explained clearly by a boy's parents discussing the early education of their son:

> BK: How old were your children when they started going to the blinds to hunt?
> P1: [Our son] was probably about five years old. Like that's when you start taking them to the blind but they don't hunt or anything. They just watch.
> P2: Learning everything. ... Like you don't actually shoot you'd just observe, but have fun ... Like he'd be outside the blind playing, looking at the little birds and all the other birds that ...come around ... He'd be watching and learning, and he'd know when to come running into the blind—like when there were geese coming, or ... we'd try to get him to see them, or we'd try to teach him as much as we can ... not necessarily starting off with shooting a gun ... That comes a little later when they're ready and when they're able to do that kind of thing, but I think there's a lot of learning that goes on when they come to the blinds.

Whether it is goose hunting, sewing, driving a snowmobile, cooking, gathering wood or other necessities of camp life, once the child becomes comfortable and familiarized with what is expected, he or she will then make attempts at the activity. However, the adults exercise discretion and determine if the child is ready for that task. In an interview, parents related the problems that occurred when their sons went to the blind when they were too young:

> P1: Well, first I'd tell them stories like how I was hunting when they were younger...like too young to hunt. But when [parents] start [children] too young, [children] don't really realize what they're hunting for or know what they're doing, but I believe that maybe eight years old is a good age to start. Nine years old that's when I started these kids. My boys ...wanted to come out and I kept telling them they're too small. I tried taking them out to the blind when they were six or seven.
> P2: ... they had to learn the hard way. [they could] go with dad and then when they [would] come back they say "oh it was so boring."
> P1: And they're shooting all over the place, get all wet. Yeah you have to learn to sit in the blind all day. They couldn't do it when

they were younger. That's one of the reasons ... Once they get the interest of actually shooting a goose they want to stay in the blind all day and get up early. So I just showed them each the way I was brought up and you know they all had to watch.

Another parent expressed the same concern with children undertaking tasks for which they are not ready, and emphasized that a child should not be forced into doing something he or she is not ready to do:

A parent has to know when they [the children] want to do it but they're not ready. You know you have to be the judge for that. Like there're too young or not safety conscious then they shouldn't be handling a gun or driving a skidoo. [My son] wouldn't drive a skidoo for a long time. ... kids younger than him were driving and I used to haul stuff and ... [I would tell him to] drive. He says 'no' you know. ... he drives the skidoo now, but I guess he drove it when he was ready, you know. ... you can't force them either and you can't let them go when they're too young. It's all judgment.

While the early education of MCFN children is gained by watching experienced practitioners, children are expected to attempt the techniques they observed once they feel they are ready. A father related the way he teaches his children bush skills:

You just let them watch and when they're ready they'll do it. When they go out by themselves they'll see how you've done it and then they'll try it. ...I think [there are a lot] of visual learners when it comes to Aboriginals. They learn by watching then they try it themselves later.

Not surprisingly, when children first attempt a task they are often unsuccessful since they have yet to master the required skills. Two interview participants recounted their children's early attempts at calling geese:

P1: When they were out there they always wanted to do things that they learned. They'd call geese but they didn't call them right at first but after awhile they got really good.
P2: Oh yeah they're really good .... Our daughter right now is really good at calling geese ...

Competence in goose hunting, like any other learned activity, is acquired through practice and development of the skills required to be successful hunters. Pálsson (2000:37) notes that "learning is not a purely cognitive or cerebral process, but is rather grounded in the contexts of practice, involvement, and personal engagement." Children often begin developing the techniques and skills required to be a goose hunter through play, and well before beginning to hunt. Children are given toy guns and they mimic the adult hunters and pretend to shoot geese. While Kuefler was at a goose camp, he noticed that a snow shovel that went missing. After a few days the shovel was found in a stand of trees behind the cabins. The young boys who had been at the camp prior to Kuefler's arrival had taken the shovel to make a snow blind which they sat in and pretended to hunt the geese that passed by.

Children are encouraged in their efforts to learn the skills required to participate in harvesting activities. For example, a teenage boy recounted what happened after he killed his first goose: "[we] had a celebration that I killed my first goose, [there was] encouragement, lots of encouragement around the camp." When a novice hunter does something that is questionable, however, attention will be drawn to his technique so that he or she may make the necessary changes to become a better hunter. Criticism will most often come from the eldest hunter in the camp. A hunter explained how elders would rebuke someone for improper hunting practices:

> [They would] give you heck that's for sure ... They'd probably start telling you about what they used to do and that you shouldn't be doing that. I used to get yelled at when I did something wrong from my grandfather. That's how I was brought up like. I learned. I won't do that again and get heck.

In goose camp, reprimands for undesirable hunting practices are less overt and take place during conversations at night. In these instances, the elder hunter at the camp would question his son on his hunting technique. He would ask questions such as "didn't you see those birds?," "were you sleeping?," "why didn't you shoot then?" Questioning of this sort allowed the younger hunter to reflect on how he hunted, to consider alternatives to what he had done, and, in general, to improve his hunting skills.

Impediments to Children Learning to Hunt Geese

Being able to go to a goose camp and interact with and learn from expert goose hunters is critically important for MCFN children to become successful hunters. However, many MCFN children do not have the opportunity to learn the necessary skills, knowledge and attitudes required to become a successful goose hunter. Members of the MCFN were aware of this and provided several reasons why some children are not getting the education needed to become successful goose hunters. The most frequently cited reason was that parents were unwilling or unable to take their children hunting. Some Moose Cree members never learned to hunt geese as children and thus goose hunting never became important to them. Other members went hunting as children but discontinued the practice as they became adults. In other instances, parents were unable to take their children hunting because they could not finance the hunting trips, had a physical ailment, or were unable to find the time to fit hunting into their lives due to job or school responsibilities.

These issues are sometimes resolved by children accompanying the families of friends or other relatives to goose camps. However, we were told that sometimes, even when these alternate means of having children participate in a goose hunt were available, parents would not allow their children to go hunting. A hunter explained this phenomena to me as:

> "there's that pride, ... "I can't take them so you can't go," you know ... it's just that they're too proud you know, they don't want to admit that they can't afford it."

This comment points to the importance of goose hunting as a cultural activity for members of the MCFN and the value that some Moose Cree place upon educating their children. Having another person instruct one's children how to hunt geese is an admission to the community that the parent/parents, is/are unable to participate in goose hunting. This not only is injurious to their pride, but it reflects negatively upon them as parents and as members of a community that highly values goose hunting as a demonstration of membership and self-expression. This comment also suggests that social and economic disparities are emerging between members of the MCFN who are able to afford the growing costs associated with goose hunting and those who struggle with meeting these costs.

Some MCFN children are also uninterested in goose hunting and do not want to learn how to hunt. These children would rather spend time in Moose Factory than at a camp. An explanation given for this ambivalence toward hunting is that, for some, the modern conveniences that are available in Moose Factory are more important and relevant to their lives than life based upon traditional hunting practices. A hunter explained the difficulties he faced with his own son:

> I've noticed there are more people into electronics. My son was saying that more kids are into electronics. It's not like [for] me; it doesn't matter. ... I can leave my electronics, but for the children, ... even [my son], we have a hard time to get him to go; but once he's there he's fine. ... he doesn't miss his games or TV or internet when he's out there.

While some children are reluctant to go hunting, once they are out in the bush they have the opportunity to learn how to hunt properly and to gain an appreciation of hunting.

Amongst the Chisasibi Cree, young men are also often uninterested in hunting activities. However, their interest and expertise in bush skills increase once they mature and gain familial responsibilities (Ohmagari 1995:329). Predicting whether or not MCFN children who are uninterested in goose hunting will ultimately grow up to be active goose hunters is difficult, but it seems probable that children who have the opportunity to learn to hunt geese as children are more likely to continue to hunt geese as they mature and teach their own children how to hunt.

Among many of the goose hunters that Kuefler had the opportunity to meet there was a sense that the local knowledge associated with goose hunting is slowly being lost. A hunter offered his opinion on this loss:

> I think ... more and more younger generations are going out on
> their own, ...they're not going with somebody that knows the
> proper way to hunt, and what to do, and what not to do, and what's
> going to happen if you do this .... Like I say, they shoot at night
> and [the geese are] not going to be there tomorrow, but some
> people don't really care.

This opinion is shared by another hunter who stated that:

> I think slowly we're, ... losing it. At least keep [up] ... what was
> taught to them by parents or grandparents ... ... some families, I
> noticed, used to go hunting when they were younger, but as they
> grew up and become a parent, they won't go out ... and then their
> kids won't go out. So you know the tradition isn't carried on ... the
> hunting tradition. You see a lot of that now.

School Programs

Many MCFN members are concerned that their children do not have adequate
opportunities to go to the bush and to learn the skills necessary to participate in
hunting and trapping. As a response to this, outdoor education programs have
been implemented in the schools in Moose Factory in which elders or
experienced hunters take children into the bush and teach them simple skills like
setting snares, catching fish or building goose blinds. A resident of Moose
Factory explained:

> The schools have these programs where the kids are taken out for a
> couple of days. It might even be one day in the fall, where they
> actually go out and into a camp. Horseshoe is one [camp] for
> DDECS [Delores D. Echum Composite School]. They have
> another one, the winter curriculum, where they're taking them out,
> I think for a couple of days … the kids are out there every fall. It's
> usually one whole class that will go out and then come back. Next
> day, another class will go out.

During Kuefler's fieldwork he had the opportunity to interview a teenage
girl (P1) and her parents about their hunting practices. The parents in this family
(P2 is the father, P3 is the mother) are active hunters and take their children with
them when they go to their camp. While conducting this interview the
discussion turned to the girl's experiences with her classmates during a school
camping trip:

> P1: What I notice is that there are only five kids in my class that go
> hunting out of 23. I could see that my teachers could teach us some
> more of our culture but ... kids are not really into it.
> BK: Do you go out on those trips that the school puts on?
> P1: Yeah. It's fun because our whole class goes, and it's a good
> thing cause the class gets along and has fun outside.

P2: That's where they go camping where we went about 12 miles upriver. I guide there too. We went on our trip there. ... I think some of the kids that had never stayed out in the bush.

P3: Even on your fishing trip there the class went fishing. She was the only one that...

P1: [interrupts] Caught fish.

P3: ...caught fish. And you knew what to do. She baited it and cleaned her fish and out of the whole class.

P1: I caught five fish...

BK: And did anyone else catch anything?

P1: There was only one person who caught one fish, but it was a baby fish. So we couldn't eat it.

P2: Same with [my son]. They have a cultural day at DDECS ... A big feast and workshop and all that and [my son] started cutting up ducks and geese and the ladies were suprised that he knew how to do that. So they got him helping out there.

P3: Yeah he was putting on his own workshop.

While this excerpt reveals the sense of pride and accomplishment of the teenager in being able to catch fish, it also stresses several recurring issues that members of the MCFN are concerned about. In this girl's class many of the children do not have the opportunity to go camping or learn how to hunt and engage in harvesting activities. The programs offered through the formal educational institutions to MCFN children, such as outdoor education and cultural days, are often the only experiences that these children will have to spend time on the land. However, despite the good intentions that surround these programs, some people expressed concerns about their efficacy in passing on the local knowledge and values of the MCFN to their younger members. A hunter related his concerns about the fall camping trip his son went on:

I'm not too sure what they teach them. 'Cause I asked my son, "what did they teach you?" He said "Oh we just got to set up a blind and we sat in the blind all day and we saw one goose." You know things like that. I just think that it should be more in-depth, where maybe it should be an overnight stay instead of just a day trip. It's a day trip they go on. Some of this stuff that, like for instance traditional knowledge and all that, as far as goose hunting is concerned is really ... I know the smaller youth are starting to lose it a little I guess. They're not taught correctly. You got to teach them correctly to pass on these traditions properly.

When questioned on the correct way to educate children about their local knowledge he replied:

The correct way would probably be basically an explanation to them that that is why it is done. It's not just like showing them something: there you made a goose blind. This is why we're making a goose blind because we have to hide from the geese. It has to be more in-depth teaching method. The methods are there but it's not so intense. It should be more of a message to them. You get them interested and then after that it's easy to teach them but you have to tell them why.

Learning and Worldview

This hunter's concerns about the formal school system's attempts to educate Moose Cree youth about traditional harvesting activities points to an important facet of goose hunting. Goose hunting, for the members of the MCFN, is more than just the activities needed to go onto the land and kill geese. Rather, hunting geese is an "empowering system of knowledge that gives life to these people… (and) informs their … cosmologies and practices, as well as the reciprocity they practice with one another and with the environment" (Riddington 1994:273). Thus, learning to hunt geese properly not only means learning how to interact with geese but also the attendant social responsibilities of being a goose hunter (e.g., sharing with those who do not have the ability to hunt for themselves.) This system of knowledge also includes individually held detailed environmental knowledge, which informs the hunter about the possible relationships held between the hunter and his environment, and provides direction in making decisions about how to act while on the land (Riddington 1982:478, 1994:281).

Hunting is a way of engaging with one's environment that is learned through hunting with experienced hunters. Learning to hunt involves developing a body of knowledge that includes knowledge about the environment as well as cultural knowledge that informs how the hunter interacts with the environment. MCFN goose hunters must learn how to successfully and respectfully interact with non-human persons and must also learn how to comprehend these elements in a way that is consistent and meaningful to the members of the MCFN. Concerned about the number of children not learning to hunt properly, one parent said to me:

> I think it's important ... just to teach them to be on the land. Not only to goose hunt, but just to be there to see the geese and know they're there. They're there to help you.

Learning to be a goose hunter "…means attending to the task at hand, being actively engaged with a social and natural environment" (Pálsson 2000). As Pálsson (2000:26-27) indicates, "this suggests a notion of enskilment [sic] that emphasizes immersion in the practical world, being caught up in the incessant flow of everyday life—and not simply … the mechanistic internalization and application of a mental script, a stock of knowledge or a 'cultural model.'"

The Moose Cree are aware that traditional modes of training their children today are not sufficient as some parents lack the resources or the desire to take their children hunting. Knowledge is possessed by individuals and not institutions (Riddington 1994:273) and thus for the local knowledge and values which informs goose hunting practice to persist, children need to be taken to goose camps and immersed in a hunting lifestyle where seasoned goose hunters and their family members can pass on their knowledge and values. The responsibility to these children rests with the entire community to ensure they have the opportunities to learn to hunt geese. This point is expressed by a hunter concerned with the number of young people not learning to hunt properly and the resultant effects on traditional Moose Cree hunting practices:

> I think we need to be more careful with the younger generation. Teach them better. Even if they're not yours.

Summary

Becoming a goose hunter is a learned activity that is conducted under the tutelage of an experienced goose hunter. Children learn to hunt geese by observing and then imitating skilled hunters; thus, having relationships with active and experienced goose hunters who are able to take them hunting is crucial for children to develop the skills and attitudes needed to become a successful goose hunter. The mentor guides the novice hunter as he or she moves from observation, to play, to actually hunting and offers encouragement, advice and, when necessary, reprimands. The child is generally allowed to learn to hunt at his or her own pace and comfort level, although a child may be restricted from participating in an activity if his or her instructor feels the child is not ready for the task.

Today, many Moose Cree children do not have the opportunity to learn to become goose hunters. Reasons for this include their parents' reluctance or inability to take them hunting themselves, and/or their parents' unwillingness to let another person teach them to hunt geese, or the child's own indifference toward hunting. To give children opportunities to experience a taste of bush life, outdoor education programs have been established in the Moose Factory schools in which an elder or an experienced hunter takes a group of children into the bush to learn simple bush skills.

However, these programs themselves may not be adequate to pass on the knowledge and values needed for children to become active and successful hunters. Becoming a goose hunter involves more than learning how to kill birds; becoming a goose hunter also means learning about the cosmologies and environmental philosophies held by members of the MCFN that give life and meaning to the relationships that exist between humans and geese. Learning to become a goose hunter means not just learning a set of activities, it involves learning how to recognize and exist within a social landscape in which hunters have obligations to both other humans and to geese. To come to embody successfully the cosmological and philosophical traditions that have been articulated by the MCFN members, a child must be immersed in a world in which these beliefs, and the practices that emerge from and reinforce these beliefs, are prevalent.

References Cited

Berkes, F. (1982). Waterfowl management and northern Native Peoples with reference to Cree Hunters of James Bay. *Musk-Ox* 30: 23-35.

Brightman, R. (1993). *Grateful Prey: Rock Cree Animal-Human Relationships.* Berkeley: University of California Press.

Cruikshank, J. (2005). *Do Glaciers Listen? Local Knowledge, Colonial Encounters, and Social Imagination.* Vancouver: UBC Press.

Environment Canada (2005). 217. *James Bay Lowland.* http://www.ec.gc.ca/soer-ree/English/Framework/NarDesc/Region.cfm?region=217 , Accessed August 10, 2009 .

Feit, H. (1994). 'Hunting and the quest for power, the James Bay Cree and whitemen in the twentieth century,' pp. 181-223 in R.B. Morrison and C.R. Wilson , eds., *Native Peoples: The Canadian Experience*, 2nd edition. Toronto: McClelland & Stewart.

Nadasdy, P. (1999). The Politics of TEK: Power and the 'integration' of knowledge. *Arctic Anthropology* 36(1-2): 1-18.

Ohmagari, K. (1995). 'Culturally sustainable development and James Bay Cree women,' pp. 322-334 in D.H. Pentland, ed., *Papers of the Twenty-sixth Algonquian Conference.* Winnipeg: University of Manitoba.

Pálsson, G. (2000). '*Finding One's Sealegs*: Learning, the process of enskilment, and integrating fishers and their knowledge into fisheries science and management,' pp. 26-40 in B. Neis and L. Felt, eds., *Finding Our Sealegs: Linking Fisher People and Their Knowledge with Science and Management.* St. John's, NL: Institute For Social and Economic Research.

Preston, R.J. (1982). 'Towards a general statement on the eastern Cree structure of knowledge,' pp. 299-306 in W. Cowen, ed., *Papers of the Thirteenth Algonquian Conference.* Ottawa: Carleton University.

Prevett, J.P., H.G. Lumsden, and F.C. Johnson (1983). Waterfowl kill by Cree hunters of the Hudson Bay Lowland, Ontario. *Arctic* 36(2): 185-192.

Riddington, R. (1982). Technology, world view, and adaptive strategy in a northern hunting society. *The Canadian Review of Sociology and Anthropology* 19(4): 469-481.

Riddington, R. (1994). 'Tools in the mind: Northern Athapaskan ecology, religion, and technology,' pp. 273-287 in T. Irimoto and T. Yamada, eds., *Circumpolar Religion and Ecology: An Anthropology of the North.* Tokyo: University of Tokyo Press.

Scott, C. (1996). 'Science for the West, myth for the rest?,' pp. 69-86 in Laura Nader, eds., *Naked Science: Anthropological Inquiry into Boundaries, Power and Knowledge.* New York: Routledge.

Tanner, A. (1979). *Bringing Home Animals: Religious Ideology and Mode of Production of the Mistassini Cree Hunters.* St. John's, NL: Institute of Social and Economic Research.

USF&W—U.S. Fish and Wildlife Service (2009). *Waterfowl Population Status, 2009.* U.S. Department of the Interior, Washington, D.C. USA.

Viveiros de Castro, E. (2005). 'Perspectivism and multinaturalism in Indigenous America,' pp. 36-74 in Alexandre Surrallés and Pedro García Hierro, eds., *The Land Within: Indigenous Territory and the Perception of the Environment.* Copenhagen: International Work Group for Indigenous Affairs.

Conclusions

Conversations We Need to Have and Spaces We Need to Create

Marc G. Stevenson and David C. Natcher

Introduction

Co-existence is not possible without reconciling the land and resource use rights and interests of Canada's Aboriginal peoples with those of other Canadians. How soon reconciliation will occur and how effective it will be when it does occur, depends on many factors, not least of which is the creation of ethical space for Aboriginal peoples in land use planning and decisions taken in respect to their lands and resources. Ethical space involves more than simply inviting Aboriginal peoples within the boundaries of existing institutions. Rather, ethical space recognizes the rights of Aboriginal peoples to define and shape new spaces for their meaningful and ethical engagement in decision-making, whether within existing institutions or through new structures that reflect or are based on their own pre-existing formations.

One of the greatest obstacles to achieving this space is the continuing assertion by provincial and federal Crowns that the historic treaties were land surrenders, rather than what the Aboriginal signatories understood them to be: agreements to co-exist and share lands and resources for the mutual benefit of both peoples. While this issue will likely need to be resolved in the courts before agreement on the spirit and intent of the treaties is reached, reconciliation must not turn on this issue. It is a process that must begin in earnest right now. However, reconciliation will undoubtedly take time, produce uneven results and require sacrifice on both sides. As the Cree scholar, Willie Ermine (2007:202-03), reminds us:

> The dimension of the dialogue might seem overwhelming because it will involve and encompass issues like language, distinct histories, knowledge traditions, values, interests, and social, economic and political realities and how these impact and influence an agreement to interact. Even so, ...the new partnership model of the ethical space, in a cooperative spirit between Indigenous peoples and Western institutions, will create new currents of thought that flow in different directions and overrun the old ways of thinking.

The task that Ermine sets out can no longer afford delay. To do otherwise, perpetuates social and political injustices, and runs counter to the foundations of peace, fairness and good government on which Canada was built (Saul 2008). In other words, meeting the challenge of this task head-on reaches to the very core of what Canada represents and what we want this country to be.

So how do we go about building a new partnership model of ethical space that Ermine and others (Wilkinson *et al.* 2007) champion, and how can we create the space needed to nurture reconciliation and co-existence? One way is to support and undertake the kinds of research reported in this volume and the earlier companion volume (Stevenson and Natcher 2009), while continuing to address the questions and issues that such analyses raise. Another is to disseminate the results of research as widely as possible and in ways accessible to the public. Still another is to engage policy makers, senior bureaucrats, politicians and other stakeholders directly in conversations that will set the stage for reconciliation and ultimately co-existence. It is likely that all of these strategies will need to be employed simultaneously in order to broaden the boundaries of political inclusion and move beyond one-off forms of consultation.

In this concluding chapter, we revisit some of the 'conversations' members of our 'community of practice' in Aboriginal forest-related research issues have had over the years concerning how best to achieve this ethical space, and consider other strategies that have come to light as we prepared this and the earlier volume. It is our hope that by having such conversations the ethical space that we aspire to create will one day be achieved.

Conversations We Choose to Engage In

Volume One
In Volume One, *Changing the Culture of Forestry in Canada...* (Stevenson and Natcher 2009) we began with three chapters that attempted to shed light on traditional Aboriginal stewardship values, land use management systems and governance institutions—in this case, the Anishinaabe of Pikangikum and the Moose Cree of Moose Factory, Ontario. These chapters also considered some of the major challenges that confront the use and application of Aboriginal values, knowledge systems and institutions today in forestry and sustainable forest management (SFM), while exploring ways to overcome them.

Chapter 5, 6 and 7 presented the results of research undertaken with the Pikangikum, Moose Cree and Little Red River Cree (Alberta) First Nations in negotiating space for their values, worldviews and management institutions in currently accepted forestry and SFM practices. In all three cases, criteria and indicator (C&I) framework methodologies play a mediating role in creating this space.

The next three chapters of the first volume sought to evaluate various institutional options available to two First Nations in British Columbia—the Kwadacha (Chapter 8) and Stellat'en Nations (Chapter 9)—and to review the experiences of a cohort of Aboriginal communities across Canada engaged in forestry and SFM practices (Chapter 10). While each option or situation had its own particular set of advantages and disadvantages, Aboriginal communities might be well-served to combine elements from several, and/or develop multiple strategies for realizing their needs, rights, and interests.

The final two chapters of Volume One examined fiduciary, legal and ethical considerations that need to be taken into account when constructing and implementing enabling policies and effective institutions. This included consideration of the scope and nature of Aboriginal and treaty rights, their interpretation by the Courts (Chapter 11) and the role of Aboriginal peoples in research (Chapter 12). It was concluded that, by taking a more nuanced, informed and ethical approach to Aboriginal rights interpretation/ accommodation and the engagement of Aboriginal peoples in research, we would far more likely than not be successful at addressing some of the most pressing environmental and social issues that we face today in Canada's forests.

Volume Two

We began this volume with an examination of cumulative impacts assessment/management (Chapter 1) and land use planning (Chapter 2) processes in Alberta's Indian Country, both of which can be considered fairly representative of provincial governments across the country. Despite substantial flaws in these processes from an Aboriginal perspective, there appears to some room for cautious optimism. Creating the space for co-existence in Alberta and other provinces, however, will depend on the extent to which Aboriginal peoples, with all the capacities and skill sets that this task demands, are involved in design and implementation of these processes.

Next, we chose to engage in a conversation about the need for, and the basis of, significant transformational change in the nature of the relationship between Canada and its Aboriginal peoples with regard to land and resource use rights. An examination of the unfulfilled Treaty obligations of Alberta and Canada to protect the livelihoods of Treaty 8 First Nations peoples in Alberta (Chapter 3) was followed by the construction of an interpretative framework for a shared understanding of Treaty, and how it may be applied in the contemporary context to achieve co-existence (Chapter 4). Next, the legal and fiduciary bases for accommodating Aboriginal land use rights and interests in British Columbia through the creation of Aboriginal forest tenures were explored (Chapter 5). This was followed by an examination of a new institutional model of co-existence developed by the Stellaquo First Nation in BC (Chapter 6), which combine elements of both traditional and modern forms of governance, while revealing contradictions that should be instructive to others. Together, these chapters help forge a path for reconciling a number of critical issues that need to be addressed on the road to co-existence.

Co-existence just doesn't happen on its own. Strategies, approaches and planning tools need to be considered, designed, and implemented that respect and accommodate both worldviews, facilitate reconciliation, and build firm foundations for co-existence. Chapters 7-11 examined several issues that are critical to achieving these goals in the context of land use planning and management. Specifically, they explored the complex issues around the delineation of Aboriginal planning areas (Chapter 7), the varied impacts of roads on Aboriginal communities and their lands (Chapter 8), the benefits and pitfalls of Aboriginal land use and occupancy studies (Chapter 9), the ethics and efficacy of Aboriginal values mapping (Chapter 10), and the development of appropriate criteria and indicators for sustainable forest management in Indian Country (Chapter 11). While these planning considerations and approaches certainly do not exhaust the range of issues that will need to be addressed to achieve reconciliation and co-existence, they do at least inform the discourse on these matters and point us in the right direction so that we can initiate the conversations we need to have.

True reconciliation and co-existence cannot be achieved in the absence of building capacities to visualize and realize these outcomes. The next two chapters in this volume suggest that current 'top down' approaches to Aboriginal capacity building must be met in equal or greater measure with 'bottom-up' approaches. Chapter 12 situates existing and future Aboriginal capacity building initiatives within a conceptual framework that involves multiple scales and dimensions, and should, if supported, facilitate Aboriginal peoples' aspirations to become true architects of their future. Chapter 13 demonstrates that, for the Moose Cree First Nation of Northern Ontario, building the capacity of community members to survive on the land and within a social context involves learning and representing the values, needs, rights and interests of the Moose Cree people, while fulfilling obligations and responsibilities to the environment and to the larger community upon which they depend. In the words of Frog Lake First Nation elder, Peter Waskahat, "culture takes practicing."[1] Both chapters suggest that Aboriginal peoples must be at the centre of determining and realizing their capacity needs. Only by building capacities in both worlds will reconciliation and co-existence, and survival of Aboriginal cultures in Canada, be realistic and achievable goals.

Conversations We Need to Have

Rights and Responsibilities

Too often, the conversations we do have get bogged down in a debate about 'rights'—who has rights of access, use and ownership, who no longer has those rights, whose rights have been infringed, what is adequate restitution for the infringement of these rights, etc. Lost in the acrimony is any discussion about the 'responsibilities' that attend those rights. As many Aboriginal elders will remind us, "with rights come responsibilities;" they are two sides of the same coin. Increasingly, Aboriginal elders lament their inability to carry out their stewardship responsibilities to the land and their social responsibilities to their communities as often as they disparage the Crown's failure to uphold its fiduciary and Treaty obligations to First Nation peoples. While pursuit of the recognition and accommodation of constitutionally protected Aboriginal and Treaty rights by Canada's Indigenous peoples needs to be more strategically coordinated, a new conversation about the 'responsibilities' that such rights confer must take place. What are these responsibilities? Do these responsibilities infringe the rights of other parties, and if so, how? What institutional and policy reforms need to occur to accommodate these responsibilities? This is the type of

---

[1] Address to Tribal Chief Ventures Inc. meeting, Cold Lake, Alberta, 25 February 2010.

discussion that might produce real results, and lead to real policy and institutional change.

Power and Authority
Closely related to any conversation about rights and responsibilities, is the discussion about power. Who has the power and the authority to wield it? How did they acquire this power? How has it been translated into existing policies and institutions? What have been the social, cultural, environmental and other effects of this inequitable distribution of power? At what cost to Aboriginal communities? At what cost to society? At what cost to the environment? Serious reflection upon these issues by both sides will underscore and support the need for significant policy and institutional reform.

Knowledge and Capacity
To say that we possess sufficient knowledge and capacity to undertake effective land use planning and environmental decision-making would be hubris. Today, in Canada, land use planning and decision-making proceed without adequate knowledge or understanding of the full environmental, social and other consequences of our actions on present and future generations. We simply do not have all the information we need to achieve the outcomes we desire. Not only does this necessitate the adoption of adaptive management approaches, it requires the consideration of alternative knowledge systems and of different ways of looking at problems. Other voices need to be heard, particularly from those most dependent on the land, and whose livelihoods are most affected by provincial government regulations. Yet, these voices and the knowledge that they might contribute are, for the most part, muted by existing provincial government legislation, policies, processes and institutions. We need to talk about ways to create room for this knowledge and we need to talk about the capacities that both sides require to realize the contributions of this knowledge.

Spaces We Need to Create

Planning co-existence necessarily entails creating the space to negotiate co-existence, to reconcile our differences and to accommodate all of our rights and interests. This space does not currently exist in Canada. It may have at one time, e.g., during the first 250 years of this country's early history when settlers depended on Canada's Aboriginal peoples for survival and safety. Yet, some stakeholders, especially those with a vested interest in maintaining existing power relationships, might argue otherwise stating that such space is created in the context of negotiating comprehensive land claims agreements within Canada. But we would disagree; the rules and terms of engagement of such agreements are determined largely by the federal Crown. This is not the space where the conversations we need to have can occur. This is not the 'new partnership model' that Ermine speaks about.

Too often, the common refrains heard from the governing and business elites of this country when it comes to contesting Aboriginal rights and interests in land include: "Why can't we just all get along?" or "Let's not focus on our differences, let's concentrate on our similarities!" (*see* Mathew Coon Come's remarks in Chapter 12 for more popular refrains). This is not the recipe for reconciliation or the path to co-existence. Reconciling similarities is easy. Reconciling differences is hard, and the challenge we must be prepared to accept. Using our commonalities as a platform, we need to focus on our differences, and the institutions and policies that contribute to the power imbalances that define our relationship and contribute to the vast economic and social gulfs that divide us. Sure, we should celebrate our similarities, but more importantly, we need to understand, respect and create the room to celebrate our differences. It is only when such space is created that we can begin to design and implement new institutions and policies to reconcile and accommodate our differences, rights and interests.

There are indications that some progress is being made on this front. British Columbia's New Relationship policy, Ontario's *Far North Act*, and even Alberta's Land Use Framework indicate that political will for reconciliation in the provinces is building despite the fact that Aboriginal peoples had little to do with their development. However, the same old posturing and acrimony on both sides may keep these policy instruments from achieving any real success. If provincial governments refuse to make real concessions to Aboriginal peoples out of fear of losing popular support or undermining business interests from which they derive most of their revenues, any 'new relationship' agreements established will almost certainly fail.

In British Columbia, the BC Union of Indian Chiefs and First Nations Summit have rejected the province's proposed *Recognition and Reconciliation Act*, which would have given 'teeth' to the New Relationship policy. The New Relationship was based on mutual respect and the accommodation of Aboriginal title and rights, coupled with the reconciliation of Aboriginal and Crown titles and jurisdictions. Despite the fact that First Nations leadership and high-ranking BC government officials worked for many months behind closed doors to develop a legal framework for this new legislation, it was ultimately rejected by BC's First Nations in September 2009 because it failed to reconcile Aboriginal and Crown title, to share in the resources of the land more fairly, and to reflect the vision espoused by the *New Relationship* (First Nations Drum 2009).

In Alberta, the Alberta Organization of Tribal Chiefs (AOTC) has rejected Alberta's Land Use Framework. The choice for Alberta's Treaty First Nations was/is inequitable: 1) either participate in the province's First Nation specific consultation process for regional land use plans, or 2) do not participate at all. While some First Nations do not endorse Alberta's *Land Stewardship Act* and Land Use Framework (ALUF), and were not involved in drafting the terms and conditions under which they are obliged to participate, others feel that they can better protect their rights and interests through participation in the development of regional land use plans. For those First Nations who choose to litigate rather than participate, they risk having their claims dismissed by the courts because they failed to participate in a consultation process that was specifically designed to address their concerns and accommodate their rights.[2]

---

[2] The courts in *Halfway River First Nation v B.C.* 1999 BCCA 470,
Mikisew Cree First Nation v. Canada 2005 SCC 69, [2005] 3 S.C.R. 388,
and other cases have said that First Nations have an obligation to 'come to

The courts have made it clear that First Nations have an obligation to 'come to the table.' The choice to participate or not in ALUF continues to divide many Alberta First Nation peoples and governments, and puts them in the unenviable position of being 'between a rock and hard place.' However, the AOTC may have a third option: continue to reject ALUF in the interim, but put forward a proactive proposal that Alberta Treaty First Nations would be willing to participate in ALUF provided that: 1) each First Nation has a role to play in determining the terms and conditions of its participation, 2) the province would ensure sufficient time and resourcing for each First Nation to complete its own land use plans and that these would be included within regional land use plans, and 3) resource royalty-sharing agreements would be an outcome of participation. This would signal, without prejudice, to the courts that the Alberta First Nations are willing to participate in ALUF on terms and conditions acceptable to them.

Many First Nations refuse to participate in these processes because they have seen similar instruments (see Frideres and Rowe in this volume) used in the past in ways that reinforce government interests and government's position of power. Rather than being transformative in nature, these frameworks merely reflect and reinforce the values and interests of those already in power. While these 'invited spaces' appear, on the surface, to offer Aboriginal peoples a more equitable role in decision-making, in reality these spaces permit only particular voices to be heard and specific versions of reality to be considered (Cornwall 2004:76). In Foucault's (1984) terms, these invited spaces are fundamental to the exercise of power. As political constructs, these invited spaces are not "neutral containers waiting to be filled, but are dynamic, humanly-constructed means of control, domination and power" (Lefebvre 1991:24). Rather than providing Aboriginal peoples an opportunity to challenge existing hierarchies, these invited spaces, which are infused with political power, reinforce inequalities. In this way, these spaces continue to be defined by those inviting others in (Cornwall 2004:80).

Yet, these invited spaces can also represent opportunity. Owing to the 'strategic reversibility' of power (Foucault 1991:5), these spaces can serve as sites of resistance where possibilities of subversion and reconstitution can be realized. Factoring in the agency of those 'invited in,' these spaces may provide opportunities for Aboriginal peoples, who admittedly remain sceptical of the efficacy of these processes, to negotiate the ethical space now being called for. In this way, hope persists that these new spaces and more respectful forms of engagement can provide for a continuous and dynamic approach to learning, allow for deliberation and compromise, manage historical tensions, and bridge the knowledge and authority gaps that have long dissuaded equitable engagement. For Palmer (2004:251) these types of spaces can facilitate the construction of a 'new deal' for Aboriginal peoples. Moreover, "these arrangements hold the potential to supersede the formal denial of Aboriginal

---

the table' and be consulted if the Crown has made a reasonable effort to 'set a table' that might lead to the accommodation of their interests and concerns.

rights by settler legal and political culture, and recognize Aboriginal polities as an integral part of the continuing nation-building dialectic" (Palmer 2004:251).

In regards to the knowledge and wisdom that Aboriginal peoples might contribute to land use planning and environmental decision-making (and certainly sustainable forest management), much as been made of the potential value of Aboriginal traditional knowledge (ATK). However, virtually no space has been created for the respectful and appropriate use of this knowledge. The promise of ATK has fallen short not because it has little to offer, but because it has had to reside in a space created by conservation bureaucrats and practitioners of western environmental science (WES). In the meantime, the currency of Aboriginal traditional knowledge declines with each passing day, and with each passing elder. We have simply not envisioned a space where ATK can reside side by side with WES, where both are required to make informed decisions about lands and resources, and the peoples who depend on them. In this regard, consideration of the different philosophical underpinnings of each knowledge system, whereby WES is designed to manage 'resources' and ATK endeavours to manage 'relationships,' both sustainably, offers some promise for reconciling their differences, while creating the space for each to co-exist on its own terms in a complementary manner (*see* Stevenson 2006).

Conclusions

If the Aboriginal Program of the SFM Network has had a shortcoming, it has been our inability to influence policy-makers to design the institutional space necessary to achieve reconciliation. Admittedly, this is a tall order for any research program. Yet, it was our belief that once disseminated, policy-makers would use our research to design more equitable policies leading to institutional reforms. In retrospect, this expectation was quite optimistic, if not altogether naive. To be sure, there have been victories, albeit largely at a local or community-specific scale. However, even these victories have been challenged through constant struggles for legitimacy, government resistance and threats of cooptation. Thus, despite our collective efforts, and the quality of research conducted, we are unable to claim success at creating the necessary ethical space for institutional and policy reforms. Through this we have learned that as long as reconciliation and co-existence are framed as threats and not opportunities, as attempts at coercion rather than mutual gain, Aboriginal peoples will likely remain the subjects, rather than co-authors, of any future policy and institutional reforms.

Despite the challenges, if changes are to occur they will likely need to originate locally, among those who are on the ground (Aboriginal and non-Aboriginal) and who are affected most directly by government policy, and whose lives and livelihoods are most at stake. Although achieving reconciliation and co-existence may seem an elusive and daunting task, change is possible, particularly if the closed spaces traditionally occupied by government can somehow be breached through the transformative engagement of Aboriginal peoples. However, for fundamental changes to occur and to be sustained, local action alone may be insufficient. Rather, new space found at the convergence of 'top-down' and 'bottom-up' approaches, where conversations occur in the absence of prejudice, in a new spirit of partnership that Ermine and others advocate, may be required. Whether there is the political will for this new type of space remains to be seen. To date the formation of this space has been avoided by most policy-makers who tiptoe around the margins, too frightened of change, and who cling to the legacies and injustices of old.

Space may also be found at the convergence of Aboriginal and industrial interests, where industry—a major source of revenue for most provinces—has a vested interest in creating an economic climate of certainty for shareholders while embracing corporate social responsibility. While the convergence of Aboriginal and industry interests in the resource development and management sectors may not result immediately in the political reforms we seek, it may be a site of institutional innovation that might foster political change, and ultimately co-existence.

Yet, by avoiding calls for social and environmental justice, and stifling institutional innovation, both government and industry further entrench the status quo, leaving yet another opportunity missed, and another generation of Aboriginal peoples left to bear the social, economic and political consequences of inaction. We cannot wait for this space to open up without identifying the conversations we need to have. Ethical space may be created by starting to engage in these conversations now. If the 27 chapters that comprise both volumes in this series play even a small role in initiating these conversations, our mission will have been accomplished.

References Cited

Cornwall, A. (2004). 'Spaces for transformation? Reflection on issues of power and difference in participation in development,' pp. 5-91 in S. Hickey and G. Mohan, eds., *Participation: From Tyranny to Transformation?* London: Zed Books.

Ermine, W. (2007). The Ethical Space of Engagement. *Indigenous Law Journal* 6(1): 193-203.

Foucault, M. (1991). 'Governmentality,' pp. 87-104 in G. Burchell, C. Gordon and P. Miller, eds., *The Foucault Effect: Studies in Governmentality*. Chicago, IL.: University of Chicago Press.

Foucault, M. (1984). 'Space, Knowledge and Power,' pp. 239-256 in P. Rabinow, ed., *The Foucault Reader*. New York: Pantheon Books.

First Nations Drum (2009.) Chiefs Reject Proposed *Recognition Act*. Sept. 2009, Vol. 19, Issue 9.

Lefebvre, H. (1991). *The Production of Space*. London: Verso.

Palmer, L. (2004). 'Agreement making, outcomes, constraints and possibilities,' pp. 251-253 in M. Langton, M. Tehan, and L. Palmer, eds., *Hounour Among Nation? Treaties and Agreements with Indigenous Peoples*. Melbourne, Australia: Melbourne University Press.

Saul, J.R. (2008). *A Fair Country: Telling Truths about Canada*. Toronto: Peguin.

Stevenson, M.G. (2006). The possibility of difference: Rethinking co-management. *Human Organization* 65(2): 167-180.

Stevenson, M.G. and D.C. Natcher, eds. (2009). *Changing the Culture of Forestry in Canada: Building Effective Institutions for Sustainable Forest Management*. ed. Occasional Publication No 60., Edmonton: CCI Press and SFM Network

Wilkinson, K.M., S.G. Clark, and W.R. Burch (2007). *Other Voices, Other Ways, Better Practices: Bridging Local and Professional Environmental Knowledge.* Yale University of Forestry and Environmental Studies Bulletin 14.

APPENDICES

Appendix A—*Notes on Authors*

Marie-Christine Adam is a PhD Candidate at the Faculty of Environmental Sciences, University of Quebec in Montreal. Her current area of research is the integration of Aboriginal values and objectives in forestry. She has written on local level criteria and indicators as a tool to assess Aboriginal values, and their use for sustainable forest management. She has also studied methods to access and integrate Aboriginal values and goals in management and the effects of culture in portraying Aboriginal forestry issues.

Sarah Allen is an MSc Forestry Candidate at Lakehead University and the Regional Fire Advisor for the Ontario Ministry of Natural Resources. Sarah's research focuses on Aboriginal forest-based economic development through developing comprehensive community frameworks for measuring success. Sarah continues to work with First Nation communities to develop partnerships for fire protection and stewardship projects around FireSmart education and awareness.

Hugo Asselin (PhD) is a Canada Research Chair in Aboriginal Forestry at the Université du Québec en Abitibi-Témiscamingue (UQAT). His research combines Aboriginal forestry, palaeoecology and forest ecology to develop sustainable forest management strategies that reflect Aboriginal cultures and worldviews.

Iain Davidson-Hunt (PhD) is an Assistant Professor at the Natural Resources Institute, University of Manitoba and associated with the Centre for Community-based Resource Management. He has undertaken research and worked professionally in northwestern Ontario, Mexico and Bolivia in the areas of community-based land-use planning, cultural landscapes and non-timber forest products.

Nathan Deutsch is a PhD candidate at the Natural Resources Institute, University of Manitoba. He is studying change in land-use institutions with Pikangikum First Nation. He holds a Masters degree in Natural Resource Management, Governance and Globalization from Stockholm University.

Jean-François Fortier is a PhD student in sociology at University Laval, Québec, and his main interests are aboriginal and social forestry and local governance of resources and lands. He has a master in sociology, has worked as a research assistant for an SFMN funded project examining collaboration between aboriginals and the forest industry and has been involved in the organisation of a number of workshops on related themes.

Jim Frideres (PhD) is Professor of Sociology, Chair of Ethnic Studies and the current director of the International Indigenous Studies program at the University of Calgary. He is the author of numerous articles on Aboriginal peoples. He also is the author of the well known *Aboriginal Peoples of Canada* (8th edition) as well a forthcoming book on *First Nations in the 21st Century*.

Daniel Kneeshaw (PhD) is a Professor in the Department of Biological Sciences, the Institute for Environmental Sciences and the Centre for Forest Studies at the University of Quebec in Montreal. He is active in research on the effects of disturbance on forest dynamics, and the effects of forest management ecologically and socially. His research has focused on forest and human communities from Northern Eurasia and Canada.

Brent Kuefler (MA) is a PhD candidate in Anthropology at Memorial University. He holds a Masters degree from Memorial University. His research explores Cree environmental ethics and the politics of animal rights movements.

Mario Larouche (MSc) is an environmental consultant for Genivar specialising in restoration ecology and land compensation agreements. He has extensive experience in landscape modelling of forest management including testing aboriginal forestry scenarios.

Eddison Lee-Johnson (MSc) is the Natural Resources Director for Stellat'en First Nation, in British Columbia. He has Master of Science in Forestry from the University of British Columbia and a Bachelor of Education from Milton Margai College of Education and Technology in Sierra Leone. Eddison's work and research interests are; resource governance, food systems and food security, rural community empowerment and participation, and community economic development.

Deborah McGregor (PhD, RPF) is an Associate Professor in the Department of Geography and Aboriginal Studies at the University of Toronto, and an Anishnabe from Whitefish River First Nation, Ontario. She has been involved in curriculum development, research and teaching at both the university and community levels for two decades. Her research focus is on Traditional Knowledge (TK) and its application in environmental management, forestry, sustainable development and water conservation. Primary themes found throughout her work include improving relations between Aboriginal and non-Aboriginal interests and ensuring appropriate consideration of Aboriginal peoples' knowledge, values and rights in environmental and resource management in Canada.

David Natcher (PhD) is an Associate Professor in the Department of Bioresource Policy, Business and Economics at the University of Saskatchewan. He also serves as Academic Chair of the Indigenous Peoples Resource Management Program and Executive Director of the Indigenous Land Management Institute.

Monique Passelac-Ross is a Research Associate at the Canadian Institute of Resources Law (CIRL), Calgary. Her research interests include forestry law, natural resources law and policy, Aboriginal law and environmental law. She has written extensively on the legal and policy framework of forest management in Canada, and has published several articles on the Crown's duty to consult and accommodate and its application in the Alberta context.

Pamela Perreault (M.Sc) is a member of Garden River First Nation (ON) and a PhD candidate in the Faculty of Forestry at the University of British Columbia. Pamela has worked extensively with First Nation communities and organizations on broad range of natural resource consultation, planning, governance, and non-timber product development issues. Her current research interest and the subject of her doctoral dissertation is community empowerment and transformative processes associated with capacity building for forest management.

Gerardo Reyes (PhD) is a post-doctoral fellow at the University of Quebec in Montreal with research interests focused on conservation biology and forest ecosystem management. He is currently involved with examining the role of ecological resilience as a management option in northern temperate and boreal forest ecosystems.

Cash Rowe is a PhD student in Interdisciplinary Graduate Studies at the University of Calgary. His work has focused on the social determinants of health for Aboriginal people and he is currently analyzing data from the Aboriginal Peoples Survey. He is a sessional lecturer in the International Indigenous Studies program at the University of Calgary. Cash is Cree whose family comes from Ahtahkakoop.

Marie Saint-Arnaud (PhD) specializes in Aboriginal forestry and, in particular, the interdisciplinary link between biology and ethnology. She has worked with anthropologists on a variety of environmental issues involving First Nations in Canada and Mexico, and, since the late 1980s, been involved with the Anicinapek of Kitcisakik (Quebec).

Peggy Smith (PhD, RPF) is a Registered Professional Forester, an Associate Professor in Lakehead University's Faculty of Natural Resources Management and a Senior Advisor for the National Aboriginal Forestry Association. Her research interests focus on the social impacts of forest management, including Aboriginal peoples' involvement in forest management, community forestry and public participation, northern development, and forest certification.

Marc Stevenson (PhD) is the former Aboriginal Program Manager for the SFMN. Marc has conducted research with Canada's Inuit, First Nations and Métis communities on social, economic, environmental and political issues directly relevant to them for the last 30 years. He currently provides research and consulting services and policy advice to Aboriginal, government and industry clients on a broad range of issues at the interface of Aboriginal and resource development.

Erin Symington (RPF) is an MSc Candidate in the Faculty of Forestry and the Forest Environment, Lakehead University. Erin's research is focussed in the area of Aboriginal initiatives to create local-level criteria and indicator frameworks for sustainable development. She is currently working as a Forest Management Planning Intern in Northern Ontario.

Adrian Tanner (PhD Anthropology) has been a Professor in the Anthropology Department at Memorial University since 1972, where he is now Honorary Research Professor. His current research interests and most recent publications focus on social suffering, community healing, forestry, land tenure and the documentation of local knowledge with the Cree and Innu of Quebec, Labrador and Ontario. He is also conducting research on regional culture, land tenure, public ritual and subsistence economics of the people of Colo Navosa, Fiji.

Ronald Trosper (PhD) is a member of the Salish and Kootenai Tribes, Montana, and is Associate Professor of Aboriginal Forestry, Faculty of Forestry, University of British Columbia. Ron teaches courses on Indigenous Peoples and forestry and is the author of many publications on these subjects.

Jimmie Webb (MA) is the Senior Policy Advisor for Little Red River Cree Nation. Jim has coordinated a range of First Nation organizational, business and resource development consultation/negotiation processes within the Treaty 8 area of Alberta and British Columbia, and represents First Nation organizations on numerous national, regional, and local organizations and committees, including the SFMN. Jim writes, and speaks often on Treaty-based approaches to sustainable resource management, cooperative management planning, and environmental impact assessment.

Stephen Wyatt (PhD) is an Assistant Professor, Faculty of Forestry, Université de Moncton at Edmundston, New Brunswick. Stephen was a general manager of the Quebec Forest Research Council, scientific coordinator for the XII World Forestry Congress at Québec in 2003, forestry program manager for an NGO in the South Pacific and a government-employed forester in Australia. He has worked with the Atikamekw Nation in Haute-Mauricie, Québec, and his current research interests include First Nations' roles in forestry, public participation and the management of private woodlots.

Appendix B—*List of Acronyms*

AAAND	Alberta Aboriginal Affairs and Northern Development
AAC	Annual Allowable Cut
ABL	Algonquins of Barrier Lake
ACWG	Aboriginal Capacity Working Group
AFLW	Alberta Forestry, Lands, and Wildlife
AFN	Assembly of First Nations
AFPP	Aboriginal Forest Planning Process
AIP	Agreement in Principle
ALUF	Alberta Land Use Framework
ALUOS	Aboriginal Land Use and Occupancy Studies
ALSA	Alberta Land Stewardship Act
ATK	Aboriginal Traditional Knowledge
AVM	Aboriginal Values Mapping
BCG	Band Council Governance
C&I	criteria and indicators
CBI	Canadian Boreal Initiative
CCFM	Canadian Council of Forest Minsters
CFSA	*Crown Forest Sustainability Act*
CEMA	Cumulative Effects Management Association
CIFOR	Center for International Forestry Research
CNR	Canadian National Railway
CPSA	Canadian Political Science Association
CSA	Canadian Standards Association
CSSP	Clayoquot Sound Scientific Panel
CSTC	Carrier Sekani Tribal Council
DFMP	Detailed Forest Management Plan
DFO	Department of Fisheries and Oceans
EAB	Environmental Assessment Board
ENGO	Environmental Non-Governmental Organization
FAO	Food and Agriculture Organization
FL	Forest Licence
FRAs	Forest and Range Agreements
FROs	Forest and Range Opportunities
FSC	Forest Stewardship Council
GFA	Gitanyow Forestry Agreement
GIS	Geographic Information System
HBC	Hudson's Bay Company
HDI	Human Development Index
HFN	Huu-Ay-Aht First Nation
HLFN	Heart Lake First Nation
IAA	Indian Association of Alberta
ICC	Indian Claims Commission
IIED	International Institute for Environment and Development
IOG	Institute of Governance
IRMP	Integrated Resource Management Plan
IUCN	International Union for the Conservation of Nature

JBACE	James Bay Advisory Committee on the Environment
JBNQA	*James Bay and Northern Quebec Agreement*
JPRF	John Prince Research Forest
JRC	Joint Resources Council
LARP	Lower Athabasca Regional Plan
LRRCN	Little Red River Cree Nation
LRMP	Land and Resource Management Planning
MAA	Ministry of Aboriginal Affairs
MCFN	Noose Cree First Nation
MNR	Ministry of Natural Resources (Ontario)
MOE	Ministry of the Environment
MOF	Ministry of Forests
NAFA	National Aboriginal Forestry Association
NBI	Northern Boreal Initiative
NRTA	*Natural Resources Transfer Agreement*
NRTEE	National Round Table on the Environment and Economy
OMNR	Ontario Ministry of Natural Resources
RACs	Regional Advisory Councils
RCAP	Royal Commission on Aboriginal Peoples
SCC	Supreme Court of Canada
SFI	Sustainable Forestry Initiative
SFMN	Sustainable Forestry Management Network
SFMP	Sustainable Forest Management Plan
SFN	Stellaquo First Nation
SMA	Special Management Area
SRC	Stellat'en Research Council
T8FNs	Treaty 8 First Nations
TEK	Traditional Ecological Knowledge
TFL	Tree Farm License
THG	Traditional Hereditary Governance
TLUS	Traditional Land Use Study
TOR	Terms of Reference
UN	United Nations
UNBC	University of Northern British Columbia
UNCED	UN Conference on Environment and Development
UNDP	United Nations Development Program
UNDSD	United Nations Division for Sustainable Development
UNESCAP	UN Economic and Social Commission for Asia and the Pacific
UNGA	United Nations General Assembly
VIU	Vancouver Island University
VOITs	Values, Objectives, Indicators and Targets
WCED	World Commission on Environment and Development
WCMF	Waswinipi Cree Model Forest
WES	Western Environmental Science
WIPO	World Intellectual Property Organization
WFMC	Whitefeather Forest Management Corporation
WMFN	West Moberly First Nation

DATE DUE